T0205418

Fungal Biology

Series Editors

Vijai Kumar Gupta
Department of Chemistry and Biotechnology
ERA Chair of Green Chemistry
School of Science
Tallinn University of Technology
Akadeemia tee, Tallinn
Estonia

Maria G. Tuohy
School of Natural Sciences
National University of Ireland Galway
Galway
Ireland

About the Series

Fungal biology has an integral role to play in the development of the biotechnology and biomedical sectors. It has become a subject of increasing importance as new fungi and their associated biomolecules are identified. The interaction between fungi and their environment is central to many natural processes that occur in the biosphere. The hosts and habitats of these eukaryotic microorganisms are very diverse; fungi are present in every ecosystem on Earth. The fungal kingdom is equally diverse, consisting of seven different known phyla. Yet detailed knowledge is limited to relatively few species. The relationship between fungi and humans has been characterized by the juxtaposed viewpoints of fungi as infectious agents of much dread and their exploitation as highly versatile systems for a range of economically important biotechnological applications. Understanding the biology of different fungi in diverse ecosystems as well as their interactions with living and non-living is essential to underpin effective and innovative technological developments. This series will provide a detailed compendium of methods and information used to investigate different aspects of mycology, including fungal biology and biochemistry, genetics, phylogenetics, genomics, proteomics, molecular enzymology, and biotechnological applications in a manner that reflects the many recent developments of relevance to researchers and scientists investigating the Kingdom Fungi. Rapid screening techniques based on screening specific regions in the DNA of fungi have been used in species comparison and identification, and are now being extended across fungal phyla. The majorities of fungi are multicellular eukaryotic systems and therefore may be excellent model systems by which to answer fundamental biological questions. A greater understanding of the cell biology of these versatile eukaryotes will underpin efforts to engineer certain fungal species to provide novel cell factories for production of proteins for pharmaceutical applications. Renewed interest in all aspects of the biology and biotechnology of fungi may also enable the development of "one pot" microbial cell factories to meet consumer energy needs in the 21st century. To realize this potential and to truly understand the diversity and biology of these eukaryotes, continued development of scientific tools and techniques is essential. As a professional reference, this series will be very helpful to all people who work with fungi and should be useful both to academic institutions and research teams, as well as to teachers, and graduate and postgraduate students with its information on the continuous developments in fungal biology with the publication of each volume.

More information about this series at http://www.springer.com/series/11224

Sachin Kumar • Pratibha Dheeran
Mohammad Taherzadeh • Samir Khanal
Editors

Fungal Biorefineries

 Springer

Editors
Sachin Kumar
Biochemical Conversion Division
Sardar Swaran Singh National Institute of
Bio-Energy
Kapurthala, Punjab, India

Mohammad Taherzadeh
Swedish Centre for Resource Recovery
University of Borås
Borås, Sweden

Pratibha Dheeran
Maharaj Singh College
Saharanpur, Uttar Pradesh, India

Samir Khanal
Department of Molecular Biosciences
and Biological Engineering
University of Hawai'i at Mānoa
Honolulu, HI, USA

ISSN 2198-7777   ISSN 2198-7785   (electronic)
Fungal Biology
ISBN 978-3-030-08002-0   ISBN 978-3-319-90379-8   (eBook)
https://doi.org/10.1007/978-3-319-90379-8

Printed on acid-free paper

This Springer imprint is published by the registered company Springer International Publishing AG part of Springer Nature.
The registered company address is: Gewerbestrasse 11, 6330 Cham, Switzerland

Preface

Fungi are eukaryotic organisms, which may be unicellular or multicellular. Fungi have unique characteristics of producing useful products including organic acids (lactic, citric, fumaric), hydrolytic enzymes (amylase, cellulases, xylanases, ligninases, lipases, pectinases, proteases), advanced biofuels (ethanol, single cell oils), polyols (xylitol), single cell protein (animal feed), secondary metabolites (antibiotics), etc. Fungi are widely used for production of cellulolytic enzymes production at industrial scale.

This book covers the fungal-based biorefinery. It is a comprehensive book dealing with different applications of fungi in advanced biofuels, organic acids, polyols, and animal feed productions, among others. The book focuses on fungal biorefinery with particular emphasis on applications in the pretreatment of lignocelluloses, and the production of advanced biofuels (ethanol, single cell oils), lactic acid, polyols (xylitol), and single cell protein (SCP).

Chapter 1 describes the current status of work done in the area and the role of fungi/fungal endophytes in helping to resolve some of the problems relating to environment and sustainability. In particular, the authors have examined the role of fungi in new microbial bioprocesses for applications in biorefineries.

Chapter 2 discusses the potential applications of fungal whole cell biocatalysts for pretreatment of lignocellulose, bioethanol fermentation, and biodiesel production. It also examines various development strategies employed to improve the performance of fungal whole cell biocatalysts.

Chapter 3 deals with the role of white-rot fungi and their xylanases in pulp and paper industry. Potential of white-rot fungal xylanases in future, with major bottlenecks, has also been highlighted. In particular, the chapter provides a major insight to the use of xylanases in pulp and paper industry.

Chapter 4 covers the composition of lignocellulosic substrates, fungal enzymes, synergism between cellulases, and industrial integrated bioprocesses for biomass utilization. Authors also explored the applications of fungal enzymes in pulp and paper industry, textile, bioethanol, wine and brewery industries, food processing, and animal feed.

Chapter 5 addresses the strategies used for economic production of lactic acid using fungi, which is still looking for its place in industry. The role of fungi, especially Rhizopus, has been discussed in detail. The chapter also addresses the various factors affecting fermentation.

Chapter 6 provides a limelight on native fungal strains and the advances made in fungal metabolic engineering to increase the production of xylitol.

Chapter 7 describes different aspects of fungal xylose reductases including their structural characteristics, sources, production, purification, characterization and immobilization, patent status, and xylitol applications.

Chapter 8 deals with the production of SCOs from various filamentous fungi as feedstock for biodiesel when grown on lignocellulosic wastes. Approaches to improve the process efficiency via optimization of fermentation conditions, one-step transesterification, and metabolic engineering, as well as the physicochemical properties of biodiesel are also discussed.

Chapter 9 covers the biorefinery concept of fungi for biofuels, which includes their culture techniques, culture medium, growth, SCO extraction, and transesterification of SCO. It also compares the first generation to third generation biofuels.

Chapter 10 presents an updated application data and treatment of vinasse for protein production and advocates the development of research to reduce environmental impacts and use this resource for human and animal consumption.

Kapurthala, India   Sachin Kumar
Saharanpur, India   Pratibha Dheeran
Borås, Sweden   Mohammad Tahezadeh
Honolulu, HI, USA   Samir Kumar Khanal

Acknowledgments

We would like to thank all the authors, who made painstaking contributions in the making of this book. Their patience and diligence in revising the initial draft of the chapters after incorporating the comments/suggestions are highly appreciated.

We would also like to acknowledge the contributions of all the reviewers for their constructive and valuable comments and suggestions to improve the quality of the contributions of various authors.

The support and encouragement of Mr. Eric Stannard, Senior Editor, Botany, Springer, are gratefully acknowledged. We also acknowledge Dr. Vijai Kumar Gupta and Dr. Maria G. Tuohy, Series Editors, for their encouragement. We are also grateful to Ms. Hemalatha Gunasekaran, Project coordinator (Books), Springer, for her consistent follow-up.

Contents

Contributors

Geiza Suzart Araujo Department of Technology, Food Engineering, State University of Feira de Santana, Feira de Santana, BA, Brazil

Sheelendra M. Bhatt Sant Baba Bhag Singh University, Jalandhar, India

Eliana Vieira Canettieri Department of Energy, São Paulo State University, Guaratinguetá, SP, Brazil

V. Capella Marisa State University (Unesp), Institute of Chemistry, Department of Biochemistry and Technological Chemistry, Araraquara, SP, Brazil

Kanika Chowdhary Centre for Rural Development and Technology, IIT-Delhi, Delhi, India

Kelly J. Dussán State University (Unesp), Institute of Chemistry, Department of Biochemistry and Technological Chemistry, Araraquara, SP, Brazil

Sílvia Maria Almeida de Souza Department of Technology, Food Engineering, State University of Feira de Santana, Feira de Santana, BA, Brazil

Jéssica Ferreira dos Santos Department of Technology, Food Engineering, State University of Feira de Santana, Feira de Santana, BA, Brazil

Baranitharan Ethiraj Department of Biotechnology, Bannari Amman Institute of Technology, Sathyamangalam, Erode District, India

Longinus I. Igbojionu IPBEN - Bioenergy Research Institute, Institute of Chemistry, São Paulo State University - UNESP, Araraquara, SP, Brazil

Shaik Jakeer International Centre for Genetic Engineering and Biotechnology, Aruna Asaf Ali Marg, New Delhi, India

Shraga Segal Department of Microbiology, Immunology, and Genetics, Ben-Gurion University of the Negev, Beer-Sheva, Israel

Gail Joseph Department of Energy and Environmental Systems, North Carolina Agricultural and Technical State University, Greensboro, NC, USA

Gouri Katre Institute of Bioinformatics and Biotechnology, Pune, India

Maksudur Rahman Khan Faculty of Chemical and Natural Resources Engineering, University Malaysia Pahang, Gambang, Malaysia

Mahesh Khot Institute of Bioinformatics and Biotechnology, Pune, India

Cecilia Laluce IPBEN - Bioenergy Research Institute, Institute of Chemistry, São Paulo State University - UNESP, Araraquara, SP, Brazil

Yogita Lugani Enzyme Biotechnology Laboratory, Department of Biotechnology, Punjabi University, Patiala, Punjab, India

Ernesto Acosta Martínez Department of Technology, Food Engineering, State University of Feira de Santana, Feira de Santana, BA, Brazil

Domenico Pirozzi Department of Chemical Engineering, Materials and Industrial Production, University Naples Federico II, Naples, Italy

Usha Prasad Gargi College, University of Delhi, Siri Fort, Delhi, India

Ameeta RaviKumar Institute of Bioinformatics and Biotechnology, Pune, India

Department of Biotechnology, Savitribai Phule Pune University, Pune, India

Rita de Cássia Lacerda Brambilla Rodrigues Department of Biotechnology, Engineering School of Lorena, São Paulo University, Lorena, SP, Brazil

Satyawati Sharma Centre for Rural Development and Technology, IIT-Delhi, Delhi, India

Shalini Singh School of Bioengineering and Biosciences, Lovely Professional University, Phagwara, Punjab, India

Balwinder Singh Sooch Enzyme Biotechnology Laboratory, Department of Biotechnology, Punjabi University, Patiala, Punjab, India

Lijun Wang Department of Natural Resources and Environmental Design, and Department of Chemical, Biological and Bioengineering, North Carolina Agricultural and Technical State University, Greensboro, NC, USA

Abu Yousuf Department of Chemical Engineering & Polymer Science, Shahjalal University of Science and Technology, Sylhet, Bangladesh

Smita Zinjarde Institute of Bioinformatics and Biotechnology, Pune, India

About the Editors

Sachin Kumar is a Deputy Director in the Biochemical Conversion Division at the Sardar Swaran Singh National Institute of Bio-Energy, Kapurthala, India. He was as a visiting professor in the Department of Chemical and Biological Engineering at South Dakota School of Mines and Technology, Rapid City, USA, for a year. He obtained his Ph.D. in Chemical Engineering from Indian Institute of Technology, Roorkee, India, and has more than 12 years of research experience in biochemical conversion of biomass to biofuels including lignocellulosic ethanol, biogas, and biohydrogen. He has completed six research projects and one consultancy project and actively engaged in two ongoing research projects. Dr. Sachin has published more than 45 papers in peer-reviewed journals, book chapters, and papers in conference proceedings, and 7 edited books. He has one US patent and one Indian patent (pending). He has delivered more than 15 invited/plenary lectures and presented over 60 papers at national and international conferences. He is a recipient of 2016 ASM-IUSSTF Indo-US Research Professorship and selected as Bioenergy-Awards for Cutting Edge Research (B-ACER) Fellow 2016 by DBT and IUSSTF.

Pratibha Dheeran is an Assistant Professor in the Department of Botany, Maharaj Singh College, Saharanpur, India. Dr. Dheeran has obtained Ph.D. degree in Biochemistry from Jiwaji University, Gwalior/ Indian Institute of Petroleum, Dehradun India. She has a master degree in Biotechnology from Indian Institute of Technology, Roorkee, and an M.Sc. in Botany from Chaudhary Charan Singh University, Meerut. She has been awarded DST Fast Track Young Scientist Grant by

Department of Science and Technology, India. She is a recipient of Innovation Postdoctoral Fellowship by Cape Peninsula University of Technology, Cape Town, South Africa. Dr. Dheeran has 8 years of research experience in biocatalysis, thermophilic microbiology, and biofuels. She has eight research papers in various peer-reviewed journals, three book chapters, and three conference proceedings.

Mohammad Taherzadeh is Professor of Bioprocess Technology since 2004 at University of Borås in Sweden and Director of the Research School at Swedish Centre for Resource Recovery. Prof. Taherzadeh has Ph.D. in Bioscience from Sweden, and M.Sc. and B.Sc. in Chemical Engineering from Iran. He is developing processes to convert wastes and residuals to value-added products such as ethanol, biogas, human food, animal feed, and biopolymers by fermentation. He has worked with filamentous fungi since 1999 and has many publications on industrial applications of the fungi and developing fungal-based biorefineries. He has more than 220 publications in scientific peer-reviewed journals, 19 book chapters, 5 patents and 3 books. More information about him is available at www.taherzadeh.se or www.hb.se/scrr.

Samir Kumar Khanal is Professor of Biological Engineering at the University of Hawai'i at Mānoa. Prof. Khanal is a leading researcher, internationally in the field of anaerobic digestion, waste-to-resources, and environmental biotechnology. Prof. Khanal obtained his M.S. in Environmental Engineering from Asian Institute of Technology, Bangkok, Thailand and Ph.D. in Civil Engineering (Environmental Biotechnology) from Hong Kong University of Science and Technology, Hong Kong. Prof. Khanal was Post-doctoral Research Associate (2 years) and Research Assistant Professor (4 years) at Iowa State University, Ames, IA. He wrote and published a book entitled *Anaerobic Biotechnology for Bioenergy Production: Principles and Applications* (Wiley-Blackwell, 2008, Bestseller). Recently, he also published a bioenergy textbook entitled *Bioenergy: Principles and Applications* (Wiley-Blackwell, 2017). Prof. Khanal is Associate Editor of *Bioresource Technology*. He is a recipient of Board of Regents' Medal for Excellence in Research (2018) and CTAHR Dean's Award for Excellence in Research (2016), University of Hawai'i at Mānoa. Please visit www.samirkkhanal.com for more details.

Chapter 1
Role of Fungi in Biorefinery: A Perspective

Kanika Chowdhary, Usha Prasad, and Satyawati Sharma

Abstract Fossil fuels and petroleum that have driven the industrial development in the modern society for the last two centuries are on the verge of decline. The usage of renewable resources and green technologies will ensure the environmental sustainability in the long run. Biorefinery includes sustainable processing of biomass into a spectrum of marketable products and energy. The main driver for the establishment of biorefineries is the holistic environmental sustainability. Biorefinery integrates upstream, midstream and downstream processing of biomass into a range of value-added products. Industrially scaled up biorefineries are expected to contribute to an increased competitiveness and wealth of the countries by responding to the need for supplying a wide range of bio-based products and energy in an economically, socially and environmentally sustainable manner in future.

Fungi play an important role in addressing major global challenges. The shift from chemical processes to biological processing achieved by using fungal (and bacterial) enzymes in industries such as textiles, leather, paper and pulp has significantly reduced negative impact on the environment. Filamentous fungi (belonging to the ascomycetes, basidiomycetes and zygomycetes classes) are of great interest as biocatalysts in biorefineries as they naturally produce and secrete a variety of different organic acids that can be used as building blocks in the chemical industry. Fungal endophytes are also the storehouse of naturally occurring bioactive compounds which are not only useful to plants but also to humans. The chapter summarises the current status of work done in the area and the role of fungi/fungal endophytes in helping to resolve some of the problems relating to environment and sustainability. In particular, we examine the role of fungi in new microbial bioprocesses for applications in biorefineries.

K. Chowdhary · S. Sharma (✉)
Centre for Rural Development and Technology, IIT-Delhi, Delhi, India

U. Prasad
Gargi College, University of Delhi, Delhi, India

© Springer International Publishing AG, part of Springer Nature 2018
S. Kumar et al. (eds.), *Fungal Biorefineries*, Fungal Biology,
https://doi.org/10.1007/978-3-319-90379-8_1

1.1   Introduction

Microorganisms are the backbone of all ecosystems. They are directly or indirectly responsible for the production of desired chemicals in industrial bioprocesses. Due to their versatility, microorganisms are known to play an important role in conversion processes employed in biorefineries. They have been used in large industrial set-ups for the production of food, beverages and pharmaceutical compounds.

Biomass is the most important source of renewable energy and the only source of carbon.

The most abundant and recalcitrant biomass available on earth is the lignocellulosic biomass, and it holds potential for production of multiple products, viz. bioethanol, biogas, biodiesel and other by-products. A biorefinery is a multiproduct system that manages its refinery products, fractions and other inputs in accordance with the chemistry and physiology of the corresponding input plant raw material as described in Fig. 1.1.

Biorefinery utilises biomass obtained from plethora of untreated sources, e.g. oil crops, starchy plants, forest residue, agriculture residue, organic waste, wet feedstocks, grass and green crops as inputs, and transformed with the aid of microbes at various processing stages. In this chapter, we primarily focus on the contribution of fungi (including fungal endophytes) in various processes and techniques of biorefineries. Fungal endophytes are regarded as a fascinating group of organisms that colonise the living internal tissues of their host – usually higher plants without causing any evident symptoms of disease in the host cells. They produce natural bioac-

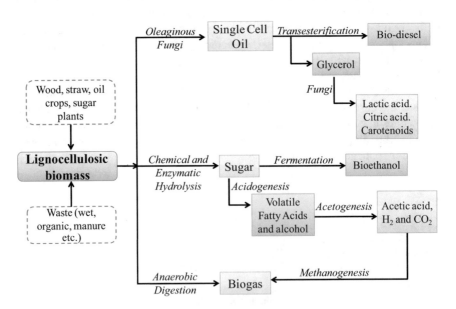

Fig. 1.1 Different approaches of biorefinery (Adapted from Naik et al. 2010)

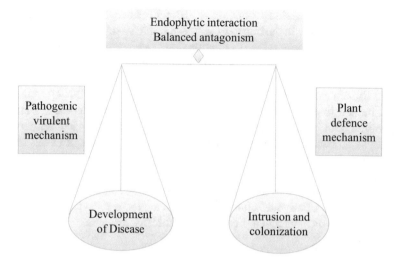

Fig. 1.2 Schematic depiction of balanced antagonism hypothesis (Adapted from Schulz et al. 1999)

tive compounds that act as elicitors for plant secondary metabolites production. Hypothesis describing the asymptomatic existence of endophytes as balanced antagonism is depicted in Fig. 1.2 (Schulz et al. 1999; Schulz and Boyle 2005).

As mentioned earlier a biorefinery aims for maximum utilisation of biomass with the aid of variable processes and techniques to produce multiple, environmentally adaptable products such as fuels (bioethanol, biogas and biodiesel) and other value-added chemicals. The approach is greener and environmentally safer than a petroleum refinery. The National Renewable Energy Laboratory of US Department of Energy defines a biorefinery as a competent facility that efficiently converts biomass to fuels, power and chemicals (http://www.nrel.gov/biomass/biorefinery. html[1]). Thus, a biorefinery system entails biomass production, its transformation and, lastly, generation of a large number of usable products.

Globally, India ranks amongst the fourth largest greenhouse gas (GHG) emitter and energy consumer countries in the world. Because of the fast depletion and uncertain availability of our natural non-renewable resources like fossil fuels and the boost in oil prices, there is a growing awareness that our natural resources will not last long. India, being the second most populous country of the world, needs to find sustainable energy generation sources to meet its evergrowing increase in energy demands. The most viable and economical alternative are biofuels. Because of their environment-friendly nature, they are widely in use all over the world. The use of biomass instead of oil as raw material for the production of chemicals and fuel leads to the development of biorefineries. In biorefineries, almost all types of feedstocks can be converted to different classes of biofuels and biochemicals

---

[1] https://www.nrel.gov/workingwithus/re-biomass.html. Last visited on 16th March, 2017.

through various conversion technologies. Biofuels have emerged as a preferred alternatives, especially in the transport sector in India as a multipronged strategy for reducing greenhouse gas (GHG) emissions, developing sustainable environment and generation of rural employment (Government of India 2008).

National biofuel policy approved by GOI in 2009 proposed 20% blending of biofuels (bioethanol and biodiesel) with petroleum by the end of 2017 (GAIN report 2016). India's first 2G ethanol biorefinery was set up by Hindustan Petroleum Corporation (HPCL) at Bathinda in Punjab in Dec 2016. The present refinery has a capacity to produce 100 kl of ethanol per day and provides employment to about 1300 people. Farmers can generate additional income (to the tune of about Rs. 20 crores per annum) by selling their agricultural wastes (http://www.news18.com/news/tech/indias-first-2g-ethanol-bio-refinery-to-be-set-up-in-punjab-1327128.html[2]). India as a country has a huge potential to tap its renewal energy available as biomass through its abundant agricultural and forestry residues.

1.2   Types of Biofuels

Biofuels include biodiesel (alkyl esters), bioethanol and biogas. There are three generations of biofuels, namely, first, second and third generation, mainly depending upon the source of biomass. Each one of them has its own limitation as a source of renewable energy. The first-generation biofuels are commercial today, with almost 50 billion litres produced annually (Naik et al. 2010). They are produced through well-known technologies like fermentation, distillation and transesterification. The main first-generation biofuels are ethanol (produced by fermentation of sugars and starches) and biodiesel (by transesterification of plant oil or animal fat). The main drawback of first-generation biofuels is that they come from biomass that is also a food source like sugar, starch or vegetable oil and their use is posing a threat to food prices. This presents a problem when there is not enough food to feed everyone, which ultimately leads to "food-versus-fuel debate". First-generation biofuels are also more expensive and economically unviable. Second-generation biofuels are derived from non-food biomass such as wood, organic waste and food waste which largely belongs to cheap and abundantly available lignocellulosic materials. They do not create the "fuels vs. food debate", as for the first-generation fuels, since they come from underutilised biomass. However, highly economical and energy-intensive procedure of breaking down of hemicelluloses and lignin layers that shield cellulose remains an issue with second-generation biofuel. Third-generation biofuel is based on energy production from photosynthetic microorganisms such as microalgae capable of fixing CO_2. The primary advantage of using microalgae is that it can be cultivated on nonarable land and it also helps in greenhouse gas mitigation

---

[2] http://www.news18.com/news/tech/indias-first-2g-ethanol-bio-refinery-to-be-set-up-in-punjab-1327128.html. Last visited on 16th March, 2017.

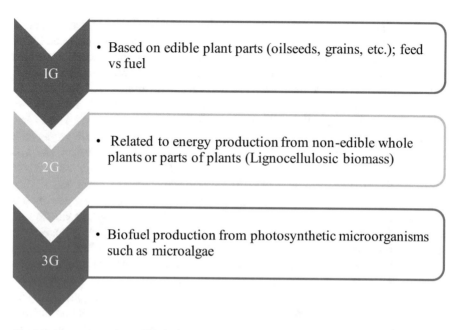

Fig. 1.3 Three generations of biofuels

(Buisson and Pettier 2013). They can be grown on variety of substrates like fresh water, salt water, brackish sea, waste water and even on sewage. Microalgae possess stupendous attributes for source of lipid-based biofuels: high growth rate, high lipid content and ease of extraction. They can store lipids in the form of triacylglycerols, which can be used to synthesise biodiesel via transesterification, and the remaining carbohydrate content can also be converted to bioethanol via fermentation. Some species of *Chlorella* have even been reported to accumulate up to 60% of oils in their biomass (Christi 2003). Nevertheless, in order to turn microalgae into commercial success, improvements through genetic and metabolic engineering have to be carried out leading to low-cost and high lipid-producing algal strains (Fig. 1.3).

1.2.1 Bioethanol

Bioethanol is produced from biomass mostly via a fermentation process using glucose derived from sugars, starch or cellulose. Conventional crops such as corn and sugarcane are unable to meet the global demand of bioethanol production due to their primary value as food and feed, whereas agricultural residues, forest residue and postharvest produce amongst others consist of stored carbohydrates in the form of lignocellulosic biomass. These agricultural wastes are attractive feedstocks for bioethanol production as they are cost effective, renewable and abundant in nature. A structural unit of lignocelluloses consists of cellulose, hemicelluloses and lignin (Huber and Corma 2007). In the case of bioethanol production, lignocellulosic biomass is converted into ethanol by following three orderly and systematic steps: physiochemical pretreatment, enzymatic saccharification and fermentation (Alrumman 2016). These are chiefly achieved by microbial enzyme arsenals, i.e. laccases, peroxidases, cellulases and hemicelluloses amongst others. Most common sources of these enzymes are different species of naturally occurring fungi and bacteria, e.g. *Saccharomyces cerevisiae*, *Zymomonas mobilis*, *Escherichia coli* and *Bacillus subtilis*, along with and engineered species *Caldicellulosiruptor bescii*. All these species are known to ferment cellulosic and hemicellulosic substrates as very well documented in literature (Cha et al. 2013). Another species *Picher pastors* has been routinely investigated as a host microbe for heterogonous production of eukaryotic proteins such as fungal biomass-degrading enzymes, mainly because of its relatively high growth rate and its ability to be cultivated on low-cost media (Várnai et al. 2014). *Picher pastoris* was examined for bioethanol production via pentose and hexose with the help of multiple functional enzyme complexes by utilising lignocellulosic biomass. It was able to efficiently degrade both xylan and carboxymethylcellulose (CMC) substrates resulting in the approximate production of 1.18 and 1.07 g/L ethanol from each substrate, respectively. Similarly, *Miscanthus sinensis* (commonly known as Chinese grass) has also been found to be a potential source of biomass to produce bioethanol. It resulted in the production of 1.08 g/L of ethanol when it was treated with recombinant strain of *P. pastoris*; the ethanol production was approximately 1.9-folds higher than that of the wild-type strain (Shin et al. 2015). *Neurospora intermedia* and *Aspergillus oryzae* were also examined for the synthesis of bioethanol and proteins using stillage as a substrate (by degrading complex substrates including arabinan, glucan, mannan and xylan). *N. intermedia* gave yield of 4.7 g/L ethanol from the stillage and were subsequently increased to 8.7 g/L with addition of 1 FPU of cellulase per g of suspended solids. *Saccharomyces cerevisiae* produced 0.4 and 5.1 g/L ethanol and protein, respectively. Under a two-stage cultivation with both fungi, up to 7.6 g/L of ethanol and 5.8 g/L of biomass containing 42% (w/w) crude protein were obtained (Bátori et al. 2015). Bioethanol and biomass as feed production were also exploited by using wheat-based thin stillage with the aid of fungal catalysts, namely, *Rhizopus* sp., *A. oryzae*, *F. venenatum*, *M. pursuers* and *N. intermedia* in submerged cultivation. *N. intermedia* was proved to be the best performing strain with production of 16 g/L biomass containing 56% crude protein and a reduction of around 34% of total solids (Ferreira et al. 2014).

White-rot fungi are major species of fungi that can selectively cleave non-phenol lignin structures found in woody biomass with high efficiency. Owing to the presence of an elaborate arsenal of lignocellulolytic enzymes, which include lignin peroxidase, Mn-oxidizing peroxidases, laccases, cellulases and hemicelluloses, white-rot fungi are commonly referred to as delignifying microbes (Stajić et al. 2016). *Phanerochaete chrysosporium* has become the model system for studying the physiology and genetics of lignin degradation since it is able to act upon on all the major components of wood including cellulose, hemicelluloses and lignin. Major components of the *P. chrysosporium* lignin depolymerisation systems include lignin peroxidase (Lip), manganese peroxidase (Mn) and a peroxide-generating enzyme, glyoxalin oxidise (GLOX) (Wymelenberg et al. 2006).

In a study, various basidiomycetes fungi, i.e. *Ganoderma lucidum, Phanerochaete chrysosporium, Pleurotus ostreatus, Pleurotus pulmonarius* and *Trametes* sp., were analysed for their efficacy in biological pretreatment of *Eucalyptus grandis* sawdust (agricultural waste). FTIR and SEM results showed that pretreatment with *P. ostreatus* and *P. pulmonarius* promoted structural changes which would lead to improved enzymatic hydrolysis of lignin such as pores formation, collapsing of cell walls and increased fibber detachment. As analysed by various parameters, the amount of reduced sugars after saccharification of sawdust increased from 2.5 to 48.0 μmol/ml in second hydrolysed fraction (Castoldi et al. 2014).

Basidiomycete's fungus *Phanerochaete sordida* YK-624 was transformed with lactate dehydrogenase-encoding gene knocked off from *Bifidobacterium longue.* The transformant strain was able to successively delignify and ferment beech wood meal substrate. The lactic acid production was greatly enhanced by supplementing the medium by exogenous addition of calcium carbonate and cellulase going up to 4 g/l (Mori et al. 2016).

Activity of cellulolytic enzymes was further enhanced by growing two species, i.e. *Euc-1 and IPEX lacteous* (basidiomycetous fungi), on wheat straw (slow-decomposing agricultural waste).These fungal strains enhanced the saccharification rate many fold by exposing the carbohydrate content (cellulose and hemicelluloses) of the straw for further conversion. The cellulose/lignin ratio increased from 2.7 (in untreated straw) to 5.9 and 4.6 in *Euc-1* and *I. lacteous* strains, respectively, after the treatment with fungi (Dias et al. 2010). Likewise, *Ascocoryne sarcoides* inhabited in *Eucryphia cord folia* stem tissues produces metabolites when cultivated on minimal medium base with cellulose (15 g/L). It produced eight-carbon volatile organic compounds, which are potential biofuel metabolites. High expression of lipoxygenase pathway genes correlated with production of these metabolites suggested linoleum acid catabolic pathway as the possible mechanism of eight-carbon metabolite synthesis (Gianoulis et al. 2012).

Emphasis is now being laid on low-cost production of bioethanol from lignocellulosic biomass by employing newer and much advanced approaches like consolidated bioprocessing (CBP). This advanced technology basically combines together four biological events (physiochemical pretreatment, enzyme hydrolysis, saccharification of cellulose and hemicelluloses and fermentation) associated with this conversion process in one reactor, thereby considerably reducing the associated capital investment in raw material input and other utilities (Xu et al. 2009). However, CBP requires specially engineered microbial workhorse developed for imparting different process-specific functions in the biorefinery. Fungi genera investigated as CBP-efficient microbes are *Aspergillus, Fusarium, Monilia, Neocallimastix, Rhizopus* and *Trichoderma. Fusarium oxysporum* was transformed with constitutively

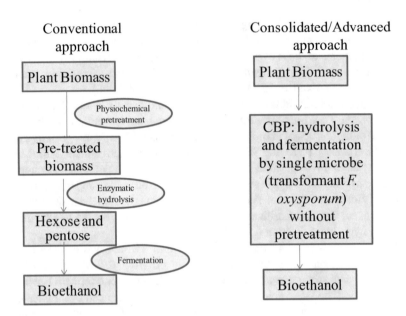

Fig. 1.4 Conventional vs. CBP/advanced technology for bioethanol production (Adapted from Anasontzis et al. 2011)

over-expressed endo-β-1,4-xylanase. When this transformant was cultivated in CBP system having corncob as fermentative medium, there was an increase in ethanol production by about 60% as compared to the wild-type strain (Anasontzis et al. 2011). In a recent study, CBP methodology improved the ethanol yield to twofold from cellulosic biomass by the action of cocultured anaerobic bacteria *Clostridium* sp. DBT-IOC-C19 and *Clostridium thermocellum* DSM 1313 (Singh et al. 2017) (Fig. 1.4 and Table 1.1).

1.2.2 Biodiesel

Biodiesel is an alternative fuel similar to conventional or fossil fuel. Technically, biodiesel is a biodegradable, non-toxic, almost sulphur-free and non-aromatic fuel which is derived from vegetable oils, waste oils, waste industrial oils, animal fats or algal oils. Biodiesel blended form of diesel is referred as BXX, where XX represents quantity of biodiesel in the blend. For instance, B100 has pure (100%) biodiesel. Generally, 20% blended biodiesel can be used in all diesel engines without any improvement. It is presently used as a major biofuel in Europe (Drozdzynska et al. 2011). Biodiesel/alkyl esters is synthesised by transesterification of free fatty acids using short-chain alcohols producing glycerol (or glycerine) as a side product (Silitonga et al. 2013).

Table 1.1 Examples of some fungi important in bioethanol production

S.No.	Fungi	Role in bioethanol production	References
1	Aspergillus, Fusarium, Monilia, Neocallimastix, Rhizopus and Trichoderma	Higher production of endo-β-1,4-xylanase could potentially enhance saccharification process leading to conversion of cellulose to bioethanol	Anasontzis et al. (2011)
2	Neurospora intermedia and Aspergillus oryzae	Higher yield of bioethanol and fungal proteins under optimised conditions	Batori et al. (2015)
3	Rhizopus sp. A. oryzae, F. venenatum, M. purpureus and N. intermedia	Higher yield of bioethanol and fungal proteins under optimised conditions	Ferreira et al. (2014)
4	Picher pastors	Recombinant strain of P. pastoris utilised Miscanthus sinensis as fermentive medium (potential biomass) for bioethanol production	Shin et al. (2015)
5	Euc-1 and Irpex lacteus	Enhanced saccharification process of wheat straw	Dias et al. (2010)
6	Phanerochaete sordida YK-624	Transformed strain delignified and fermented beech wood	Mori et al. (2016)

Glycerine recovered is purified and commonly used in chemical, pharmaceutical and cosmetic industry. Biodiesel industry generates multiple residue products (Dobson et al. 2012). Approximately 10% of glycerol is released in biodiesel production leading to expensive cost tag on the biodiesel price. Glycerol can be exploited as carbon source for microbial growth media in order to produce valuable organic chemicals (Yang et al. 2012). Lactic acid and its derivate chemicals are widely applied in the food, pharmaceutical and polymer industries (Okano et al. 2010). Vodnar et al. (2013) carried out a study for L(+)-lactic acid production by using pelletized *Rhizopus oryzae NRRL 395*, fermented on biodiesel crude glycerol media supplemented with inorganic nutrients and lucerne green juice. The results showed 3.72 g/g yield of L (+)-lactic acid. Fungus *Y. lipolytica* has been documented to produce various organic acids from glycerol. Similarly, citric acid and storage lipids were obtained from crude glycerol as a substrate in a submerged fermentation by *Yarrowia lipolytica* ACA-DC (Papanikolaou et al. 2008). In yet another study, mutant of *Y. lipolytica Wratislavia* AWG7 strain produced citric acid at 131.5 g/L in fed-batch fermentation. *Y. lipolytica Wratislavia* K1 produced citric acid (about 87–89 g/L) and polyols, i.e. mannitol and erythritol, upon cultivation on crude glycerol as carbon source (Rymowicz et al. 2009).

Surface active agents produced on living biosurfaces of microbes are called bio-surfactants. Bio-surfactants have numerous industrial applications. *Ustilago maydis* efficiently converted crude glycerol to glycolipid-type bio-surfactant in a synthetic medium using crude glycerol as the sole carbon source (Liu et al. 2011). For *Pithier irregular*, when fermented on medium containing 30 g/L crude glycerol, the EPA yield and productivity reached 90 mg/L and 14.9 mg/L per day, respectively. The study provides eicosapentaenoic acid (EPA)-fortified feeds

through fungal biomass (Athalye et al. 2009). Another value-added glycerol derivative is glycerol carbonate (Gly C). High lipase (biocatalyst) producing *Aspergillus
niger* strain was selected, and reaction condition was optimised by raw glycerol
(Gly) with dimethyl carbonate under solvent-free conditions. Reported optimised
conditions are 12% (w/w) *Aspergillus niger* lipase, to a glycerol-DMC molar ratio
of 1:10 with an incubation time of 4 h and temperature of 60 °C, respectively
(Tudorache et al. 2012).

Saenge et al. (2011) described the potential of oleaginous red yeast *Rhodotorula
glutinis* TISTR 5159 for utilisation of crude glycerol. In fed-batch stirred fermentation, this yeast strain produced 6.05 g/L of lipid content and 135.25 mg/L of carotenoids, respectively. A recent study elaborated on exploitation of biodiesel-derived
crude glycerol on fungal technology. An animal feed with balanced amino acid
profile and high contents of crude protein and lipids derived from fungal biomass is
an outstanding protein source for monogastrics. This objective was feasibly
achieved by cultivation of *Rhizopus microsporus* var. *oligosporus* on non-sterile
crude glycerol at a concentration of 75% (w/v) with nutrient supplementation
resulting in the production of prolific biomass along with elevated amounts of different amino acids like histamine, leonine, isoleucine, threonine, phenylalanine and
valine (Nitayavardhana and Khanal 2011).

Oleaginous microorganisms have the ability to accumulate lipids ~ 20% of their
biomass. Several filamentous fungi and single-celled yeasts have previously been
reported as oleaginous, having capability to synthesise and accumulate high amounts
of triacylglycerols within their cells, up to 70% of the biomass weight. The most
intensely examined oleaginous yeasts belong to the genera *Yarrowia*, *Candida*,
Rhodotorula, *Rhodosporidium*, *Cryptococcus* and *Lipomyces* as an alternative
source of biodiesel source (Rossi et al. 2011). *Mortierella* sp. being an oleaginous
microbe has been documented to produce 80% of intracellular lipid content.
Similarly, endophyte *Ascocoryne sarcoides* produced biofuel metabolites both
straight-chain and branched-chain alkenes upon fermentation on cellulose-based
medium (Gianoulis et al. 2012). Likewise, endophytic *Aspergillus* sp. upon cultivation for biodiesel production on Sabouraud dextrose broth medium and corncob
waste liquor as substrates displayed improvement in biomass production (13.6 g dry
weight/ 1000 ml) as well as lipid productivity (23.3%) (Subhash and Mohan 2011).
Single cell oil production of yeast strains *Rhodosporidium toruloides* and *Lipomyces
starkeyi* resulted in 42% and 48% lipid extractability, respectively (Bonturi et al.
2015). These yeasts could be preferably exploited for high lipid production on lignocellulosic biomass (Fig. 1.5).

1.2.3   Biogas

Biogas is a biofuel produced from the anaerobic fermentation of carbohydrates in
plant material or waste (both agricultural and municipal) by complex of microorganisms, the main player of which being the acidogenic and methanogenic bacteria.

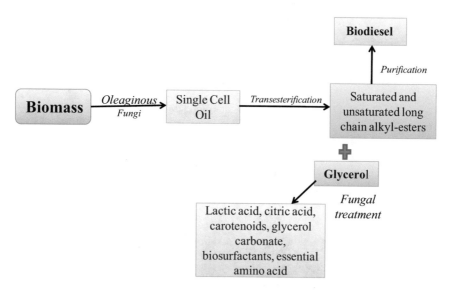

Fig. 1.5 Different steps involved in the production of biodiesel

Biogas is mainly composed of methane (65%) followed by carbon dioxide (35–40%) and other gases in traces, of which hydrogen sulphide is 0.5–1% and remaining water vapours, etc. Biogas spent slurry (by-product of biogas plant) is an excellent source of plant essential micro- and macronutrients and can be used for enriching soil health. Plant biomass is the largest reservoir of environment-friendly renewable energy on earth. Biogas is a renewable source of energy and can be utilised for cooking purposes particularly in rural areas. Anaerobic fungi belonging to phylum *Neocallimastigomycota* which are mainly housed within the digestive tracts of ruminants (larger mammalian herbivore) play an important role in the degradation of plant material resulting in biogas production. Their cellulolytic machinery consists of both secreted enzymes and high molecular weight multienzyme complexes called cellulosomes (containing cellulolytic and hemicellulolytic enzymes) capable of hydrolysing plant carbohydrates. Anaerobic fungi are the only fungi known to produce cellulosomes. Furthermore, these fungi are known to form cocultures with ruminant methanogenic archaic as a result of their high hydrogen production. They adhere onto plant surfaces and penetrate the plant tissues through the cell walls, thus opening the cells to allow the entry of cellulolytic bacteria. This makes the entire decomposition process in the rumen much faster. The same process could hold true for a biogas plant. Mucoromycotina, Pezizomycotina, Agaricomycotina, Saccharomycotina and Pucciniomycotina are some of the important groups of such fungi (Kazda et al. 2014). Anaerobic fungi have a great potential in the production of biogas from lignocellulosic waste. The production of biomethane and biohydrogen was studied on two different substrates, i.e. corn silage and cattail, using anaerobic fungus *Piromyces rhizinflata* YM600 (Nkemka et al. 2015) (Fig. 1.6 and Table 1.2).

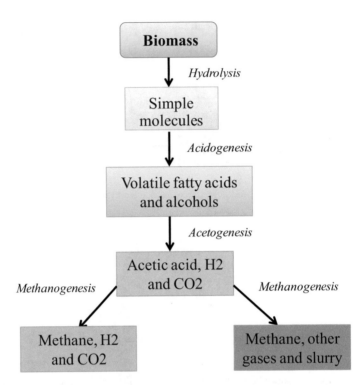

Fig. 1.6 Different steps involved in the production of biogas (Adapted from Vasudevan et al. 2015)

Table 1.2 Examples of some fungi important in biodiesel and biogas production

S. No.	Fungi	Role in biodiesel and biogas production	References
1	*Piromyces rhizinflata* YM600	Biogas formation utilising corn silage and cattail as substrates	Nkemka et al. (2015)
2	*Ascocoryne sarcoides*	High lipid yield to be utilised as a feedstock in biodiesel production	Gianoulis et al. (2012)
3	*Rhodosporidium toruloides* and *Lipomyces starkeyi*	High lipid extractability to be utilised as a feedstock in biodiesel production	Bonturi et al. (2015)
4	*Rhizopus oryzae* NRRL 395	Lactic acid production by bioconversion of glycerol	Vodnar et al. (2013)
5	*Yarrowia lipolytica* ACA-DC	Citric acid production by bioconversion of glycerol	
6	*Y. lipolytica Wratislavia* K1	Citric acid production by bioconversion of glycerol	Rymowicz et al. (2009)
7	*Ustilago maydis*	Bio-surfactant production by bioconversion of glycerol	Liu et al. (2011)
8	*Rhodotorula glutinis* TISTR 5159	Considerable lipids yield for biodiesel synthesis and carotenoids from bioconversion of glycerol	Saenge et al. (2011)
9	*Pithier irregular*	Eicosapentaenoic acid (EPA)-fortified feeds	Athalye et al. (2009)
10	*Aspergillus Niger*	Glycerol carbonate by bioconversion of glycerol	Tudorache et al. (2012)
12	*Rhizopus oryzae*	Lactic acid production by bioconversion of glycerol	Zamani (2010)

1.2.4 Other Usable Products from Fungi in Biorefinery

Many fungi are useful to humans and have been exploited both industrially and commercially. Efficient use of fungal by-products has been in use since time immemorial. Agro-industrial wastes are organic matters which can be recycled by utilising clean technology. Clean technology supports environment sustainability and maximises energy efficient use of waste.

A by-product of the sugar-alcohol industry, vinasse, is acidic dark brown slurry and is largely composed of water, organic matter and mineral elements. Because of the problems of large quantities of vinasse production in sugar industry and its potential effects as an environmental pollutant, several uses of vinasse have been developed by the scientists. For example, it has been utilised as a nutritive medium (Christofoletti et al. 2013). Vinasse has also been used as a fermentation substrate for cultivating non-edible ascomycetes fungi, e.g. *Neurospora intermedia* CBS 131.92 and *Aspergillus oryzae* var. *oryzae* CBS 819.72. High yields of dry fungal biomass of *N. intermedia* and *A. oryzae*, i.e. 202.4 and 222.8 g per litre of vinasse, respectively, were reported (Nair and Taherzadeh 2016). The study of Nitayavardhana et al. (2013) emphasised on potential use of fungal biomass cultivated on vinasse in fish feed industry and simultaneous 80% organic content reduction (as soluble chemical oxygen demand) leading to recycling of treated effluent as reclaimed water for plant use or for land applications. Likewise, another study documented production of ~50% crude protein and the essential amino acids similar to industrial aquatic feed from fungal biomass of edible fungus, *Rhizopus oligosporus*, cultivated on airlift bioreactor with 75% (v/v) vinasse and other supplements along with reclaiming water for land use (Nitayavardhana and Khanal 2010).

Fungi have been explored for biosynthesis of a vast array of low molecular weight molecules with a targeted biological function. *Aspergillus* species are utilised industrially in a number of ways. *Aspergillus terreus* has been known to be a microbial producer of HMG-CoA reductase enzymes which are marketed as statins (pharmaceutical drugs recommended to lower cholesterol and in cardiovascular diseases (Subhan et al. 2016). In a breakthrough study, *A. terreus* was cultivated on crude glycerol as a substrate producing (+) geodin and other secondary metabolites (Bizukojc and Pecyna 2011).

Different species of *Aspergillus* like *A. niger*, *A. oryzae* and *A. terreus* are high commercial producers of organic acids, i.e. citric acid, gluconic acid, itaconic acid, malic acid and oxalic acid. In addition to these organic acids, *Aspergillus* sp. also produces chitosan by hydrolysing chitin (a cell wall polysaccharide) present in their cell walls. Chitosan is a superabsorbent and commercially used in the industry. *Rhizopus oryzae*, a zygomycete member, is a good commercial producer of lactic acid (Zamani 2010). *Monascus* sp. has been an important source of polyketide pigments which are used as colourants in food industry and are derived from *Monascus* sp. Likewise, the other species of fungi like *Eurotium* sp., *Aspergillus* sp. and *Fusarium* sp. are sources of hydroxyanthraquinoid pigments used in different human food products (Dufosse et al. 2014; Ferreira et al. 2015). Heterogonous strains of *Saccharomyces* sp. have been able to synthesise various antioxidant polyphenolic compounds through MAGE (multiplex automated genome engineering) technology. Nevertheless, yeasts (unicellular ascomycetes) like *Saccharomyces* sp., *Picher* sp. and *Yarrowia* sp. have been widely used as prospective sources of organic acids, non-gelling agent, antifreeze proteins, collagen polysaccharides, polyunsaturated fatty acids, steroids and lipids (Pandey et al. 2015). The substrates for industrial production of value-added products by ascomycetes include mainly refined sugars (Silva da et al. 2012; Pandey et al. 2015).

The ascomycete's filamentous fungus *Trichoderma reesei* has been studied primarily for its benchmark production of cellulases enzymes; earlier this strain was designated as *Trichoderma* QM6a strain collected from Massachusetts, USA (Mandels and Reese 1957). Since the 1990s genetic-improved mutant *T. reesei* has progressively been applied as a host microbe for production of industrially significant homologous and heterogonous enzymes for feed, textile and other industries (Puranen et al. 2014).

Kluyveromyces marxianus, Aspergillus niger, Aspergillus sydowii and *Aspergillus fumigatus* have been isolated from bagasse piles in Brazilian alcohol plant and examined for presence of contributory enzymes involved in the saccharification of lignocellulosic material. Amongst the different species of fungi isolated, the three strains of *Aspergillus* (i.e. *A. niger* (SBCM 3), *A. sydowii* (SBCM7) and *A. fumigatus* (SBC4)) were found to produce good quantity of hemicellulolytic enzymes. *A. niger* SBCM3 strain tops the list in β-xylosidase production (Santos et al. 2015) (Table 1.3).

Table 1.3 Some of the usable products from fungi in biorefinery

S. No.	Fungi	Products	References
1	*Neurospora intermedia* CBS 131.92 and *Aspergillus oryzae* var. *oryzae* CBS 819.72	Protein-rich biomass for fish feed industry	Nitayavardhana et al. (2013)
2	*Rhizopus oligosporus*	Protein-rich biomass for fish feed industry	Nitayavardhana and Khanal (2010)
3	*Rhizopus microsporus* var. *oligosporus*	Fungal biomass with high protein content specially essential amino acid threonine	Nitayavardhana and Khanal (2011)
4	*A. niger* SBCM3 strain	β-xylosidase production can be utilised in improved saccharification of lignocellulosic matter	Santos et al.(2015)
5	*Trichoderma* QM6a strain	Transformed strain secretes high amounts of cellulolytic enzymes	Puranen et al. (2014)
6	*Aspergillus terreus*	(+)-geodin production by bioconversion of crude glycerol	Bizukojc and Pecyna (2011)

1.3   Fungal Endophytes: Rich Source of Enzymes

Fungal endophytes have been recognised as a potential depository of highly useful and functional enzymes that can be exploited in various processes and techniques in biorefineries. For example, enzymes like cellulases, lactases and lipases which are known to be secreted from different endophytic fungi may find a key role in various important industrial processes. Cellulose, a linear polymer of β-(1, 4)-D-glucopyranose unit, is the most vastly available biomass. Biologically, cellulose can be hydrolysed by cellulase which is a complex of three enzymes, viz. endo-β-1, 4-glucanase, β-1, 4-glucosidase and exo-β-1, 4-glucanase. Three different endophytic fungi, namely, *Trichoderma* sp., *Aspergillus* sp. and *Penicillium* sp., were recombined together through the process of intergeneric protoplast fusion and mutagenesis to produce Tahrir-25 – a recombinant strain showing high cellulase activity. These three endophytes were isolated from marine sponge *Latrunculia corticata*. Tahrir-25 was able to completely hydrolyse cellulose present in bagasse (agricultural waste). This pretreated bagasse was then further fermented to ethanol by *Saccharomyces cerevisiae* NRC2. These hypercellulase-producing endophytic fungal strains may be further tapped for their bioethanol production potential from lignocellulosic biomass (El-Bondkly and El-Gendy 2012).

Biopulping in wood and paper industry requires pretreatment of wood or its residues for removal of lignin to facilitate cellulose accessibility for saccharification (Young and Akhtar 1997). Different endophytic fungi (*Neofusicoccum luteum*, *Ulocladium* sp., *Hormonema* sp. and *Neofusicoccum austral*) isolated from *Eucalyptus* sp. were identified as potent producers of laccase enzymes and were used to pretreat *Eucalyptus globulus* wood chips.

Laccases are oxidoreductase enzymes mainly involved in lignin degradation. Fungal laccases have higher redo potential as compared to the ones isolated from bacteria or plants (Kunamneni et al. 2007). The concentrated extracts of *Hormonema* sp. CECT-13092 showed high levels of laccase activity. This fungus was further used for the pretreatment of wood for pulping and bleaching of unbleached pulps followed by mechanical pulping, which resulted in the enhancement of saccharification capacity of this fungus. In fact, this fungal strain possessed ~ 27% higher delignification capacity than the reference fungus *Teammates* sp. I-62. *Ulocladium* sp. showed better mechanical properties in post-pulp treatment (Martín-Sampedro et al. 2015). Lignocellulosic biorefineries must depend on pretreatments that consume high energy and investment for lignin removal. To overcome this problem, one can exploit lignin-degrading microorganisms which not only increase accessibility of the cellulose to enzymatic hydrolysis but also have minimum chemical requirements under mild conditions (Fillat et al. 2017). Laccase activity in *Hormonema* sp. and *N. luteum* was significantly enhanced by addition of certain inducers like ethanol and copper sulphate. The increase in activity was about 85% for *Hormonema* sp. and 95% for *N. luteum* (Fillat et al. 2016). Similarly, *Monotospora* sp. (another endophyte) isolated from *Cynodon dactylon* also showed lactase enzyme activity of 13.55 ± 1.38 U/ml under optimal growth conditions (Wang et al. 2006).

Table 1.4 Enzymes secreted by endophytic fungi being utilised in biorefinery

S.No.	Endophytic Fungi	Enzymes secreted	References
1	*Hormonema* sp. and *Neofusicoccum luteum*	Laccase	Fillat et al. (2016)
2	*Ulocladium* sp. and *Hormonema* sp.	Laccase	Martín-Sampedro et al. (2015)
3	*Trichoderma* sp., *Aspergillus* sp. and *Penicillium* sp.	Cellulase	El-Bondkly and El-Gendy (2012)
4	*Aspergillus Niger*	Cellulase	Mrudula and Murugammal (2011)
5	*Monotospora* sp.	Laccase	Wang et al. (2006)

Yeasts, particularly genus *Candida*, have been commercially used for the production of lipases (Salihu et al. 2011). Endophytic yeast identified as *Candida guilliermondii* recovered from castor beans yielded18 ± 1.4 U mL^{-1}of lipase enzyme when grown in 14 L of fermentative medium. The enzyme exhibited 81% of esterification rate in hexane. Higher ester conversion rate was obtained using hexane (Oliveira et al. 2014). Thus, enzymatic transesterification using lipases as catalysts has been considered as greener and safer approach than chemical-based production of biodiesel.

Solid-state fermentation is most favourably utilised for enzyme production in filamentous fungi. SSF has several advantages over liquid fermentation because of its technical simplicity, low energy and water input, less expensive substrates (agro-industrial wastes), lesser effluent, better extraction and high volumetric production (Zeng and Chen 2009).For example, endophytic fungus *Aspergillus niger* isolated from coir retting industry showed higher production of cellulase enzyme in SSF (14.6 fold higher) as compared to liquid fermentation under optimal conditions (Mrudula and Murugammal 2011). The study concludes that the coir waste can be utilised for the production of these enzymes at industrial scale using SSF technology (Table 1.4).

1.4 Conclusions and Future Directions

Microorganisms form the backbone of all ecosystems and are the main players in industrial bioprocesses, being directly or indirectly responsible for the production of desired chemicals. Due to their versatility, fungi are known to play an essential role in conversion processes deployed in biorefineries. The role of anaerobic fungi in the complex degradation of cellulose-rich substances is likely to receive more attention in future biogas research. The use of fungal processes and products can lead to increased sustainability through more efficient use of natural resources. In the arena of greener technologies, progress of work will increasingly focus on the commercial feasibility and viability of value-added products derived from biorefinery.

Additionally, both genetic modification of potent strains and utilisation of mixed consortia of competent fungi that can jointly perform complex processes efficiently, yielding the desired product at an amplified rate, need to be taken up and studied further. Also, coproducts such as protein biomass can be used to offset the processing cost of biorefinery and generate more revenue. The potential of the endophytic filamentous fungi and yeast strains have been realised to some extent and continued to be explored vigorously in fields of development of high-value products required in biorefinery processes. Most importantly, fungal endophytes can be effectively utilised for agriculture wastes, particularly lignocellulosic biomass, as a substrate.

Acknowledgement Kanika Chowdhary would like to acknowledge the financial assistance provided by DST-NPDF scheme (Grant no. PDF/2016/000317) and CRDT, IIT Delhi for providing infrastructural support. Usha Prasad would like to acknowledge the grant of sabbatical leave to her parent institution, Gargi College, University of Delhi.

References

Alrumman SA (2016) Enzymatic saccharification and fermentation of cellulosic date palm wastes to glucose and lactic acid. Brazilian J Microbiol 47(1):110–119

Anasontzis GE, Zerva A, Stathopoulou PM, Haralampidis K, Diallinas G, Karagouni AD et al (2011) Homologous overexpression of xylanase in Fusarium oxysporum increases ethanol productivity during consolidated bioprocessing (CBP) of lignocellulosics. J Biotechnol 152:16–23

Athalye SK, Garcia RA, Wen ZY (2009) Use of biodiesel-derived crude glycerol for producing Eicosapentaenoic Acid (EPA) by the fungus Pythium irregular. J Agric Food Chem 57:2739–2744

Batori V, Ferreira JA, Taherzadeh MJ, Lennartsson PR (2015) Ethanol and protein from ethanol plant by-products using edible fungi *Neurospora intermedia* and *Aspergillus oryzae*. Biomed Res Int 2015:10

Bizukojc M, Pecyna (2011) Lovastatin and (+)-geodin formation by Aspergillus terreus ATCC 20542 in a batch culture with the simultaneous use of lactose and glycerol as carbon sources. Eng Life Sci 11(3):272–282

Bonturi N, Matsakas L, Nilsson R, Christakopoulos P, Miranda EA, Berglund KA, Rova U (2015) Single cell oil producing yeasts *Lipomyces starkeyi* and *Rhodosporidium toruloides*: selection of extraction strategies and biodiesel property prediction. Energies 8(6):5040–5052

Castoldi R, Bracht A, de Morais GR, Baesso ML, Correa RCG, Peralta RA et al (2014) Biological pretreatment of *Eucalyptus grandis* sawdust with white-rot fungi: study of degradation patterns and saccharification kinetics. Chemical Engn J 258:240–246

Cha M, Chung D, Elkins JG, Guss AM, Westpheling J (2013) Metabolic engineering of *Caldicellulosiruptor bescii* yields increased hydrogen production from lignocellulosic biomass. Biotechnol Biofuels 6(1):85

Christi Y (2003) Biodiesel from microalgae beats bioethanol. Trends Biotechnol 26(3):126–131

Christofoletti CA, Esher JP, Correia JE, Marinho JFU, Fontanetti CS (2013) Sugarcane vinasse: environmental implications of its use. Waste Manang 33:2753–2761

Dias AA, Freitas GS, Marques GS, Sampaio A, Fraga IS, Rodrigues MA, Bezerra RM (2010) Enzymatic saccharification of biologically pre-treated wheat straw with white-rot fungi. Bioresour Technol 101(15):6045–6050

Dobson R, Gray V, Rumbold K (2012) Microbial utilization of crude glycerol for the production of value-added products. J Ind Microbiol Biotechnol 39:217–226. https://doi.org/10.1007/s10295-011-1038-0

Drozdzynska A, Leja K, Czaczyk K (2011) Biotechnological production of 1,3-propanediol from crude glycerol. J Biotechnol Comp Bio Bionanotechnol 92:92–100

Dufosse L, Fouillaud M, Caro Y, Mapari SA, Sutthiwong N (2014) Filamentous fungi are large-scale producers of pigments and colorants for the food industry. Curr opinion biotechnol 26:56–61

El-Bondkly AM, El-Gendy MM (2012) Cellulase production from agricultural residues by recombinant fusant strain of a fungal endophyte of the marine sponge *Latrunculia corticata* for production of ethanol. Ant Van Leeu 101(2):331–346

Ferreira JA, Lennartsson PR, Taherzadeh MJ (2014) Production of ethanol and biomass from thin stillage using food-grade Zygomycetes and Ascomycetes filamentous fungi. Energies 7(6):3872–3885

Ferreira JA, Lennartsson PR, Taherzadeh MJ (2015) Production of ethanol and biomass from thin stillage by Neurospora intermedia: a pilot study for process diversification. Engn Life Sci 15(8):751–759

Fillat Ú, Martín-Sampedro R, Macaya-Sanz D, Martín JA, Ibarra D, Martínez MJ, Eugenio ME (2016) Screening of eucalyptus wood endophytes for laccase activity. Process Biochem 51(5):589–598

Fillat Ú, Martín-Sampedro R, Ibarra D, Macaya D, Martín JA, Eugenio ME (2017) Potential of the new endophytic fungus *Hormonema* sp. CECT-13092 for improving processes in lignocellulosic biorefineries: biofuel production and cellulosic pulp manufacture. J Chem Technol Biotechnol 92(5):997–1005

GAIN Report (2016) Number: IN6088- India Biofuels Annual-2016, Global Agricultural Information Network, USDA Foreign Agricultural Service

Gianoulis TA, Griffin MA, Spakowicz DJ, Dunican BF, Sboner A, Sismour AM, Strobel SA (2012) Genomic analysis of the hydrocarbon-producing, cellulolytic, endophytic fungus *Ascocoryne* barcodes. Plops Genet 8(3):e1002558

Government of India. 2008. National Policy on Biofuels, Ministry of New & Renewable Energy. Government of India

Huber GW, Corma A (2007) Synergies between bio-and oil refineries for the production of fuels from biomass. Angew Chem Int Edit 46(38):7184–7201

Kazda M, Langer S, Bengelsdorf FR (2014) Fungi open new possibilities for anaerobic fermentation of organic residues. Ener Sustainability Soc 4(1):6

Kunamneni A, Ballesteros A, Plou FJ, Alcalde M (2007) Fungal laccase – a versatile enzyme for biotechnological applications. In: Méndez-Vilas A (ed) Communicating current research and educational topics and trends in applied microbiology. Badajoz, Spain, pp 233–245

Liu Y, Koh CMJ, Ji L (2011) Bioconversion of crude glycerol to glycolipids in *Ustilagomaydis*. Bioresour Technol 102:3927–3933

Mandels M, Reese ET (1957) Induction of cellulase in Trichoderma *viride* as influenced by carbon sources and metals. J Bacteriol 73(2):269–278

Martín-Sampedro R, Fillat Ú, Ibarra D, Eugenio ME (2015) Use of new endophytic fungi as pretreatment to enhance enzymatic saccharification of Eucalyptus globulus. Bioresour Technol 1196:383–390

Mori T, Kako H, Sumiya T, Kawagishi H, Hirai H (2016) Direct lactic acid production from beech wood by transgenic white-rot fungus *Phanerochaete sordida* YK-624. J Biotechnol 239:83–89

Mrudula S, Murugammal R (2011) Production of cellulase by *Aspergillus niger* under submerged and solid state fermentation using coir waste as a substrate. Brazilian J Microbiol 42(3):1119–1127

Naik SN, Goud VV, Rout PK, Dalai AK (2010) Production of first and second generation biofuels: a comprehensive review. Renewable Sustainable Ener Rev 14(2):578–597

Nair RB, Taherzadeh MJ (2016) Valorization of sugar-to-ethanol process waste vinasse: a novel biorefinery approach using edible ascomycetes filamentous fungi. Bioresour Technol 221:469–476

Nitayavardhana S, Khanal SK (2010) Innovative biorefinery concept for sugar-based ethanol industries: production of protein-rich fungal biomass on vinasse as an aquaculture feed ingredient. Bioresour Technol 101(23):9078–9085

Nitayavardhana S, Khanal SK (2011) Biodiesel-derived crude glycerol bioconversion to animal feed: a sustainable option for a biodiesel refinery. Bioresour Technol 102(10):5808–5814

Nitayavardhana S, Issarapayup K, Pavasant P, Khanal SK (2013) Production of protein-rich fungal biomass in an airlift bioreactor using vinasse as substrate. Bioresour Technol 133:301–306

Nkemka VN, Gilroyed B, Yanke J, Gruninger R, Vedres D, McAllister T, Hao X (2015) Biooaugmentation with an anaerobic fungus in a two-stage process for biohydrogen and biogas production using corn silage and cattail. Bioresour Technol 185:79–88

Okano K, Tanaka T, Ogino C, Fukuda H, Kondo A (2010) Biotechnological production of enantiomeric pure lactic acid from renewable resources: recent achievements, perspectives, and limits. Appl Microbiol Biotechnol 85:413–423. https://doi.org/10.1007/s00253-009-2280-5

Oliveira ACD, Fernandez ML, Mariano AB (2014) Production and characterization of an extracellular lipase from *Candida guilliermondii*. Brazilian J Microbiol 45(4):1503–1511

Pandey A, Höfer R, Taherzadeh M, Nampoothiri M, Larroche C (2015) Industrial biorefineries and white biotechnology. Elsevier, Amsterdam

Papanikolaou S, Fakas S, Fick M, Chevalot I, Galiotou-Panayotou M, Komaitis M, Marc I, Aggelis G (2008) Biotechnological valorisation of raw glycerol discharged after bio-diesel (fatty acid methyl esters) manufacturing process: production of 1, 3-propanediol, citric acid and single cell oil. Biomass Bioenergy 32:60–71

Puranen T, Alapuranen M, Vehmaanperä J (2014) Trichoderma enzymes for textile industries. In: Gupta VK, Schmoll M, Herrera-Estrella A et al (eds) Biotechnology and biology of *Trichoderma*. Elsevier, Waltham, pp 351–362

Rossi M, Amaretti A, Raimondi S, Leonardi A (2011) Getting lipids for biodiesel production from oleaginous fungi. Biodiesel-Feedstocks Processing Technol 1:72–74

Rymowicz W, Rywińska A, Marcinkiewicz M High-yield production of erythritol from raw glycerol in fed-batch cultures of *Yarrowia lipolytica*. Biotechnol Lett 2009, 2009(31): 377–380

Saenge C, Cheirsilp B, Suksaroge TT, Bourtoom T (2011) Potential use of oleaginous red yeast Rhodotorula glutini for the bioconversion of crude glycerol from biodiesel plant to lipids and carotenoids. Process Biochem 46:210–218

Santos BSLD, Gomes AFS, Franciscon EG, Oliveira JMD, Baffi MA (2015) Thermotolerant and mesophylic fungi from sugarcane bagasse and their prospection for biomass-degrading enzyme production. Brazilian J Microbiol 46(3):903–910

Schulz B, Boyle C (2005) The endophytic continuum. Mycol Res 109:661–687

Schulz B, Römmert AK, Dammann U, Aust HJ, Strack D (1999) The endophyte host interactions: a balanced antagonism? Mycological Res 103:1275–1283

Shin SK, Hyeon JE, Kim YI, Kang DH, Kim SW, Park C, Han SO (2015) Enhanced hydrolysis of lignocellulosic biomass: Bi-functional enzyme complexes expressed in *Pichia pastoris* improve bioethanol production from *Miscanthus sinensis*. Biotechnol J10(12):1912–1919

Silitonga AS, Masjuki HH, Mahlia TMI, Ong HC, Chong WT, Boosroh MH (2013) Overview properties of biodiesel diesel blends from edible and non-edible feedstock. Renew Sust Energ Rev 22:346–360

Silva da SS, Chandel AK, Wickramasinghe SR, Domínguez JM (2012) Fermentative production of value-added products from lignocellulosic biomass. BioMed Res Intern 826162:1–2

Singh N, Mathur AS, Tuli DK, Gupta RP, Barrow CJ, Puri M (2017) Cellulosic ethanol production via consolidated bioprocessing by a novel thermophilic anaerobic bacterium isolated from a Himalayan hot spring. Biotechnol Biofuels 10(1):73

Stajić M, Vukojević J, Milovanović I, Ćilerdžić J, Knežević A (2016) Role of mushroom Mn-oxidizing peroxidases in biomass conversion. In: Microbial enzymes in bioconversions of biomass. Springer International Publishing, Cham, pp 251–269

Subhan M, Faryal R, Macreadie I (2016) Exploitation of *Aspergillus terreus* for the production of natural statins. J Fungi 2(2):13

Subhash GV, Mohan SV (2011) Biodiesel production from isolated oleaginous fungi Aspergillus sp. using corncob waste liquor as a substrate. Bioresour Technol 102(19):9286–9290

Tudorache M, Protesescu L, Coman S, Parvulescu VI (2012) Efficient bio-conversion of glycerol to glycerol carbonate catalyzed by lipase extracted from *Aspergillus niger*. Green Chem 14(2):478–482

Várnai A, Tang C, Bengtsson O, Atterton A, Mathiesen G, Eijsink VG (2014) Expression of endoglucanases in *Pichia pastoris* under control of the GAP promoter. Microb Cell Factories 13(1):57

Vasudevan P, Sharma S, Sharma VP and Verma M Editors: (2015) Women, technology and development. published by Narosa Publishers, Daryaganj, New Delhi

Vodnar DC, Dulf FV, Pop OL, Socaciu C (2013) L (+)-lactic acid production by pellet-form *Rhizopus oryzae* NRRL 395 on biodiesel crude glycerol. Microb Cell Factories 12(1):92

Wang JW, Wu JH, Huang WY, Tan RX (2006) Laccase production by Monotospora sp., an endophytic fungus in *Cynodon dactylon*. Bioresour Technol 97(5):786–789

Wymelenberg AV, Minges P, Sabat G, Martinez D, Aerts A, Salamov A, Dosoretz C (2006) Computational analysis of the *Phanerochaete chrysosporium* v2. 0 genome database and mass spectrometry identification of peptides in ligninolytic cultures reveal complex mixtures of secreted proteins. Fungal Genetics Biol 43(5):343–356

Xu Q, Singh A, Himmel ME (2009) Perspectives and new directions for the production of bioethanol using consolidated bioprocessing of lignocellulose. Curr Opin Biotechnol 20:364–371

Yang F, Hanna MA, Sun R (2012) Value-added uses for crude glycerol-a byproduct of biodiesel production. Biotechnol Biofuels 5:13. https://doi.org/10.1186/1754-6834-5-13

Young R, Akhtar M (1997) Environmentally friendly Technologies for the Pulp and Paper Industry. Wiley, Hoboken, p 592

Zamani A (2010) Superabsorbent polymers from the cell wall of zygomycetes fungi. Doctoral dissertation, Chalmers University of Technology

Zeng W, Chen HZ (2009) Air pressure pulsation solid state fermentation of feruloyl esterase by *Aspergillus niger*. Bioresour Technol 100:1371–1375

Chapter 2
Production of Biofuels from Biomass by Fungi

Gail Joseph and Lijun Wang

Abstract Fungal whole-cell biocatalysts have been used for the pre-treatment of lignocellulosic biomass, lignocellulosic ethanol fermentation, and enzymatic bio-diesel production. Fungal whole cells or enzymes bound to whole-cell membrane have significantly reduced the production costs of biocatalysts while also increasing the reusability of biocatalysts. Various strategies based on the advances in genomics and genetic engineering have been developed to improve the biocatalysis efficiency. Genetic engineering of the fungal cells and enzymes via directed evolution has significantly increased the yields and productivity of biofuels in biological processes. Future research on the use of fungal whole-cell biocatalysts can lead to a more sustainable production of biofuels, biodiesel, and other value-added bioproducts.

2.1 Introduction

The energy, water, land, and natural resources have been under great pressure to support increasing world population. The availability of fossil energy has been declining worldwide. It is essential to change the global fossil-dependent economy into a sustainable bio-based economy. Environmental concern associated with fossil-based energy is another key driver for the emerging bioenergy industry. There is an increasing interest in converting various biomass resources into biofuels. Biomass materials with a high moisture content (e.g., >50%) are usually better suited to biological processes such as anaerobic digestion, while biomass materials

G. Joseph
Department of Energy and Environmental Systems, North Carolina Agricultural and Technical State University, Greensboro, NC, USA
e-mail: gjoseph@aggies.ncat.edu

L. Wang (✉)
Department of Natural Resources and Environmental Design and Department of Chemical, Biological and Bioengineering, North Carolina Agricultural and Technical State University, Greensboro, NC, USA
e-mail: lwang@ncat.edu

© Springer International Publishing AG, part of Springer Nature 2018   21
S. Kumar et al. (eds.), *Fungal Biorefineries*, Fungal Biology,
https://doi.org/10.1007/978-3-319-90379-8_2

with a low moisture content are better for thermochemical conversion processes such as combustion, gasification, and pyrolysis. In many cases, it is the form of marketable bioenergy products that determines the process selection. Fermentation, transesterification, pyrolysis, and liquefaction produce liquid fuels including ethanol, biodiesel, and bio-oil suitable for the use as transportation fuels. Combustion, gasification, and anaerobic digestion produce gaseous energy products such as hot gas for steam generation, syngas, biogas, and hydrogen, which are suitable to be used at the production location (Yan et al. 2014).

Biocatalysts are a competent and sometimes superior alternative to traditional metallo-catalysts and organo-catalysts (Bornscheuer et al. 2012). Enzymes are extensively used in the synthesis of industrial chemicals as enzymes offer several advantages over traditional catalysts such as no extra tedious step for separation, high temperature and pressure conditions, low costs, and environmental sustainability. As enzymes have high reactivity even at an ambient condition and excellent stereoselectivity, they have been widely used in synthesis of organic chemicals and asymmetric transformation. Both isolated enzymes and whole cells have been used as biocatalysts in industries (Ishige et al. 2005). Several studies have reported the utilization of microorganisms such as bacteria, yeast, and fungi as whole-cell biocatalysts can reduce the cost of bioconversion processes (Ban et al. 2001; Narita et al. 2006). Filamentous fungi have been identified as the most robust whole-cell biocatalyst for industrial applications. Filamentous fungi have been further immobilized on biomass support particles (BSPs) made of polyurethane foam for the separation of the whole-cell catalysts from a reaction mixture and facilitating their repeated use in a bioconversion process (Fukuda et al. 2008).

One of the applications of whole-cell catalysts for biofuel production is the pretreatment of lignocellulosic biomass and ethanol fermentation. Lignocellulosic biomass is abundant renewable resource for producing biofuels. Lignin in the lignocellulosic biomass is one of the major barriers to enzymatic hydrolysis, and the removal of lignin can increase the accessibility of cellulose enzymes to cellulose (Taniguchi et al. 2005; Yu et al. 2009a). Pre-treatment is an essential step for biological conversion of lignocellulose into biofuels. Fungi produce enzymes that can degrade hemicellulose, lignin, and polyaromatic phenols. White-rot and soft-rot fungi have been used to successfully degrade lignocellulosic materials (Sun and Cheng 2002; Anderson and Akin 2008). Filamentous fungi have attracted a lot of research interest due to its ability to produce a variety of lignocellulolytic enzymes and ferment both hexoses and pentoses to ethanol. The fungi of ascomycetes are a potential biocatalyst for conversion of organic wastes to value-added products. They have the ability to synthesize enzymes for degradation of lignocellulosic materials for the production of organic acids, ethanol, and enzymes (Ferreira et al. 2016).

The conversion of triglycerides (vegetable oil and animal fats) into biodiesel has been a promising method to supply an alternative fuel for diesel engines (Fukuda et al. 2008). The enzymatic transesterification of triglycerides into biodiesel is another application of whole-cell biocatalysts. As vegetable oils and animal fats have high viscosity, acid composition, and free fatty acid contents, it is impractical to directly use them in diesel engines (Fukuda et al. 2008). Alkali-catalyzed transesterification can achieve high conversion rates of triglycerides to their corresponding alkyl esters within a short reaction time. However, it has several drawbacks such as high energy consumption and difficulty in glycerol recovery and alkaline removal from the product and a need for the treatment of highly alkaline wastewater generated. The enzymatic production of biodiesel has substantial advantages over conventional alkali-catalyzed transesterification processes such as easy recovery of glycerol byproduct. However, although the lipase immobilized on acrylic resins provides high conversion rates and capacity to recycle the biocatalyst, the high production cost of lipase enzyme makes the process very costly (Fukuda et al. 2008). The use of whole cell instead is a promising methodology for biodiesel production (Aguieiras et al. 2015).

This chapter discusses the potential applications of fungal whole-cell biocatalysts for pre-treatment of lignocellulose, bioethanol fermentation, and biodiesel production. It also examines various development strategies employed to improve the performance of fungal whole-cell biocatalysts.

2.2 Fungi as Biocatalysts

2.2.1 Bioreactors for the Cultivation of Fungi

Cellulase is the enzyme that plays a key role in hydrolyzing cellulose that is a dominant component in plant cell wall (Singhania et al. 2010). Cellulase can be produce by a wide range of microorganisms such as aerobic and anaerobic bacteria, anaerobic fungi, soft-rot fungi, white-rot fungi (WRF), and brown-rot fungi (BRF) by means of bacterial and fungal fermentation. Two strains of soft-rot fungi *Trichoderma reesei* and *Aspergillus niger* produce cellulase via submerged fermentation. However, submerged fermentation leads to a low concentration of end products, which needs further purification. The additional downstream step increases the production cost of cellulase. Due to the short coming, solid-state fermentation (SSF) has gained its importance in cultivation of fungi (Couto and Sanromán 2005).

Cellulase production by white-rot fungi and brown-rot fungi has been carried out on a lab-scale batch process using solid-state fermentation (SSF) (Raghavarao et al. 2003). When the fungi are cultivated by SSF, effective diffusion of nutrients and oxygen is the key factor to ensure optimal fungal growth. As the fermentation progresses, the depletion of nutrients occurs, and heat is generated due to the metabolic activity of fungus (Durand 2003). A bioreactor with appropriate designs will enable in improving the cellulase production by overcoming the heat and mass transfer problems that become more prevalent in large-scale production. Tray bioreactor, packed bed bioreactor, rotary drum bioreactor, and fluidized bed bioreactor are the most commonly used for SSF (Mitchell et al. 2006).

Tray bioreactor is the most widely employed in lab, pilot, and even industrial scales. Tray bioreactors are simple to operate. There is no need to agitate the substrate in a tray bioreactor, which can reduce the energy consumption. The reaction temperature of a tray bioreactor can be maintained by incubating the trays in a temperature-controlled chamber. Tray bioreactors are widely used for the selection of fungal strains and the optimization of fermentation conditions. However, tray bioreactors require a large space and huge manpower to be used in industrial production plants. Furthermore, the temperature gradient may occur in a tray bioreactor due to poor heat transfer (Chen et al. 2005; Brijwani et al. 2011).

A packed bed bioreactor is suitable for cultivation of fungi that are sensitive to shear force. It has a simple design and does not require any mechanical agitation. It is also flexible for modification to include a water jacket or cooling plates. There may be a pressure drop and temperature gradient across the substrate bed in a packed bed bioreactor, which can be overcome by proper mixing of the substrate across the bed. Mixing the substrate bed intermittently might reduce the mass transfer problem (Couto and Sanromán 2006). Evaporation cooling can be used to enhance heat transfer in a packed bed bioreactor (Durand 2003; Mitchell et al. 2006; Chen 2013).

A rotary drum bioreactor is also commonly used to cultivate fungi. The temperature of a rotary drum bioreactor is usually controlled by an external water jacket. The substrate particle in the bioreactor can be exposed to air that in turn would enhance the heat transfer. For the continuous rotation speed of more than 10% of the critical speed, higher power is required. As the aggregation of substrate bed due to rotation may also lead to fungal growth inhibition, glass beads may be embedded in the substrate bed to decrease substrate agglomeration. Intermittent agitation can also be adopted to minimize the effect of shear force on fungal growth (Chen 2013).

A fluidized bed bioreactor provides complete homogeneity of the substrate bed. It is simple to add water, nutrient, and a pH buffer solution to the substrate bed in a fluidized bed bioreactor. The heat exchange between solid particles and the surroundings can be easily facilitated. A fluidized bed bioreactor also has some limitations: the reactor might impose shear damage to fungal mycelia; and moist lignocellulosic substrate particles might aggregate to form larger particles in the reactor (Shrestha et al. 2008). Some substrates could be sticky to form particles agglomerate that prevents effective fluidization and fungal penetration into the substrate. The fungal morphology might also experience some changes during fluidization as mixing can damage the fungal mycelium. The efficiency of fluidization might change as the structure and size of lignocellulosic substrate particles changes as fermentation progress (Chen 2013).

2.2.2 Enzymes and Whole-Cell Biocatalysts

Enzymes as biocatalysts have been used for producing ethanol from lignocellulosic biomass and biodiesel from vegetable oil and animal fats. Recent developments in functional genomics and molecular biology have led to the discovery of many new enzymes and various approaches for the enhancement of enzymatic performance

(Fernandez-Arrojo et al. 2010). One of the examples was the improvement of the thermostability of chimeric fungal family 6 cellobiohydrolase (HJPlus) using random mutagenesis and recombination, which could produce more thermostable laccases and cellobiohydrolase. However, the production costs of enzymes such as lipases and cellulase are usually high because of the requirement of expensive purification. Many studies have been focused on immobilizing enzymes onto a solid structure to increase the stability and recycle the enzymes (Wu and Arnold 2013). In comparison with the isolated enzymes, the production of whole-cell biocatalysts is inexpensive and fast. Enzymes in whole-cell applications are well protected from external environment and are generally more stable than free enzymes. The recent advances in the life sciences have increased the availability of whole-cell biocatalysts. The recombinant DNA technique has enabled the overproduction of a desired enzyme and modification of metabolic pathways in various hosts such as microorganism, which enables the use of whole-cell biocatalysts (Ishige et al. 2005).

2.2.3 Immobilization of Whole-Cell Biocatalysts

Whole-cell biocatalysts are frequently immobilized onto porous biomass support particles (BSPs) and membranes for their reuse. Research has particularly been focused on improving the stability of membrane-bound enzymes as they are critical in industrial biocatalysis for the production of fuels and chemicals. Ban et al. (2002) observed that glutaraldehyde cross-linking of membrane bound lipase showed high lipase activity. The high activity of the cross-linked whole-cell biocatalysts could be maintained over several cycles (>70 yield) while the lipase activity in the cells that did not receive the glutaraldehyde treatment gradually deceased (Ban et al. 2002). Oda et al. (2005) reported that the durability of the whole-cell biocatalysts depended on cultivation methods. The *Rhizopus oryzae* cells immobilized in BSPs cultivated in an air-lift bioreactor showed higher conversion efficiency than those obtained from a shake flask. Furthermore, their conversion rate in an air-lift bioreactor could be maintained over many repeated cycles. These findings indicate that the use of whole-cell biocatalysts immobilized within BSPs offers a promising means for the industrial production of biodiesel as it provides a simplified and cost-effective approach to supply lipase that exhibits high enzymatic activity over a long period (Fukuda et al. 2008). Although the immobilization technique has been widely used, sometimes biocatalysts are directly used without immobilization to eliminate the time and expense of immobilization (Shiraga et al. 2005). Furthermore, Hama et al. (2004) studied the lipase activity of whole-cell biocatalyst of *R. oryzae* and found that culturing in suspension facilitated the enzymes to be secreted out while the immobilized fungal mycelium retained much of its enzymes within its cells.

2.2.4 Directed Evolution of Whole-Cell Biocatalysts

A number of approaches have been applied toward modifying the activity and physiology of enzymes using a method called directed evolution (DE). DE is similar to a natural selection process to evolve proteins or nucleic acids but toward a user-defined goal (Lutz 2010). DE consists of repeated cycles of (a) using random mutagenesis and/or DNA recombination to generate a diversity of a target gene sequence and (b) using high-throughput screening or selection to identify the desired protein variants (Cobb et al. 2012). DE experiments have four main steps (Denard et al. 2015):

1. Choice of parent protein.
2. Create a mutant library based on the parent protein.
3. Identify variants with improved target properties.
4. Repeat the entire process until the desired function is achieved or until no further improvement is possible.

DE has become an important tool for improving traits of biocatalysts such as thermostability, activity, selectivity, and tolerance to organic solvents for industrial applications. Asial et al. (2013) developed a new screening technique to improve protein thermostability. It was hypothesized that the unfolding and aggregation quality of the recombinant protein above a critical temperature would be similar to purified proteins. This technique allowed physical separation of folded and aggregated proteins after incubation at a high temperature where only thermostable variant was able to diffuse through (Asial et al. 2013).

DE represents the Darwinian evolution process in the test tube (Cobb et al. 2012). The use of in silico models to engineer microorganisms enables in utilizing new substrates to produce biofuels more efficiently. The ability to use a wide variety of biomass as feedstocks would decrease the cost of biofuel production. The enzymes of cellulases that exist in many microorganisms are responsible for the degradation of the cellulose in lignocellulosic biomass that is the most abundant biopolymer on earth. Both glucose and xylose can be produced from lignocellulosic biomass. *S. cerevisiae* was engineered with the genes encoding xylose reductase and xylitol dehydrogenase from a fungus *Pichia stipitis* to enable utilization of xylose. However, the overexpression of the genes led to a slow growth and low fermentation rate due to the redox imbalance. Further modifications in the genes led to an increase in ethanol production and a reduction in byproduct generation (Johannes and Zhao 2006). Cellulase produced by thermophilic fungi is promising at optimizing industrial processes, such as biomass degradation and biofuel production. Thermophilic cellulases are vital enzymes for efficient biomass degradation. Two endochitinase genes from *Melanocarpus albomyces* when transformed to fungus *Trichoderma reesei* increased the cellulase activity several times higher than that of the parental *M. albomyces* strain (Haakana et al. 2004). DE of *T. reesei* was reported to augment its stability and specific activity. The clone that harbored a total of six mutation demonstrated a broad range of pH stability (4.4–8.0) and also an increase in the thermotolerance as compared to the wild type (100% activity at 55 °C for 30 min) (Nakazawa et al. 2009).

In vitro DE was used to harness the power of natural selection to obtain desirable proteins that are not found in nature. The wild-type cbh2 gene of a thermophilic fungus *Chaetomium thermophilum* which encodes for cellobiohydrolase II (CBHII) was mutagenized. Cellobiohydrolase (CBH) is a cellulase that hydrolyses the 1,4-β-D-glycosidic bonds in cellulose and are of two types: *CBHI* that cleaves gradually from the reducing end and *CBHII* that cleaves progressively from the nonreducing end of cellulose (Divne et al. 1998). The mutant library of the gene was established in the yeast of *Pichia pastoris* and was screened for enhanced CBHII activities. After purification from the yeast, the selected mutant CBHII proteins were analyzed for their physical and chemical properties. The mutant CBHII showed an increase in proteins activities, optimum reaction temperature, thermal stability, and optimum reaction pH compared to the wild-type CBHII. When kept at 80 °C for 1 h, the wild-type lost all activity, whereas the mutant gene retained more than 50% of their activities (Wang et al. 2012a). The advantage of DE is that the enzyme characteristics can be improved even with limited knowledge of protein structure and accurate predictions on the active site or the binding pocket (Leisola and Turunen 2007).

2.3 Fungal Pre-treatment of Lignocellulosic Biomass

2.3.1 Fungal Whole-Cell Biocatalysts for Delignification

There is an increasing interest in producing biofuels such as alcohols from lignocellulosic biomass. Pre-treatment is an important step in the process of conversion of lignocellulosic biomass into fermentable sugars and then biofuels. Delignification is essential to reduce the biomass recalcitrance prior to the enzymatic hydrolysis (Mosier et al. 2005). Wood-decay fungi can be classified according to the type of decay caused such as brown-rot, soft-rot, and white-rot. Table 2.1 summarizes the different types of wood-rotting fungi.

Fungal pre-treatment using lignin-degrading microorganism has been used an alternative to thermal/chemical pre-treatment for biofuel production due to the economic and environmental benefits (Mosier et al. 2005). Microorganisms such as white-rot fungi, brown-rot fungi, and soft-rot as shown in Table 2.2 can degrade lignocellulosic biomass.

White-rot fungi have a unique ligninolytic system which makes them the most efficient for delignification. White-rot fungi have high selectivity for the degradation of lignin over cellulose, which can reduce cellulose loss to make fungal pre-treatment practical for biofuel production (Wan and Li 2012). White-rot fungi secrete oxidative enzymes such as lignin peroxidase (Forti et al. 2015), manganese peroxidase (MnP), and laccase which are responsible for the degradation of lignin (Wan and Li 2012). White-rot fungi such as *Phanerochaete chrysosporium, Phlebia*

Table 2.1 Characteristics of wood attack by several types of fungi

	White-rot	Brown-rot	Soft-rot
Decay attributes	Whitened appearance, soft, moist, loss in strength loss after advanced decay	Brown, dry, brittle consistency, drastic strength loss at initial stage of decay	Soft consistency in wet environments, brown and crumbly in dry environments
Host	Hardwood and softwood	Softwood, seldom hardwoods	Hardwood and minor degradation in softwood
Degradation of cell wall constituent	Cellulose, hemicellulose and lignin	Cellulose, hemicellulose, mild modification of lignin	Cellulose, hemicellulose, and mild modification of lignin
Anatomical features	Attack of cell wall gradually from lumen, lignin degradation in middle lamella and secondary wall	Degradation at a great distance from hyphae (diffusion mechanism), entire cell wall rapidly attacked	Cell wall attacked in the proximity of hyphae from cell lumen
Agents	*Basidiomycetes* (e.g., *T. versicolor, Irpex lacteus, P. chrysosporium*, and *Heterobasidium annosum*), *Ascomycetes* (e.g., *Xylaria hypoxylon*), and *Basidiomycetes* (e.g., *Ganoderma australe, Phlebia tremellosa, C. subvermispora, Pleurotus* spp., and *Phellinus pini*)	*Basidiomycetes* exclusively (e.g., *C. puteana, Gloeophyllum trabeum, Laetiporus sulphureus, Piptoporus betulinus, Postia placenta*, and *Serpula lacrymans*)	*Ascomycetes* (*Chaetomium globosum, Ustulina deusta*), *Deuteromycetes* (*Alternaria alternata, Thielavia terrestris, Paecilomyces* spp.), some white (*Inonotus hispidus*) and brown-rot (*Rigidoporus crocatus*), *Basidiomycetes* cause facultative soft-rot decay

Source: Martinez et al. (2005)

radiate, Dichomitus squalens, Rigidoporus lignosus, and *Jungua separabilima* have been used to successfully delignify wood and wheat straw (Itoh et al. 2003). The delignification is typically fungi and feedstock specific. The *P. ostreatus* was found to have higher effectiveness for the delignification of straw materials than other fungi (Taniguchi et al. 2005). Wheat straw pre-treated with *P. ostreatus* yielded 27–33% cellulose digestibility during 72-h enzymatic hydrolysis with cellulase loading of about 10 FPU/g solid (Taniguchi et al. 2005). The fungi, however, was less selective in lignin and cellulose degradation when pre-treatment time was extended to several weeks. Therefore, the digestibility of cellulose stabilized during the later stage of cultivation (Taniguchi et al. 2005; Yu et al. 2009b). Keller et al. 2003 reported that the 29-day pre-treatment of corn stover with *P. chrysosporium* only modestly increased saccharification yield compared to the control but the cellulose loss was as high as 40.49% and lignin degradation was in the range of 19.38–35.53%. Longer fungal pre-treatment time has been more fruitful in the digestibility of woody biomass such as aspen and birch. Softwood required a longer pre-treatment as compared to hardwood (Yu et al. 2009a). Chinese willow pre-treated with *E. taxodii* 2538 for 120 d exhibits 37% conversion of the polysaccharides when enzymatic hydrolysis was conducted at a cellulase loading of 20 FPU/g solid (Zhang

Table 2.2 Effect of fungal pre-treatment on lignocellulosic biomass

Fungus	Type of decay	Class	Substrate	Sugar/ethanol yield
Phanerochaete chrysosporium	White rot	*Agaricomycetes*	Cotton stalk	Reduced glucose yield
Phanerochaete chrysosporium	White rot	*Agaricomycetes*	Rice straw	50% glucose yield 50% ethanol yield
Phanerochaete chrysosporium	White rot	*Agaricomycetes*	Beechwood	9.5% total sugar yield
Phanerochaete chrysosporium	White rot	*Agaricomycetes*	Corn fiber	Reduced sugar yield or no significant yield
Pleurotus ostreatus	White rot	*Agaricomycetes*	Rice straw	33% glucose yield
Pleurotus ostreatus	White rot	*Agaricomycetes*	Rice hull	38.9% glucose yield
Pleurotus ostreatus, Pycnoporus cinnabarinus 115	White rot	*Agaricomycetes*	Wheat straw	27–28% glucose yield
Irpex lacteus	White-rot	*Hymenomycetes*	Corn stover	66.4% total sugar yield
Ceriporiopsis subvermispora	White-rot	*Agaricomycetes*	Corn stover	56–66% glucose yield and 57.8% ethanol yield
Stereum hirsutum		*Agaricomycetes*	Japanese red pine	13.7% glucose yield
Echinodontium taxodii 2538	White-rot	*Agaricomycetes*	Chinese willow, China fir	5–35% glucose yield
Echinodontium taxodii 2538	White-rot	*Agaricomycetes*	Bamboo culm	37% total sugar yield
Cyathus stercoreus	Brown-rot	*Agaricomycetes*	Corn stover	36% glucose yield

Source: Wan and Li (2012)

et al. 2007). Complete decontamination of feedstocks may not be necessary for fungal pre-treatment since white-rot fungi can survive in contamination and actively act on degradation. Long pre-treatment time is a major and common barrier for the application of fungal pre-treatment. Combining fungal pre-treatment concurrently with on-farm wet storage is a promising option to solve the long pre-treatment time issue. Another option is to apply fungal pre-treatment prior to physical or thermo-chemical pre-treatment. As short-term fungal pre-treatment can modify the cell walls before evident degradation takes place, the required level of thermochemical pre-treatment can be substantially reduced (Wan and Li 2012).

Brown-rot fungi pre-treatment technologies would offer tangible energy and cost benefits to the whole biofuel process. Brown-rot fungi causes rapid and extensive depolymerization of the holocellulose (both cellulose and hemicellulose) components of lignocellulosic biomass with only minimal changes to the lignin (Ray et al. 2010). The hyphae of the brown-rot fungi grow in the lumina of the plant cell that leads to disruption of the carbohydrate polymers at some distance from the site of contact. Reports have suggested that brown-rot fungi secrete oxalic and other organic acids which reduce the pH of lignocellulose and depolymerize hemicellulose and cellulose by acid-catalyzed hydrolysis and thus further increases the porosity of the plant cell wall (Kaneko et al. 2005). Brown-rot fungi principally colonize softwoods, and only 6% of *Basidiomycotina* are brown-rot fungi (Tuomela et al. 2000). The pre-treatment of lignocellulosic biomass using brown-rot fungi offers advantages by significantly reducing the energy and material requirements for the process of bioethanol production from lignocellulosic biomass. Brown-rot fungal species, such as *Coniophora puteana* and *Postia placenta*, could significantly enhance the yield of glucose after enzymatic saccharification of pine without the generation of any fermentation inhibitors. This method could potentially substitute some thermomechanical pre-treatments (Ray et al. 2010). Brown-rot fungi mineralize methoxyl groups of lignin, but the mineralization of other parts is lower. Brown-rot fungi are members of the *Basidiomycota*, including common species such as *Schizophyllum commune*, *Fomes fomentarius*, and *Serpula lacrymans*. The brown-rot fungi derive the name from the color of the decayed wood, because most of the cellulose and hemicelluloses are degraded, leaving the lignin more or less intact which is brown (Deacon 2005).

Soft-rot fungi belong to the subdivision of *Ascomycotina* or *Deuteromycotina* which degrade both hardwood and softwood. However, the degradation rate is low as compared to white-rot and brown-rot fungi. Soft-rot fungi can degrade wood in a wet environment and decompose plant litter in soil (Tuomela et al. 2000). Soft-rot fungi can degrade wood under extreme environmental conditions (high or low water potential) where the activity of other fungi is limited. Soft-rot fungi can withstand a wide range of humidity, temperature, and pH conditions and can attack a variety of wood substrates. These fungi are most commonly found in soils, but can also appear in other environments. Some soft-rot fungi produce cracks within the secondary walls, while some may cause complete erosion of the secondary wall leaving a relatively intact middle lamella. Soft-rot is more commonly observed in hardwood than in softwood because of the difference of lignin in the hardwood and softwood (Hamed 2013). Soft-rot fungi secrete cellulase from their hyphae, an enzyme that breaks down cellulose in the wood. The soft-rot fungi belong to Phylum *Ascomycota*, and species *Chaetomium* and *Ceratocystis* are found in terrestrial environments and species of *Lulworthia*, *Halosphaeria*, and *Pleospora* in marine and estuarine environments (Deacon 2005).

Many groups of viable microorganisms have demonstrated the ability to degrade lignocellulosic biomass for biofuel production. However, the lignin-degrading fungi are particularly adept at degrading specific polymers within lignocellulose. Brown-rot fungi produces enzymes that only target and degrade cellulose and hemicellulose with negligible degradation of lignin. Soft-rot fungi specifically degrade plant polysaccharides. White rot-fungi has gained significant consideration as they can degrade all three parts of lignocellulose and typically degrade lignin before cellulose (Martinez et al. 2008). White rot-fungi have gained increasing attention for liquid biofuel production as the degradation of lignin is an essential factor for cost-effective operations (Martinez et al. 2008; Kumar et al. 2009). Fungal pre-treatment prior to mild physical and chemical pre-treatment has shown synergism on the improvement of cellulose digestibility with advantages similar to that of the bio-pulping process (Wan and Li 2012). Compared to current leading thermal or chemical pre-treatment processes, pre-treatment using fungal whole-cell biocatalyst is an environmental-friendly and energy-efficient process. However, fungal pre-treatment is usually less effective than thermochemical pre-treatment (Yu et al. 2009a).

2.3.2   Fungal Whole-Cell Biocatalysts for Enzymatic Hydrolysis of Lignocellulosic Biomass

Cellulase is required to breakdown cellulose to fermentable sugars for the subsequent fermentation to produce alcohols such as ethanol and butanol. The high cost of cellulases is a major challenge for the production of lignocellulosic ethanol (Merino and Cherry 2007). Wen et al. (Wen et al. 2009) proposed two protein engineering strategies to overcome these challenges (a) improving cellulases properties and (b) optimizing the enzyme cocktail for maximized synergy. Cellulose hydrolysis involved three major types of cellulases of exoglucanases, endoglucanases (EG), and β-glucosidases (BGL). Engineering efforts are mainly focused on the endoglucanases (EG) and β-glucosidases (BGL), and their activities could be assayed in a high-throughput manner with the help of artificial substrates that were soluble or chromogenic. Carboxymethyl cellulose (CMC) was widely used as a model substrate for endoglucanases (Wang et al. 2012b). Chromogenic analogues of the natural substrate cellobiose were used for screening β-glucosidases by fluorescence-activated cell sorting (FACS) (Hardiman et al. 2010). Cellulase has been isolated from fungi ascomycetes and used in textile, paper and pulp, food and animal feed, fuel, chemical, waste management, and pharmaceutical industries (Jabasingh and Nachiyar 2011). The industrial production of cellulases is presently dominated by *Trichoderma* sp. and *Aspergillus* sp. (Pandey et al. 2015). For optimal hydrolysis of lignocellulosic materials, a synergistic action of a collection of complementary cellulases is necessary. The ratios and combinations of cellulases greatly affect the hydrolysis efficiency. Substrates from different sources or with different pre-treatment methods require distinct enzyme cocktails (Merino and Cherry 2007). Recombinant *Saccharomyces cerevisiae* strains incorporating three cellulase genes, BGL, EG, and CBH, using δ-integration method with one marker gene showed that the ratio between the cellulases was more important than the total copies integrated. This suggests the underlying importance of synergy engineering to reduce the amount and cost of enzymes needed for cellulose hydrolysis (Yamada et al. 2010).

Directed evolution (DE) was used to improve the cellulase properties. Moderate success was achieved due to the difficulties in developing high-throughput screening methods on activities toward the insoluble cellulosic substrates (Zhang et al. 2006). The most challenging problem associated with DE experiments on cellulases is the use of artificial substrates to facilitate the high-throughput assay. The positive hits from the screening using artificial substrates may not show higher activity on natural substrates (Chakiath et al. 2009). High-throughput screening on natural substrates is more desirable than artificial substrates but is more difficult to be carried out (Zhang et al. 2006). Dinitrosalicylic acid (DNS) assay has been used to monitor xylanase-mediated hydrolysis of wheat straw and has been successful in the screening of mutated endo-b-1,4-xylanase library from *Thermobacillus xylanilyticus* with 74% increase in hydrolytic activity (Song et al. 2010). Filamentous ascomycetes can grow on both untreated and treated lignocellulosic materials. The ascomycetes can also be used within the concept of consolidated bioprocessing (CBP) where the fungus is responsible for the hydrolysis of the substrate and conversion of the simple sugars to value-added products (Salehi Jouzani and Taherzadeh 2015).

2.4 Ethanol Production Using Fungal Whole-Cell Biocatalysts

Fuel ethanol, or ethyl alcohol, is a volatile, flammable, and colorless liquid that can be produced from a mixture of sugars hydrolyzed from lignocellulosic biomass. The current two major ethanol producers are the USA and Brazil with an annual production of more than 15.5 billion gallons and 7295 million gallons, respectively (EIA 2017; RFA 2017). Employing robust microorganisms with higher conversion rates and yields would simplify process steps and reduce capital and operating costs. Current production methods include widely used solid-state fermentation (SSF) or submerged fermentation (SmF). SSF which involves the growth of microorganisms on moist solid substrates similar to their natural habitat in the absence of free-flowing water has gained more importance (Barrington et al. 2009). Substrate degradation will be catalyzed by free extracellular enzymes or enzymes organized in cellosomes produced by the growing microorganism. SSF is influenced by the strain of microorganism, the substrate, and numerous process parameters such as carbon and nutrient composition, moisture content, particle sizes, incubation temperature, pH, and inoculum density (Bari et al. 2009; Pandey et al. 2015). SSF offers several advantages such as lower energy requirements, lower operation costs and less wastewater produced (Pandey 2001), and higher productivity (Sreedharan 2016).

SmF is a method which utilizes enzymes and other reactive compounds submerged in a free-flowing liquid substrate such as oil, nutrient broth, or alcohol to produce biomolecules. It can be operated in a continuous mode. As a continuous process, substrates are consumed rapidly and need to be replaced/supplemented with nutrients. The SmF technique is best suited for microorganisms that require a high moisture content. Another advantage of the SmF technique is that it is easier to purify products than SSF. SmF is primarily used in the extraction of secondary metabolites that need to be used in a liquid form (Subramaniyam and Vimala 2012). SmF has a wider range of biotechnological applications due to its better heat and mass transfer and culture homogeneity, which renders it to be more reliable,

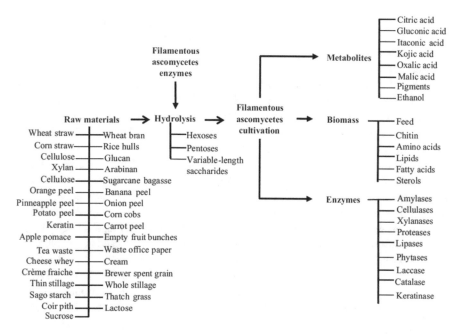

Fig. 2.1 Potential substrates and value-added products in filamentous ascomycetes based waste biorefineries (Ferreira et al. 2016)

reproducible, flexible, and easier to monitor (Fazenda et al. 2008). However, SmF has some disadvantages compared to SSF, which include high cost of equipment and high electricity consumption, high sensitivity to interruptions in aeration, and high susceptibility to contamination that would lead to yield loss and even complete failure (Betiku et al. 2016). Filamentous fungi are generally used for their wide-ranging metabolic activities and secondary metabolites. Ascomycetes are commonly used for ethanol production, but they can also be used to produce various value-added products such enzymes, organic acids, and biomass from waste materials. The range of possible substrates and value-added products in filamentous ascomycetes-based "waste biorefineries" is presented in Fig. 2.1 (Ferreira et al. 2016).

However, the morphology of the filamentous fungi poses a constraint during a biological process because the long filaments can get entangled into clumps or pellets (Gibbs et al. 2000). When fungi grow as pellets, the medium is less viscous due to the lower impact of the pellets in the bulk medium. Another issue arises when the fungal mycelium growth becomes dense and the medium becomes more viscous, which causes resistance in oxygen and other mass transfer. The growth of fungi as pellets is mostly ideal for biotechnological applications. However, pellets can suffer shear stress which depends on their sizes (Gibbs et al. 2000). The most advantageous fungal growth morphology for enhanced production of citric acid by *Aspergillus niger* is the pellet form (Dhillon et al. 2013). Filamentous fungi have advantage over unicellular microorganisms such as bacteria and yeasts due to their extracellular enzymatic system coupled with hyphal penetration (Sharma and Arora

2015). *Aspergillus* spp. and *Fusarium* spp. have been successfully used to hydrolyze lignocellulosic biomass, and yeast could convert the released sugars to ethanol (Salehi Jouzani and Taherzadeh 2015).

Recent research has been carried out using *Fusarium* spp. and *Neurospora* spp. for ethanol production (Ferreira et al. 2016). Bacteria can also be used for the production of ethanol or organic acids from lignocellulosic biomass, but the capacity of producing the required concentrations of enzymes for complete saccharification of pre-treated substrates is still limited to fungi (Amore and Faraco 2012). Baker's yeast which is the most widely used biocatalyst for industrial production of ethanol is incapable of fermenting pentose sugars (Ferreira et al. 2015). Among the genus of *Fusarium*, *F. oxysporum* is the most widely used for ethanol production, but few studies have investigated *F. verticillioides*, *F. equiseti*, and *F. acuminatum* for ethanol production by consolidated bioprocessing (Xiros and Christakopoulos 2009). Ethanol production is influenced by the source of nitrogen and alcohol dehydrogenase activity where higher values were reported when urea was used instead of yeast extract with *F. equiseti* and *F. acuminatum* (Anasontzis et al. 2011). *Neurospora crassa* is the most widely used *Neurospora* genus for ethanol production from sweet sorghum bagasse (Dogaris et al. 2012). *N. crassa* was also used to produce ethanol from alkali treated brewer spent grain with a theoretical yield of 36% (Dogaris et al. 2012). The intracellular enzyme activities of xylose reductase, xylitol dehydrogenase, and xylulokinase were found to be dependent on oxygen concentration which affected xylose uptake (Zhang et al. 2008). The xylose fermenting capability of *N. crassa* has a direct effect on ethanol yield. Pre-treating the substrate such as wheat bran increased the fermentation and ethanol yield by up to 51% (Nair et al. 2015).

2.5 Biodiesel Production Using Fungal Whole-Cell Biocatalysts

2.5.1 Enzymatic Biodiesel Production

Biodiesel is an environmentally friendly alternative fuel for diesel engines (Fukuda et al. 2008). Current industrial biodiesel production uses homogeneous alkaline transesterification of edible oils or animal fat. This process, however, has several disadvantages such as difficulty in catalyst recovery, a large amount of highly alkaline wastewater produced, and required high-quality and high-price raw materials with low contents of free fatty acids (FFAs) (acidity less than 0.5%) and water to avoid soap formation. The refined oils are comparatively expensive and account for 70–80% of overall biodiesel production costs which makes biodiesel economically incompetent with petro-diesel (Aguieiras et al. 2015). The use of biocatalysts (enzymes) for biodiesel production is a promising substitute to overcome these problems (Villeneuve et al. 2000; Ranganathan et al. 2008). An enzymatic biodiesel production plant has several economic advantages over the chemical process plant:

low-quality oils with high moisture content can be utilized; the process requires little to no heat; it can use simpler and cheaper raw materials pre-treatment and post-reaction catalyst separation methods; no salts are produced; and the by-product glycerol can be sold at a higher value (Hass et al. 2002; Fukuda et al. 2008; Aguieiras et al. 2015).

Lipases are enzymes that catalyze the hydrolysis of esters, particularly long-chain triglycerides (TAGs) to produce free fatty acids, di- and monoglycerides, and glycerol. In a nonaqueous medium, lipases catalyze esterification, acidolysis, aminolysis, alcoholysis, and interesterification reactions with high selectivity and under mild conditions which make them the most employed groups of enzymes in biotechnological applications (Villeneuve et al. 2000; Ranganathan et al. 2008). Lipases used in biodiesel production are mainly obtained from fungi (yeasts and filamentous fungi) and are commercially available in the market. There are four main forms of biocatalysts used in biodiesel synthesis: free lipases, immobilized lipases, whole cells, and fermented solids enzymatic preparation (SEP) (Aguieiras et al. 2015). Many of these enzymes are commercially available in immobilized forms. Lipase B is the most widely used lipase in biodiesel synthesis which is obtained from the yeast *Candida antarctica* (CALB) (Aarthy et al. 2014). Other yeast lipases including *Candida rugosa* (CRL), *C. antarctica* A (CALA), and *Candida* sp. 99–125 have been immobilized on textile and used for biodiesel production in China (Aguieiras et al. 2015). Lipases obtained from filamentous fungi, such as *Rhizomucor miehei* lipase (RML) (Rodrigues and Fernandez-Lafuente 2010), *Rhizopus oryzae* lipase (ROL) (Canet et al. 2014), *Thermomyces lanuginosus* lipase (TLL) (Fernandez-Lafuente 2010), *Aspergillus niger* lipase (ANL) (Contesini et al. 2010), and *Penicillium expansum* lipase (PEL) (Li and Zong 2010), have also been studied.

Free lipases are usually in liquid formulations that contain stabilizers to inhibit the denaturation of enzymes (e.g., glycerol, sorbitol) and antimicrobials (e.g., benzoate) or in powder forms that are obtained by freeze-drying (Kaieda et al. 2001; Nielsen et al. 2008). Free enzymes are comparatively cheaper than immobilized enzymes. However, in many cases, they have demonstrated poor operational stability and could be used only once as they are inactivated (Yan et al. 2014). A few reports are available on soluble lipase-catalyzed biodiesel production (Zhao et al. 2007; Chen et al. 2008), and the methanolysis of low-cost crude soybean oil using Callera Trans L (lipase from *T. lanuginosus)* (Cesarini et al. 2013).

Lipases are immobilized on solid supports to improve their catalytic properties. Immobilized biocatalysts offer several advantages. The enzymes are distributed on the surface of the support which increases the number of available active sites. Enzymes offer structural rigidity by forming enzyme-support bonds which increases thermal stability and resistance to solvents (Aguieiras et al. 2015). The recovery of immobilized lipases is relatively simple and can be recovered by filtration or centrifugation which makes it reusable or carry out continuous processes. High-purity glycerol by-product can also be easily recovered without requiring any complex separation process (Tan et al. 2010; Yan et al. 2014). The immobilized fungal lipases most widely used for biodiesel production are Novozym 435, Lipozyme RM IM, and Lipozyme TL IM (Aguieiras et al. 2015).

Adsorption is the most common immobilization method used due to its ease, no requirement of toxic and expensive reagents, and ease of biocatalyst regeneration (Al-Zuhair 2007). Immobilization surface materials include acrylic resins, macro- and microporous resins, silica gels, hydrotalcite, Celite, zeolite, porous kaolinite, diatomaceous earth, and textile membranes. Hydrophobic membranes are preferred to eliminate nonspecific binding of glycerol onto the membrane (Séverac et al. 2011; Garcia-Galan et al. 2014). Lipases immobilized on textile membranes have shown to increase fatty acid methyl ester (FAME) yield by about 80% (Aguieiras et al. 2015). Immobilization of *Trichosporon asahii* lipase on basic alumina yielded similar FAME production (Kumari and Gupta 2012). Kuo et al. (2013) reported that the immobilization of *C. rugosa* lipase onto polyvinylidene fluoride membrane in methanolysis of soybean oil obtained a FAME yield of 97% and offered good reusability for sustained periods. Lipase from *T. lanuginosus* immobilized by covalent attachment onto polyglutaraldehyde-activated styrene-divinylbenzene copolymer yielded a 97% conversion in methanolysis of canola oil (Dizge et al. 2009).

2.5.2   Fungal Whole-Cell Biocatalysts for Biodiesel Production

The main obstacle for a commercially feasible enzymatic production of biodiesel fuels is the high cost of lipase enzyme. Despite the advantages offered by immobilized lipases, these biocatalysts are expensive since their preparation steps of enzyme extraction and immobilization are costly (Bajaj et al. 2010). Furthermore, the immobilized biocatalysts have mass transfer limitation due to the molecular sizes of substrates and products and the effect of alcohol in contact (Fjerbaek et al. 2009). In order to reduce enzyme-associated process costs, immobilization of fungal mycelium within biomass support particles (BSPs) and expression of the lipase enzyme on the surface of yeast cells have been developed to generate whole-cell biocatalysts for industrial applications (Fukuda et al. 2008). A relatively new technology of using the entire microorganism containing the enzymes as whole-cell biocatalysts offers advantages in production cost and simplification of operation (Yan et al. 2014). Three different forms of whole-cell catalysts have been used in biodiesel production, which are presented in Table 2.3.

The first type consists of filamentous fungi, mainly *Rhizopus* sp. and *Aspergillus* sp., which produce cell-bound or intracellular lipases. Filamentous fungi are immobilized on biomass support particles (BSPs) or glutaraldehyde to improve the stability and activity of the biocatalyst, enable the separation of the whole-cell biocatalyst, and make its reuse to be easier (Aguieiras et al. 2015). Ban et al. (2001) was the first to use fungal whole cell for biodiesel production. The mycelium of *R. oryzae* was immobilized in biomass support particles (BSPs) which were made of polyurethane foam and methanolysis of soybean oil was investigated. Hatzinikolaou et al. (1999) used cell-bound lipase produced by the yeast *Rhodotorula glutinis* to study the

Table 2.3 Use of different whole-cell biocatalysts for biodiesel fuel production

Whole-cell biocatalyst	Oil	Alcohol	Solvent	Methyl ester (%)	Time (h)	Temp (°C)
BSPs with *R. oryzae*	Soybean	Methanol	None	80–90	72	32
BSPs with *R. oryzae*	Soybean	Methanol	None	90	48	35
BSPs with *R. oryzae*	Soybean	Methanol	t-butanol	72	NA	35
BSPs with *R. oryzae*	Jatropha	Methanol	None	89	60	30
BSPs with *R. oryzae*	Rapeseed (refined)	Methanol	t-butanol	60	24	35
BSPs with *R. oryzae*	Rapeseed (crude)	Methanol	t-butanol	60	24	35
BSPs with *R. oryzae*	Rapeseed (acidified)	Methanol	t-butanol	70	24	35
Mycelium of *R. chinensis*	Soybean	Methanol	None	86	72	NA
S. cerevisiae (intracellular ROL)	Soybean	Methanol	None	71	165	37
S. cerevisiae (cell surface ROL)	Soybean	Methanol	None	78	72	37

Source: Fukuda (2008)

esterification of palmitic and oleic acid with butanol and obtained about 60% of conversion. Srimhan et al. (2011) reported the use of strain of *Rhodotorula muci-laginosa* whole cell for transesterification of refined palm oil and methanol, and a FAME yield of 83.3% was achieved in 72 h. The lipase-producing whole cells of *Rhizopus oryzae* (ROL) were immobilized onto biomass support particles (BSPs) and were used for the production of biodiesel from relatively low-cost nonedible Jatropha oil. The activity of ROL was compared with a commercially available lipase (Novozym 435). The presence of water had a significant effect on the rate of methanolysis and the whole-cell ROL performed better than Novozym 435 whose activity was severely inhibited by the presence of added water. The results suggested that whole-cell ROL immobilized on BSP was a promising biocatalyst for producing biodiesel from oil as the expensive downstream processing steps for biodiesel production from Jatropha oil can be avoided with whole-cell biocatalysts (Tamalampudi et al. 2008).

Second approach of surface display technology is the use of yeast strains that naturally produce cell-bound lipases and could be cultivated to a high density in a low-cost medium (Aarthy et al. 2014). The enzymes are immobilized on the surface of yeast cells which allows the expression of target proteins or peptides on the microbial cell surface by fusing an appropriate protein as an anchoring motif (Huang et al. 2012; Liu et al. 2014). The expression level and enzymatic activity of the anchored lipases are usually low but with the advances in genetic engineering, it is very likely that more lipases will be displayed on yeast cell surfaces with high expression levels and activity in the near future (Liu et al. 2014; Yan et al. 2014). Four structural units are crucial in cell-surface display development, i.e., the target enzyme, anchor protein, linker sequence, and the host microbial cell. Yeasts such as *Pichia pastoris, Yarrowia lipolytica*, and *Saccharomyces cerevisiae* are used as host for the surface display technique due to the simplicity in their genetic manipulation and posttranslational modification of expressed heterologous proteins (Liu et al. 2014).

The third strategy of preparing lipase whole-cell biocatalysts for biodiesel production is the use of recombinant DNA techniques to produce recombinant *Aspergillus oryzae* strains, *Escherichia coli* cells, or yeast cells (Aguieiras et al. 2015). Yan et al. (2012) reported that recombinant *E. coli* cells with intracellular co-expression of CALB and TLL showed five times more reusability and maintained 75% of its original productivity. Intracellular lipases have limited substrate accessibility due to the shielding effect of cell walls and cell membranes, and hence techniques like air-drying and combination of isopropyl alcohol treatment and freeze-drying have been employed to improve the permeability of the cells (Ban et al. 2001; Yan et al. 2012).

Another type of biocatalyst used for biodiesel production is the solid enzymatic preparation (SEP), which is obtained by solid-state fermentation (SSF). In SSF, fungal cells are immobilized on particles of the solid fermentation medium, and lipases are emitted from the cells which remain adhered in fermented matrix. This approach can reduce the production cost of the biocatalysts significantly. Agro-industrial residues can be used as substrates for cell growth in SSF. Also, dry SEP obtained after fermentation process avoids extraction, purification, and enzyme immobilization steps which reduce the costs of the biocatalyst considerably (Esvelt et al. 2011). Aguieiras et al. (2014) reported that the use of dry SEP with the lipase emitted from *R. miehei* as a biocatalyst for biodiesel production achieved a conversion of FFAs to ester of 91% in 8 h.

2.6   Challenges and Perspectives of Fungal Whole-Cell Biocatalysts in Biofuel Production

Biocatalysis has made significant strides in the production of biofuels, bulk chemicals, and even pharmaceuticals owing to their high efficiency, high selectivity, and green footprint. One of the challenges associated with using fungal whole-cell biocatalysts for biofuel production is their relatively slow reaction rates, which leads to the increase in cost and maintenance. A possible solution to overcome this limitation is to use packed bed reactors that contain whole-cell biocatalysts, operated under continuous operation conditions.

Directed evolution (DE) has enabled the engineering and improvement of known properties of enzymes and led to enhanced production of desired products. The creation of mutant library and high-throughput screening techniques has greatly facilitated to the engineering of novel and improved biocatalysts (Wang et al. 2012b). DE can be further developed by focusing on designing, generating, and screening of smart (sequence-space restricting) libraries, which can significantly reduce the library size despite it relies on current data of target proteins. Another strategy for developing DE is the use of ultrahigh-throughput screening methods based on drop-based microfluidics and FACS. Novel ultrahigh-throughput methods enable us to perform DE of a target protein with limited knowledge in more efficient way, which will in turn provides more insights of the target protein and "hot spots" for designing smart libraries to further improve the desired property for industrial applications (Khalil and Collins 2010; Liang et al. 2011). Both these strategies complement with each other and can significantly enhance DE-based experiments.

A hurdle in the DE experiments is that it takes relatively long time to screen variants. Phage-assisted continuous evolution (PACE) system has been designed to overcome this issue. The system makes the use of a fixed-volume vessel that is continuously supplied with uninfected *Escherichia coli* cells along with infected cells. During the process, the uninfected *E. coli* cells are continuously removed. A phage population encoding the target gene is also grown in the same vessel. The infectivity of the phage is directly linked to the function of the target gene and hence is also dependent on the improved function of the target gene. Therefore, one replication/infection cycle of the phage will be equal to one cycle of traditional mutation/ selection that takes place without any human intervention. PACE can complete 200 rounds of protein evolution over the course of 8 days, and this method has been used to identify the T7 RNA polymerase variant that could recognize the T3 promoter which is a novel DNA sequence (Esvelt et al. 2011).

The combination of rational design and DE could help create breakthrough in the development of completely novel enzymes. DE can be used as a critical tool to provide tailor-made individual enzymes for single-step transformations and multienzyme pathways for multistep catalysis which will lead to development of high-functioning cell factories for synthesizing biofuels genomes (Khalil and Collins 2010; Liang et al. 2011).A key step in lignocellulose-based biorefineries is employing cellulosic biomass pre-treatment with naturally robust microorganisms. The successful conversion of monomeric sugars present in lignocellulosic materials relies on a pre-treatment step in the absence of a potentially potent microorganism which eventually increases the overall process costs. Thus, a feasible lignocellulose-based industrial process should represent a true biorefinery producing as many as possible different value-added products. Currently biological treatment of lignocellulosic waste material is carried out by solid-state fermentation (SSF). However, SSF is restricted by its limitations in scaling up due to space requirements and other physical constraints such as sufficient aeration and dissipation of heat and insufficient organic loading. New bioreactor designs for SSF that circumvents these limitations like using submerged fermentation can significantly boost research in the industry (Ferreira et al. 2016).

2.7   Conclusions

Whole-cell biocatalysis can provide several advantages over conventional processes by providing an inexpensive method of producing catalysts and excellent operational stability for the production of fuel ethanol, biodiesel, and a variety of bioproducts. Fungal whole cells or enzymes bound to whole-cell membrane have significantly reduced the costs of purification and operation required for enzymatic bioprocesses while also increasing the reusability of biocatalysts. Recent development in molecular biology, functional genomics, and structural biology has made it possible to modify enzyme-catalyzed reaction rate and activity, and genetic engineering of the fungal cells and enzymes via directed evolution has significantly increased the yields of biofuels and bioproducts while at the same time overcoming many limitations of biological processes. Future research on the use of whole-cell biocatalysts can lead to a more sustainable production of biofuels, biodiesel, and other value-added bioproducts.

Acknowledgments This publication was made possible by Grant Number NC.X2013-38821-21141 and NC.X-294-5-15-130-1 from the US Department of Agriculture (USDA-NIFA). Its contents are solely the responsibility of the authors and do not necessarily represent the official views of the National Institute of Food and Agriculture.

References

Aarthy M, Saravanan P, Gowthaman MK, Rose C, Kamini NR (2014) Enzymatic transesterification for production of biodiesel using yeast lipases: an overview. Chem Eng Res Des 92(8):1591–1601

Aguieiras ECG, Cavalcanti-Oliveira ED, de Castro AM, Langone MAP, Freire DMG (2014) Biodiesel production from Acrocomia aculeata acid oil by (enzyme/enzyme) hydroesterification process: Use of vegetable lipase and fermented solid as low-cost biocatalysts. Fuel 135:315–321

Aguieiras ECG, Cavalcanti-Oliveira ED, Freire DMG (2015) Current status and new developments of biodiesel production using fungal lipases. Fuel 159:52–67

Al-Zuhair S (2007) Production of biodiesel: possibilities and challenges. Biofuels Bioprod Biorefin 1(1):57–66

Amore A, Faraco V (2012) Potential of fungi as category I consolidated BioProcessing organisms for cellulosic ethanol production. Renew Sust Energ Rev 16(5):3286–3301

Anasontzis GE, Zerva A, Stathopoulou PM, Haralampidis K, Diallinas G, Karagouni AD, Hatzinikolaou DG (2011) Homologous overexpression of xylanase in Fusarium oxysporum increases ethanol productivity during consolidated bioprocessing (CBP) of lignocellulosics. J Biotechnol 152(1):16–23

Anderson WF, Akin DE (2008) Structural and chemical properties of grass lignocelluloses related to conversion for biofuels. J Ind Microbiol Biotechnol 35(5):355–366

Asial I, Cheng YX, Engman H, Dollhopf M, Wu B, Nordlund P, Cornvik T (2013) Engineering protein thermostability using a generic activity-independent biophysical screen inside the cell. Nat Commun 4:2901

Bajaj A, Lohan P, Jha PN, Mehrotra R (2010) Biodiesel production through lipase catalyzed transesterification: an overview. J Mol Catal B Enzym 62(1):9–14

Ban K, Kaieda M, Matsumoto T, Kondo A, Fukuda H (2001) Whole cell biocatalyst for biodiesel fuel production utilizing Rhizopus oryzae cells immobilized within biomass support particles. Biochem Eng J 8(1):39–43

Ban K, Hama S, Nishizuka K, Kaieda M, Matsumoto T, Kondo A, Noda H, Fukuda H (2002) Repeated use of whole-cell biocatalysts immobilized within biomass support particles for biodiesel fuel production. J Mol Catal B Enzym 17(3–5):157–165

Bari MN, Alam MZ, Muyibi SA, Jamal P (2009) Improvement of production of citric acid from oil palm empty fruit bunches: optimization of media by statistical experimental designs. Bioresour Technol 100(12):3113–3120

Barrington S, Kim JS, Wang L, Kim J-W (2009) Optimization of citric acid production by Aspergillus niger NRRL 567 grown in a column bioreactor. Korean J Chem Eng 26(2):422–427

Betiku E, Emeko HA, Solomon BO (2016) Fermentation parameter optimization of microbial oxalic acid production from cashew apple juice. Heliyon 2(2):e00082

Bornscheuer UT, Huisman GW, Kazlauskas RJ, Lutz S, Moore JC, Robins K (2012) Engineering the third wave of biocatalysis. Nature 485(7397):185–194

Brijwani K, Vadlani PV, Hohn KL, Maier DE (2011) Experimental and theoretical analysis of a novel deep-bed solid-state bioreactor for cellulolytic enzymes production. Biochem Eng J 58:110–123

Canet A, Dolors Benaiges M, Valero F (2014) Biodiesel synthesis in a solvent-free system by recombinant Rhizopus oryzae lipase. Study of the catalytic reaction progress. J Am Oil Chem Soc 91(9):1499–1506

Cesarini S, Diaz P, Nielsen PM (2013) Exploring a new, soluble lipase for FAMEs production in water-containing systems using crude soybean oil as a feedstock. Process Biochem 48(3):484–487

Chakiath C, Lyons MJ, Kozak RE, Laufer CS (2009) Thermal stabilization of Erwinia chrysanthemi pectin Methylesterase a for application in a sugar beet pulp biorefinery. Appl Environ Microbiol 75(23):7343

Chen H (2013) Modern solid state fermentation. Springer, Netherlands

Chen H-Z, Xu J, Li Z-H (2005) Temperature control at different bed depths in a novel solid-state fermentation system with two dynamic changes of air. Biochem Eng J 23(2):117–122

Chen X, Du W, Liu D (2008) Effect of several factors on soluble lipase-mediated biodiesel preparation in the biphasic aqueous-oil systems. World J Microbiol Biotechnol 24(10):2097–2102

Cobb RE, Si T, Zhao H (2012) Directed evolution: an evolving and enabling synthetic biology tool. Curr Opin Chem Biol 16(3–4):285–291

Contesini FJ, Lopes DB, Macedo GA, Nascimento M d G, Carvalho P d O (2010) Aspergillus sp. lipase: potential biocatalyst for industrial use. J Mol Catal B Enzym 67(3–4):163–171

Couto SR, Sanromán MA (2005) Application of solid-state fermentation to ligninolytic enzyme production. Biochem Eng J 22(3):211–219

Couto SR, Sanromán MA (2006) Application of solid-state fermentation to food industry—a review. J Food Eng 76(3):291–302

Deacon J (2005) Fungal Biology. Blackwell Publishing, Oxford

Denard CA, Ren H, Zhao H (2015) Improving and repurposing biocatalysts via directed evolution. Curr Opin Chem Biol 25:55–64

Dhillon GS, Brar SK, Kaur S, Verma M (2013) Rheological studies during submerged citric acid fermentation by Aspergillus Niger in stirred fermentor using apple pomace ultrafiltration sludge. Food Bioprocess Technol 6(5):1240–1250

Divne C, Ståhlberg J, Teeri TT, Jones TA (1998) High-resolution crystal structures reveal how a cellulose chain is bound in the 50 Å long tunnel of cellobiohydrolase I from Trichoderma reesei11Edited by K. Nagai. J Mol Biol 275(2):309–325

Dizge N, Keskinler B, Tanriseven A (2009) Biodiesel production from canola oil by using lipase immobilized onto hydrophobic microporous styrene–divinylbenzene copolymer. Biochem Eng J 44(2–3):220–225

Dogaris I, Gkounta O, Mamma D, Kekos D (2012) Bioconversion of dilute-acid pretreated sorghum bagasse to ethanol by Neurospora crassa. Appl Microbiol Biotechnol 95(2):541–550

Durand A (2003) Bioreactor designs for solid state fermentation. Biochem Eng J 13(2):113–125

EIA, U. S. E. I. A (2017) EIA U.S. Energy Information Administration, Online: https://www.eia.gov/

Esvelt KM, Carlson JC, Liu DR (2011) A system for the continuous directed evolution of biomolecules. Nature 472(7344):499–503

Fazenda ML, Seviour R, McNeil B, Harvey LM (2008) Submerged culture fermentation of "higher fungi": the macrofungi. Adv Appl Microbiol 63:33–103

Fernandez-Arrojo L, Guazzaroni ME, Lopez-Cortes N, Beloqui A, Ferrer M (2010) Metagenomic era for biocatalyst identification. Curr Opin Biotechnol 21(6):725–733

Fernandez-Lafuente R (2010) Lipase from Thermomyces lanuginosus: uses and prospects as an industrial biocatalyst. J Mol Catal B Enzym 62(3–4):197–212

Ferreira JA, Lennartsson PR, Taherzadeh MJ (2015) Production of ethanol and biomass from thin stillage by Neurospora intermedia: a pilot study for process diversification. Eng Life Sci 15(8):751–759

Ferreira JA, Mahboubi A, Lennartsson PR, Taherzadeh MJ (2016) Waste biorefineries using filamentous ascomycetes fungi: present status and future prospects. Bioresour Technol 215:334–345

Fjerbaek L, Christensen KV, Norddahl B (2009) A review of the current state of biodiesel production using enzymatic transesterification. Biotechnol Bioeng 102(5):1298–1315

Forti L, Di Mauro S, Cramarossa MR, Filippucci S, Turchetti B, Buzzini P (2015) Non-conventional yeasts whole cells as efficient biocatalysts for the production of flavors and fragrances. Molecules 20(6):10377–10398

Fukuda H, Hama S, Tamalampudi S, Noda H (2008) Whole-cell biocatalysts for biodiesel fuel production. Trends Biotechnol 26(12):668–673

Garcia-Galan C, Barbosa O, Hernandez K, Santos CJ, Rodrigues CR, Fernandez-Lafuente R (2014) Evaluation of styrene-divinylbenzene beads as a support to immobilize lipases. Molecules 19(6):7629

Gibbs PA, Seviour RJ, Schmid F (2000) Growth of filamentous fungi in submerged culture: problems and possible solutions. Crit Rev Biotechnol 20(1):17–48

Haakana H, Miettinen-Oinonen A, Joutsjoki V, Mäntylä A, Suominen P, Vehmaanperä J (2004) Cloning of cellulase genes from Melanocarpus albomyces and their efficient expression in Trichoderma reesei. Enzym Microb Technol 34(2):159–167

Haas M, Foglia T, Piazza G (2002) Enzymatic approaches to the production of biodiesel fuels. Lipid Biotechnology, CRC Press

Hama S, Yamaji H, Kaieda M, Oda M, Kondo A, Fukuda H (2004) Effect of fatty acid membrane composition on whole-cell biocatalysts for biodiesel-fuel production. Biochem Eng J 21(2):155–160

Hamed SAM (2013) In-vitro studies on wood degradation in soil by soft-rot fungi: Aspergillus niger and Penicillium chrysogenum. Int Biodeter Biodegr 78(Supplement C):98–102

Hardiman E, Gibbs M, Reeves R, Bergquist P (2010) Directed evolution of a thermophilic [beta]-glucosidase for cellulosic bioethanol production. Appl Biochem Biotechnol 161(1–8):301–312

Hatzinikolaou DG, Kourentzi E, Stamatis H, Christakopoulos P, Kolisis FN, Kekos D, Macris BJ (1999) A novel lipolytic activity of Rhodotorula glutinis cells: Production, partial characterization and application in the synthesis of esters. J Biosci Bioeng 88(1):53–56

Huang D, Han S, Han Z, Lin Y (2012) Biodiesel production catalyzed by Rhizomucor miehei lipase-displaying Pichia pastoris whole cells in an isooctane system. Biochem Eng J 63:10–14

Ishige T, Honda K, Shimizu S (2005) Whole organism biocatalysis. Curr Opin Chem Biol 9(2):174–180

Itoh H, Wada M, Honda Y, Kuwahara M, Watanabe T (2003) Bioorganosolve pretreatments for simultaneous saccharification and fermentation of beech wood by ethanolysis and white rot fungi. J Biotechnol 103(3):273–280

Jabasingh SA, Nachiyar CV (2011) Utilization of pretreated bagasse for the sustainable bioproduction of cellulase by Aspergillus nidulans MTCC344 using response surface methodology. Ind Crop Prod 34(3):1564–1571

Johannes TW, Zhao H (2006) Directed evolution of enzymes and biosynthetic pathways. Curr Opin Microbiol 9(3):261–267

Kaieda M, Samukawa T, Kondo A, Fukuda H (2001) Effect of methanol and water contents on production of biodiesel fuel from plant oil catalyzed by various lipases in a solvent-free system. J Biosci Bioeng 91(1):12–15

Kaneko S, Yoshitake K, Itakura S, Tanaka H, Enoki A (2005) Relationship between production of hydroxyl radicals and degradation of wood, crystalline cellulose, and a lignin-related compound or accumulation of oxalic acid in cultures of brown-rot fungi. J Wood Sci 51(3):262–269

Keller FA, Hamilton JE, Nguyen QA (2003) Microbial pretreatment of biomass. Appl Biochem Biotechnol 105(1–3):27–41

Khalil AS, Collins JJ (2010) Synthetic biology: applications come of age. Nat Rev Genet 11(5):367–379

Kumar P, Barrett DM, Delwiche MJ, Stroeve P (2009) Methods for pretreatment of lignocellulosic biomass for efficient hydrolysis and biofuel production. Ind Eng Chem Res 48(8):3713–3729

Kumari A, Gupta R (2012) Purification and biochemical characterization of a novel magnesium dependent lipase from Trichosporon asahii MSR 54 and its application in biodiesel production. Asian J Biotechnol 4(2):70–82

Kuo C-H, Peng L-T, Kan S-C, Liu Y-C, Shieh C-J (2013) Lipase-immobilized biocatalytic membranes for biodiesel production. Bioresour Technol 145:229–232

Leisola M, Turunen O (2007) Protein engineering: opportunities and challenges. Applied Microbiology & Biotechnology 75(6):1225–1232

Li N, Zong M-H (2010) Lipases from the genus penicillium: production, purification, characterization and applications. J Mol Catal B Enzym 66(1–2):43–54

Liang J, Luo Y, Zhao H (2011) Synthetic biology: putting synthesis into biology. Wiley Interdiscip Rev Syst Biol Med 3(1):7–20

Liu Y, Xiong Y, Huang W, Jia B (2014) Recurrent paratyphoid fever a co-infected with hepatitis a reactivated chronic hepatitis B. Ann Clin Microbiol Antimicrob 13(1):17

Lutz S (2010) Beyond directed evolution – semi-rational protein engineering and design. Curr Opin Biotechnol 21(6):734–743

Martinez D, Berka RM, Henrissat B, Saloheimo M, Arvas M, Baker SE, Chapman J, Chertkov O, Coutinho PM, Cullen D (2008) Genome sequencing and analysis of the biomass-degrading fungus Trichoderma reesei (syn. Hypocrea jecorina). Nat Biotechnol 26(5):553

Martinez AT, Speranza M, Ruiz-Dueñas FJ, Ferreira P, Camarero S, Guillén F, Martínez MJ, Gutiérrez A, del Río JC (2005) Biodegradation of lignocellulosics: microbial, chemical, and enzymatic aspects of the fungal attack of lignin. Intern Microbiol 8:195–204

Merino ST, Cherry J (2007) Progress and challenges in enzyme development for biomass utilization. Adv Biochem Eng Biotechnol 108:95–120

Mitchell DA, Krieger N, Beroviˇc M (2006) Solid-state fermentation bioreactors. Springer, Heidelberg, p 19

Mosier NS, Hendrickson R, Brewer M, Ho N, Sedlak M, Dreshel R, Welch G, Dien BS, Aden A, Ladisch MR (2005) Industrial scale-up of pH-controlled liquid hot water pretreatment of corn fiber for fuel ethanol production. Appl Biochem Biotechnol 125(2):77–97

Nair RB, Lundin M, Brandberg T, Lennartsson PR, Taherzadeh MJ (2015) Dilute phosphoric acid pretreatment of wheat bran for enzymatic hydrolysis and subsequent ethanol production by edible fungi Neurospora intermedia. Ind Crop Prod 69:314–323

Nakazawa H, Okada K, Onodera T, Ogasawara W, Okada H, Morikawa Y (2009) Directed evolution of endoglucanase III (Cel12A) from Trichoderma reesei. Appl Microbiol Biotechnol 83(4):649–657

Narita J, Okano K, Tateno T, Tanino T, Sewaki T, Sung M-h, Fukuda H, Kondo A (2006) Display of active enzymes on the cell surface of Escherichia coli using PgsA anchor protein and their application to bioconversion. Appl Microbiol Biotechnol 70(5):564–572

Nielsen PM, Brask J, Fjerbaek L (2008) Enzymatic biodiesel production: technical and economical considerations. Eur J Lipid Sci Technol 110(8):692–700

Oda M, Kaieda M, Hama S, Yamaji H, Kondo A, Izumoto E, Fukuda H (2005) Facilitatory effect of immobilized lipase-producing Rhizopus oryzae cells on acyl migration in biodiesel-fuel production. Biochem Eng J 23(1):45–51

Pandey A (2001) Solid-state fermentation in biotechnology: fundamentals and applications. Asiatech Publishers, New Delhi

Pandey A, Höfer R, Taherzadeh M, Nampoothiri M, Larroche C (2015) Industrial Biorefineries & White Biotechnology. Elsevier, Amsterdam

Raghavarao KSMS, Ranganathan TV, Karanth NG (2003) Some engineering aspects of solid-state fermentation. Biochem Eng J 13(2):127–135

Ranganathan SV, Narasimhan SL, Muthukumar K (2008) An overview of enzymatic production of biodiesel. Bioresour Technol 99(10):3975–3981

Ray MJ, Leak DJ, Spanu PD, Murphy RJ (2010) Brown rot fungal early stage decay mechanism as a biological pretreatment for softwood biomass in biofuel production. Biomass Bioenergy 34(8):1257–1262

RFA (2017) Renewable Fuels Association, Online: http://www.ethanolrfa.org.

Rodrigues RC, Fernandez-Lafuente R (2010) Lipase from Rhizomucor miehei as a biocatalyst in fats and oils modification. J Mol Catal B Enzym 66(1–2):15–32

Salehi Jouzani G, Taherzadeh MJ (2015) Advances in consolidated bioprocessing systems for bioethanol and butanol production from biomass: a comprehensive review. Biofuel Research Journal 2(1):152–195

Séverac E, Galy O, Turon F, Pantel CA, Condoret J-S, Monsan P, Marty A (2011) Selection of CalB immobilization method to be used in continuous oil transesterification: analysis of the economical impact. Enzym Microb Technol 48(1):61–70

Sharma RK, Arora DS (2015) Fungal degradation of lignocellulosic residues: an aspect of improved nutritive quality. Crit Rev Microbiol 41(1):52–60

Shiraga S, Kawakami M, Ishiguro M, Ueda M (2005) Enhanced reactivity of Rhizopus oryzae lipase displayed on yeast cell surfaces in organic solvents: potential as a whole-cell biocatalyst in organic solvents. Appl Environ Microbiol 71(8):4335–4338

Shrestha P, Rasmussen M, Khanal SK, Pometto AL 3rd, van Leeuwen JH (2008) Solid-substrate fermentation of corn fiber by Phanerochaete chrysosporium and subsequent fermentation of hydrolysate into ethanol. J Agric Food Chem 56(11):3918–3924

Singhania RR, Sukumaran RK, Patel AK, Larroche C, Pandey A (2010) Advancement and comparative profiles in the production technologies using solid-state and submerged fermentation for microbial cellulases. Enzym Microb Technol 46(7):541–549

Song L, Laguerre S, Dumon C, Bozonnet S, O'Donohue MJ (2010) A high-throughput screening system for the evaluation of biomass-hydrolyzing glycoside hydrolases. Bioresour Technol 101(21):8237–8243

Sreedharan (2016) An overview on fungal cellulases with an industrial perspective. J Nutr Food Sci 06(01)

Srimhan P, Kongnum K, Taweerodjanakarn S, Hongpattarakere T (2011) Selection of lipase producing yeasts for methanol-tolerant biocatalyst as whole cell application for palm-oil transesterification. Enzym Microb Technol 48(3):293–298

Subramaniyam R, Vimala R (2012) Solid state and submerged fermentation for the production of bioactive substances: a comparative study. Int J Sci Nat 3:480–486

Sun Y, Cheng J (2002) Hydrolysis of lignocellulosic materials for ethanol production: a review. Bioresour Technol 83(1):1–11

Tamalampudi S, Talukder MR, Hama S, Numata T, Kondo A, Fukuda H (2008) Enzymatic production of biodiesel from Jatropha oil: a comparative study of immobilized-whole cell and commercial lipases as a biocatalyst. Biochem Eng J 39(1):185–189

Tan KT, Gui MM, Lee KT, Mohamed AR (2010) An optimized study of methanol and ethanol in supercritical alcohol technology for biodiesel production. J Supercrit Fluids 53(1):82–87

Taniguchi M, Suzuki H, Watanabe D, Sakai K, Hoshino K, Tanaka T (2005) Evaluation of pretreatment with Pleurotus ostreatus for enzymatic hydrolysis of rice straw. J Biosci Bioeng 100(6):637–643

Tuomela M, Vikman M, Hatakka A, Itävaara M (2000) Biodegradation of lignin in a compost environment: a review. Bioresour Technol 72(2):169–183

Villeneuve P, Muderhwa JM, Graille J, Haas MJ (2000) Customizing lipases for biocatalysis: a survey of chemical, physical and molecular biological approaches. J Mol Catal B Enzym 9(4–6):113–148

Wan C, Li Y (2012) Fungal pretreatment of lignocellulosic biomass. Biotechnol Adv 30(6):1447–1457

Wang X-J, Peng Y-J, Zhang L-Q, Li A-N, Li D-C (2012a) Directed evolution and structural prediction of cellobiohydrolase II from the thermophilic fungus Chaetomium thermophilum. Appl Microbiol Biotechnol 95(6):1469–1478

Wang M, Si T, Zhao H (2012b) Biocatalyst development by directed evolution. Bioresour Technol 115C:117–125

Wen F, Nair NU, Zhao H (2009) Protein engineering in designing tailored enzymes and microorganisms for biofuels production. Curr Opin Biotechnol 20(4):412–419

Wu I, Arnold FH (2013) Engineered thermostable fungal Cel6A and Cel7A cellobiohydrolases hydrolyze cellulose efficiently at elevated temperatures. Biotechnol Bioeng 110(7):1874

Xiros C, Christakopoulos P (2009) Enhanced ethanol production from brewer's spent grain by a Fusarium oxysporum consolidated system. Biotechnol Biofuels 2(1):1

Yamada R, Taniguchi N, Tanaka T, Ogino C, Fukuda H, Kondo A (2010) Cocktail delta-integration: a novel method to construct cellulolytic enzyme expression ratio-optimized yeast strains. Microb Cell Factories 9:32

Yan J, Li A, Xu Y, Ngo TPN, Phua S, Li Z (2012) Efficient production of biodiesel from waste grease: one-pot esterification and transesterification with tandem lipases. Bioresour Technol 123:332–337

Yan Y, Li X, Wang G, Gui X, Li G, Su F, Wang X, Liu T (2014) Biotechnological preparation of biodiesel and its high-valued derivatives: a review. Appl Energy 113:1614–1631

Yu H, Guo G, Zhang X, Yan K, Xu C (2009a) The effect of biological pretreatment with the selective white-rot fungus Echinodontium taxodii on enzymatic hydrolysis of softwoods and hardwoods. Bioresour Technol 100(21):5170–5175

Yu J, Zhang J, He J, Liu Z, Yu Z (2009b) Combinations of mild physical or chemical pretreatment with biological pretreatment for enzymatic hydrolysis of rice hull. Bioresour Technol 100(2):903–908

Zhang Y-HP, Himmel ME, Mielenz JR (2006) Outlook for cellulase improvement: screening and selection strategies. Biotechnol Adv 24(5):452–481

Zhang X, Yu H, Huang H, Liu Y (2007) Evaluation of biological pretreatment with white rot fungi for the enzymatic hydrolysis of bamboo culms. Int Biodeter Biodegr 60(3):159–164

Zhang Z, Qu Y, Zhang X, Lin J (2008) Effects of oxygen limitation on xylose fermentation, intracellular metabolites, and key enzymes of Neurospora crassa AS3. 1602. Appl Biochem Biotechnol 145(1–3):39–51

Zhao X, El-Zahab B, Brosnahan R, Perry J, Wang P (2007) An organic soluble lipase for water-free synthesis of biodiesel. Appl Biochem Biotechnol 143(3):236–243

Chapter 3
White-Rot Fungal Xylanases for Applications in Pulp and Paper Industry

Shalini Singh

Abstract Pulp and paper industry plays an important part in the economy of any nation, but it is also considered to be one of the most polluted industries, as it is one of the biggest consumers of natural resources and at the same time, discharges a large amount of harmful waste into the environment especially, water bodies. Stringent regulations have, thus, forced the pulp and paper industry to opt for environment-friendly products and processes. Microbial enzymes have provided the solution to the problem with a large number of them being applied in various sections of pulp and paper processing. Xylanases are one of the best among them, with distinct applications in different sections of pulp and paper processing including, pulping, bleaching, and even effluent treatment. White-rot fungi are considered to be the best candidates for production of xylanases for industrial purposes, with very promising results observed with white-rot fungal xylanases being applied in pulp and paper industry. The chapter explains the current status of pulp and paper industries worldwide and emphasizes the role of environment-friendly technologies for the paper industry. The contribution of microorganisms and their powerful enzyme systems in various applications of paper production has been explained and compared with the conventional processes being used in the industry. The chapter specifically discusses the role of white-rot fungi and their xylanases in paper industry, with details of significant studies carried out in the same section. Potential of white-rot fungal xylanases in future, with major bottlenecks, has also been highlighted. Thus, it provides a major insight to the use of xylanases in pulp and paper industry.

S. Singh (✉)
School of Bioengineering and Biosciences, Lovely Professional University,
Phagwara, Punjab, India

© Springer International Publishing AG, part of Springer Nature 2018
S. Kumar et al. (eds.), *Fungal Biorefineries*, Fungal Biology,
https://doi.org/10.1007/978-3-319-90379-8_3

3.1 Introduction

Pulp and paper industry is considered as a vital and core industry. It has been considered to be a major consumer of natural resources (wood, water) and energy (fossil fuels, electricity) and a significant contributor of pollutant discharges to the environment (Thompson et al. 2001). Pulp and paper are made of cellulosic fibers as well as certain synthetic materials (for specialty papermaking). The raw materials mainly include wood fibers; rags; flax; cotton linters; agricultural residues like wheat straw, sugarcane bagasse, etc.; perennial grasses. Recycled paper is also produced while blending used paper with virgin fibers for desirable paper properties (Jiminez and Lopez 1993).

Pulp and paper industry is faced by multifarious problems, which significantly hinder the development of this prime industrial sector. It is one of the most important industries but, per capita consumption of paper in India, is globally one of the lowest (Sood 2007). Faulty demand projections, where factors like level of national income, level of industrial production, level of literacy and education, size of population, price of paper, etc. should be considered for making real-time demand projections, unfortunately do not receive due considerations (Schumacher and Sathaye 1999). These industries also face many other serious issues. These tough competition from imports, obsolete technology, soaring environmental cost, depreciation of money, and inadequate supply of low-cost fiber. The requirement to rapidly adapt to more efficient and environmentally compatible technologies is also threatening the very existence of a good number of paper mills, worldwide.

As indicated above, the scarcity of paper raw materials has grown in parallel with environmental pollution problems where pulping and bleaching processes cause serious contamination problems. Chlorine-based bleaching especially generates and releases a large number of adsorbable organic halogens in the environment. These are toxic and persistent in the environment and pose serious health risks too. The strict legislation and the environmental concerns regarding imposition of effluent discharge norms are, thus, forcing mills to look for alternative techniques that are cost efficient as well as environment compatible (Ali and Sreekrishnan 2001).

3.2 Pulp and Papermaking

The pulp and papermaking process is an elaborate, largely a chemical, and a mechanical process, where the major steps include (1) pulping, to separate cellulose fibers; (2) beating and refining the fibers, to develop desirable paper properties; (3) diluting the fiber suspension, to form a thin slurry; (4) forming a web of fibers on a thin screen; (5) pressing the web, to increase the density of the material; (6) drying, to remove the remaining moisture; and (7) finishing, to provide suitable surface for required end use (Nissan 1981). Broadly, papermaking is broadly divided into three sections which are pulping, bleaching, and papermaking.

3.3 Conventional Processing in Pulp and Paper Industry

The raw material, rich in lignocelluloses, is first prepared for pulping. This mainly includes removal of pith and debarking, if wood is the raw material used. The raw material is then subjected to pulping using strong alkalis.

In the process of paper production, pulping is a step in which pulp fibers are separated by chemical means and most of the lignin that is associated with the middle lamella is removed. The residual lignin is then removed by a multistage bleaching process (Bajpai 1999). Chemical pulping, which is frequently used around the world, is mainly achieved by sulfate (kraft) pulping process. Non-woods are generally delignified by soda or soda-AQ pulping. These processes remove most of the lignin and dissolve and partially degrade hemicelluloses. Residual lignin is covalently bound to carbohydrate moieties in the pulp (Yamasaki et al. 1981). As the cook progresses, xylan precipitates onto the surfaces of the cellulosic fibers (Gierer and Wännström 1984; Yamasaki et al. 1981), trapping degradation products in the matrix. These degradation products impart a characteristic brown color to the pulp.

Pulping is followed by pulp washing with water to separate pulping liquor (black liquor), containing cooking chemicals and dissolved raw materials, from the pulp. This operation is used to separate pulp fibers from the black liquor.

The residual lignin is then removed by a multistage bleaching process (Bajpai 1999). Usually, one or more bleaching sequences are needed to remove the dark brown color caused by the deposition of lignin. These conventionally include chlorine and chlorine-containing compounds. This step is required to bleach the pulp fibers, and the bleaching requirements vary with the type of paper to be produced. Usually, one or more bleaching sequences are needed to remove the dark brown color caused by the deposition of lignin. Most of the Indian pulp and paper industries bleach the pulp by conventional bleaching sequence where chlorination stage has traditionally been considered as the "workhorse" of post-pulping processing steps. Chlorination followed by extraction stage can selectively remove 75–90% of the lignin remaining on the fiber after pulping. Further, hypochlorite or chlorine dioxide is used to oxidize rest of the lignin. Since the pulp produced corresponds to only approximately 40–45% of the original weight of the wood, the effluents are heavily loaded with organic matter (Peck and Daley 1994). The major modified bleaching sequences include elemental chlorine-free (ECF) bleaching, where oxygen delignification followed by chlorine dioxide and other chemical agents achieves brightness, and total chlorine-free (TCF) bleaching, where combination of oxygen delignification with ozone/peroxide brightening leads to TCF bleaching.

Further, stock preparation, which involves additions of filters, retention aids, sizing agents, and various other additives, required for paper development, are added to the pulp slurry, which is then pumped into paper machine for production of finished paper.

Biological Processes in Pulp and Paper Industry The possibilities for employing biotechnology in the pulp and paper industries are numerous as it has the potential to increase the quality and supply of feedstocks for pulp and paper, reduce manufacturing costs, and create novel high-value products. Novel enzyme technologies can reduce environmental problems and alter fiber properties (Jeffries 1992; William and Jeffries 2003). Consequently, the use of microorganisms and their enzymes to replace or supplement older chemical methods in the pulp and paper industry is gaining utmost interest. Enzymes are usually very specific as to which reactions they catalyze and the substrates that are involved in these reactions (Jaeger and Eggert 2004).

A number of biological processes involving microorganisms, as well as their enzymes, are being explored nowadays for enhancing the quality and environment compatibility of these processes. The major applications of enzymes in paper industry include the following:

1. Enzymes for wood debarking: Polygalacturonases in wood barking (Rättö et al. 1992) are used to hydrolyze cambial layer,
2. Biopulping: Pretreatment of wood/cellulose fiber with microbial cells especially fungal cultures can modify lignin and extractive content. The microbial processes are less energy intensive and provide better quality of paper than conventional chemical pulping. Paper strength is especially improved by biopulping. This is mainly due to selective nature of enzymes to specifically remove lignin during pulping The major enzymes include ligninases (Johnsrud et al. 1987; Arbeloa et al. 1992).
3. Retention in papermaking: Pectinases are used for hydrolyzing pectins present in cationic demand. This helps in reduction of cationic demand in peroxide brightened mechanical pulps (Thorton et al. 1992).
4. Control of extractives: Lipases and ligninases are used for control of extractives as lipases hydrolyze triglycerides and esters present in extractives. Fungi have been predominantly used for such purposes. Also, laccases (ligninases) are used for extractive control as they can oxidize fatty acids and glycerides present in these extractives (Blanchette et al. 1992; Xia et al. 1996).
5. Stickies control: Esterases can be used for control of stickies (Patrick 2005), as they are able to hydrolyze polyvinyl acetate (a major constituent of stickies), to less sticky polyvinyl alcohol.
6. Control of boil-outs and slime formation: Amylases, in conjunction with lipases and proteases are very effective for boil-outs and slime control, in comparison to the conventional caustic treatment. These enzymes efficiently control the slime forming bacterial growth in paper machine systems (Paice and Zhang 2005)
7. Control of starch viscosity: Amylases are commonly employed for targeting starch viscosity in tissue mills (Paice and Zhang 2005) with promising results.
8. Reduction in usage of strong acids: Catalases are being used for deactivation of residual peroxide after mechanical pulp bleaching, which helps to avoid the use of strong acids for the same purpose (Thompson et al. 2003).

9. Deinking: Cellulases are extensively being used for deinking, which is increasing the recycling potential of waste paper worldwide (Xia et al. 1996).
10. Effluent treatment: A large number of microbial processes, viz., trickling filters, oxygen-activated sludge, activated sludge, rotating biological contactors, anaerobic fluidized beds, etc., are being used for treatment of waste water of pulp and paper industry. The powerful ligninases (Ali and Sreekrishnan 2001) has an important role to play in using microbial systems for effluent treatment.
11. Pulp drainage improvement: Cellulases and xylanases are being explored for improvement of pulp drainage on pulp or paper machine (Paice and Zhang 2005).
12. Pulp bleaching: Hemicellulases (especially xylanases) and ligninases (especially laccases) have been extensively used for pulp bleaching nowadays (Bajpai and Bajpai 1992; Viikari et al. 1994a, b; Jimenez et al. 1997).

3.4 Xylanases for Paper Industry

Xylanases (endo-1,4-β-xylanase) have greatly attracted the attention of researchers worldwide (Prade 1995) for their immense biotechnological potential in various industrial processes like in manufacturing of bread, food, and drinks, in the improvement of nutritional properties of agricultural silage and grain feed, in the textile industry to process plant fibers, in pharmaceutical and chemical applications, and in cellulose pulp and paper (Kuhad and Singh 1993).

Xylan, the substrate of xylanases, constitutes the major component of hemicelluloses (Shallom and Shoham 2003), which together with cellulose and lignin make up the major polymeric constituents of plant cell walls (Kulkarni et al. 1999). It is a complex, highly branched heteropolysaccharide and varies in structure between different plant species. The homopolymeric backbone chain of 1,4-linked β-D-xylopyranosyl units can be substituted to varying degrees with glucuronopyranosyl, 4-O-methyl-D-glucuronopyranosyl, α-L-arabinofuranosyl, acetyl, feruloyl, and/or p-coumaroyl side-chain groups (Chanda et al. 1950; Eda et al. 1976; Kulkarni et al. 1999). Due to its heterogeneity and complexity, the complete hydrolysis of xylan requires a cocktail of cooperatively acting enzymes (Subramaniyan and Prema 2002).

Xylanases are inducible enzymes, with rare examples of constitutive xylanase expression (Srivastava and Srivastava 1993). Xylanolytic enzymes are generally induced by xylan, xylobiose, and xylose and also by lignocellulosic residues that contain xylan (Flores et al. 1996). The low-molecular-mass fragments of xylan (including xylose, xylobiose, xylooligosaccharides, heterodisaccharides of xylose and glucose, and their positional isomers) that are liberated from xylan by the action of small amounts of constitutively produced enzyme play a key role in the regulation of xylanase biosynthesis (Kulkarni et al. 1999). Though xylanases have been found to be produced by a variety of sources, including microorganisms, plant seeds, snails, and crustaceans, fungi and bacteria are considered to be the best sources. Xylanases from bacteria and actinomycetes usually exhibit higher stability to a broader pH (5–9) and temperature range (35–60 °C) (Beg et al. 2000; Mandal 2015; Motta et al. 2013), while fungal xylanases are usually more effective at a pH range of 4–6 and temperature below 50 °C (Mandal 2015). However, fungi exhibit higher xylanase activity than bacteria or yeast (Mandal 2015; Motta et al. 2013) and show high yields and extracellular release of the enzymes. Also, fungal xylanases have been linked without cellulase activity (Subramaniyan and Prema 2002).

Bacteria like *Bacillus* spp., *Pseudomonas* spp., and *Streptomyces* spp. are efficient sources of xylanases (Mandal 2015; Motta et al. 2013). On the other hand, *Aspergillus* sp., *Fusarium* sp., and *Penicillium* sp. are effective fungal xylanase producers (Mandal 2015). Xylan-induced xylanases have been found to be produced by *Trametes trogii* (Levin and Forschiassin 1998), *Aspergillus awamori* (Siedenberg et al. 1998), and *Streptomyces* sp. (Beg et al. 2000) where xylan is found to be a strong inducer. *Cellulomonas flavigena*, an important xylanases source, in contrast, exhibits poor induction by xylan (Avalos et al. 1996). L-Sorbose in medium induces the xylanase production in *Sclerotium rolfsii* (Sachslehner et al. 1998) and *Trichoderma reesei* (Xu et al. 1998). In *Bacillus* sp. (Lopez et al. 1998) and *Trichosporon cutaneum* (Liu et al. 1998), xylanase is induced by xylose. Several reports have shown xylanase induction by lignocelluloses (Gupta et al. 2001; Kuhad et al. 1998; Puchart et al. 1999). The enhancement of xylanase production in the presence of amino acids has also been shown in *Bacillus* sp. (Ikura and Horikoshi 1987), *Bacillus* sp. (Balakrishnan et al. 1997), *Trametes trogii* (Levin and Forschiassin 1998), *Staphylococcus* sp. (Gupta et al. 1999), and *Streptomyces* sp. (Beg et al. 2000). Synthetic compounds like calcium-containing zeolite (CaA) have also been found to enhance xylanase production up as reported for *Bacillus* sp. Xylanase induction in *Trichosporon cutaneum* (Hrmova et al. 1984) by positional isomers has been reported.

White-Rot Fungal Xylanases for Paper Industry The wood-rot fungi are the most potent producers of xylanases as well as other extracellular polysaccharide-degrading enzymes since they excrete the enzymes into the medium and their enzyme levels are much higher than those of yeast and bacteria (Qinnghe et al. 2004). The most prominent among wood-rotters are white-rot fungi, as they are able to degrade all the components of the wood cell wall including the highly recalcitrant polymer, lignin (Kuhad and Singh 1993). They efficiently degrade even the aromatic polymer lignin, causing a characteristic white appearance on degraded wood (Hatakka and Hammel 2011). They have strong hemicellulolytic as well as significant cellulolytic enzymes, which facilitate colonization of various plants. White-rot fungi exist primarily as branching hyphae, usually 1 to 2 μ in diameter. Originating from spores or from nearby colonies, hyphae rapidly invade wood cells and lie along the lumen walls. From that vantage, they secrete the battery of enzymes and metabolites that bring about the depolymerization of the hemicelluloses and cellulose and fragmentation of the lignin (Kirk and Cullen 1998). The polymer-fragmenting enzymes and required metabolites are secreted into a fungal polysaccharide matrix (a β-1-3 glucan) that extends from the hyphae onto the lumen wall surface, sometimes over the entire inner surface (Blanchette 1991). This matrix is thought to direct the enzymes to the site of action – the exposed surfaces of the wall polymer composite (Flournoy et al. 1993). White-rot fungi have a large set of putative hemicellulose-degrading enzymes, having multiple copies of genes from GH families 10, 11, 43, and 74 that are xylan/xyloglucan related. Also, multiple copies of genes from carbohydrate esterase (CE) families 1 (xylan related) are also present in white-rot fungi, thus making them one of the most significant producers of xylanases (Rytioja et al. 2014).

The genetic incompatibility systems found that white-rot fungi are highly variable homothallic species. Instead, most of the white-rot fungi possess a heterothallic incompatibility system. Interestingly, one of the most important white-rot fungi used in industry, *Phanerochaete chrysosporium* Burds, is reported to be homothallic as well as heterothallic.

Pleurotus ostreatus is found to preferentially degrade lignin instead of polysaccharides, and *P. chrysosporium* has been one of the most studied white-rot fungi for bioremediation of environmental pollutants. *Armillaria* spp. are notorious for attacking living trees, while *P. ostreatus* and other oyster mushrooms are commonly cultivated white-rot fungi. A few other important white-rot fungi include *Coprinus* sp., *Coprinellus* sp., *Trametes* sp., *Ganoderma* sp., *Fomes* sp., *Lentinula* sp., and *Schizophyllum* sp.

As indicated by their ability to degrade all the structural components of wood, no doubt exists to prove that white-rot fungi have effective hemicellulase systems (Tenkanen et al. 1996). Microscopic examination of white-rot fungi reveals the presence of basidium, basidiospores, pileocystidea, hymenial layers and chlamydospores (mitotic submerged spores), and clamp connections. This is a characteristic feature of class *Hymenomycetes* of *Basidiomycotina* (James et al. 2006).

White-rot fungi, apart from hydrolases, produce various isoforms of extracellular oxidases which are involved in the degradation of lignin in their natural lignocellulosic substrates and directly involved in the degradation of various xenobiotic compounds and dyes (Wesenberg et al. 2003).

A large number of substrates have been explored for production of xylanases using white-rot fungi, but the most attractive alternatives have been lignocellulosic agricultural waste (Sanghi et al. 2008; Singh et al. 2009).

Xylanases are optimally expressed at the end of the exponential phase (Kulkarni et al. 1999). Studies have shown that most of the white-rot fungi grow best at slightly acidic pH but deviations from such observations have also been reported. Most of the white-rot fungi are mesophiles, yet many authors have suggested different temperature ranges (40–70 °C) for xylanase production (Reid 1989; Garg 1996). Xylanase repression has been reported in the presence of easily metabolizable sugars in different microorganisms (Gessesse and Mamo 1999; Sanghi et al. 2008). Xylanase induction, in general, is a complex phenomenon, and the level of response to an individual inducer varies with the organisms (Hrmova et al. 1989). Similarly, the nature and concentration of nitrogen sources are powerful nutrition factors regulating xylanase production by wood-rotting basidiomycetes (Sun et al. 2004; Zakariashvili and Elisashvili 1993). The mechanisms that govern the formation of extracellular enzymes are influenced by the availability of precursors for protein synthesis (Qinnghe et al. 2004). The xylanase production has been successfully carried out using both liquid-state fermentation and solid-state fermentation but, in general, greater xylanase production under SSF compared to LSF (Silva da R. et al. 2005; Aguilar et al. 2001)

As indicated earlier, xylanases are the most potent enzymes for pulp and paper industry, and a large number of investigations have already established their significance in the given industrial sector. Xylanases, along with cellulases, are the major enzymes involved in pulp drainage on a pulp or paper machine (Paice and Zhang 2005). Cellulases and xylanases are widely reported to facilitate deinking of mixed office waste but appear to be less effective for ONP (old newsprint) deinking (Xia et al. 1996). Biological processes like activated sludge, trickling filters, rotating biological contactors, and anaerobic fluidized beds are usually employed in secondary or polishing treatments that follow sedimentation or other primary treatment (Ali and Sreekrishnan 2001). Products like xylitol, an artificial sweetener, and those having therapeutic value can be obtained as by-products (Paice and Zhang 2005), thereby, improving profitability of kraft mills. Biotechnology has a high potential in the pulp bleaching sector as it allows the development of more sustainable and environment-friendly products and processes by reducing chemical usage, saving energy and water, and minimizing waste products (Pawar et al. 2002). Several workers have approached biological bleaching by use of lignolytic enzymes (Bajpai and Bajpai 1992) and hemicellulolytic enzymes (Jimenez et al. 1997), with the focus being mainly on the xylanases (Roncero et al. 2003). Gallardo et al. (2010) characterized GH5 xylanase from *Bacillus* sp. for its potential in pulp bleaching and found it to be a promising bleaching aid for eucalyptus pulp. *Ganoderma australe* and *Ceriporiopsis subvermispora* have shown to selective wood biodelignification in hardwoods and hence are attractive candidates for lignin degradation in biopulping (Mendonca et al. 2008).

There are many white-rot fungi like *Coriolus versicolor* that are not selective for lignin degradation, and large losses of polysaccharides also are removed along with lignin. Blanchette and Burnes (1988) evaluated different white-rot fungal strains, viz., *C. versicolor*, *Dichomitus squalens*, *Phellinus pini*, *Phlebia tremellosus*, *Poria medulla-panis*, and *Scytinostroma galactinum*, for their potential use in biopulping where all the species tested showed promising results. Xylanases and endoglucanases from *Trichoderma reesei* and *Humicola insolens* were used by Kibblewhite and Wong (1999) to improve drainage of radiate pine kraft pulp. They reported an improvement in tear index and tensile index of the hand sheets prepared with the treated pulp.

Ganoderma australe was evaluated for production of ligninolytic enzymes as well as hydrolytic enzymes (cellulases and xylanases) using oil palm empty fruit bunches in SSF. Significant xylanase production was observed, and the yield of empty fruit bunches increased up to 18% during biopulping with satisfactory paper properties (Hasnul et al. 2015). *Trametes trogii* (MYA 28-11) was found to be an outstanding producer of laccase and a promising producer of endoxylanase under solid-state fermentation (Levin et al. 2008), where a wood-based solid medium supplemented with malt extract was used. The culture medium yielded crude extracellular extracts with high laccase (510 U g^{-1}) and endoxylanase (780 U g^{-1}) yields indicating its promising use in pulp bleaching. Dowarah et al. (2015) used *P. chrysosporium* for xylanase production using statistical approach, and xylanase activity of 7.2 U/mL in 53 h was attained in the bioreactor. *C. subvermispora* was also found to produce high amount of xylanase (Milagres et al. 2005). Govumoni et al. (2015) also reported that *P. chrysosporium* produced significant xylanase activity (28.5 IU/mL) on the 9th day of incubation, using corn cobs as substrate. *P. ostreatus* and *P. sajor-caju* were investigated for their ability to produce various lignolytic and cellulolytic enzymes including xylanases, on banana agricultural waste using solid substrate fermentation where both strains showed promising enzyme activity with exception of cellulases, which were recorded very low in amount. Montoya et al. 2015, reported the production of xylanases from *P. ostreatus*, *C. versicolor*, and *Lentinus edodes* on different low-cost lignocellulosic substrates and found that *C. versicolor* exhibited the highest ability to degrade lignocellulosic polymers. Carabajal et al. (2012) and Elisashvili et al. (2008) also reported significant production of lignocellulolytic enzymes including xylanases, by *P. ostreatus* and *L. edodes*. Similarly, *Coprinellus disseminatus* SW-1 NTCC1165 was also reported to produce significant amount of xylanase under solid-state fermentation (Agnihotri et al. 2010). Qinnghe et al. (2004) also reported high xylanase production on the 7th day of incubation under solid-state fermentation with prominent potential in pulp and paper industry. Kachlishvili et al. (2005) reported high xylanase activity with peptone as nitrogen source using *L. edodes* IBB363 under SSF. *Coprinopsis cinerea* HK-1 NFCCI-2032 has also been reported to produce xylanase (693.6 IU/mL), with no cellulase activity under SSF using wheat bran as substrate, and shows promise for commercial application in paper industry (Kaur et al. 2011). Lal et al. (2015) investigated strains of *C. disseminatus* (MLK01 NTCC1180 and MLK-07 NTCC1181) for xylanase production under submerged fermentation for potential use in pulp biobleaching. Xylanolytic hydrolysis of xylans and mannans was carried out using hemicellulolytic enzymes of *Pleurotus* sp. BCCB068 and *Pleurotus tailandia* where both the strains recorded more than 73% degradation of xylan, which can effectively be utilized for bleaching of kraft pulps (Ragagnin et al. 2010).

As mentioned earlier, xylanases have a great impact on bleaching processes (Viikari et al.1994a, b), where they improve pulp fibrillation and water retention reduces beating time in virgin pulps, restores bonding and increased freeness in recycled fibers, and selectively removes xylans from dissolving pulps. Xylanases are also useful in yielding cellulose from dissolving pulps for rayon production and biobleaching of cellulose pulps (Srinivasan and Rele 1999). The use of chlorine as a chemical bleach in the pulp and paper industry generates toxic chlorinated organic by-products (Swanson et al. 1988). Pollutants, like polychlorinated dibenzodioxins and dibenzofurans (dioxins and furans), are recalcitrant to degradation and tend to persist in nature (Ali and Sreekrishnan 2001). Many of these contaminants are acute or even chronic toxins and can induce genetic alterations in exposed organisms (Easton et al. 1997). Thus, in response to the growing concerns for the environment, regulations on the release of waste bleach waters from the pulp and paper industry are becoming stringent (McGeorge et al. 1985). Faced with market, environmental, and legislative pressures, the pulp and paper industry is modifying its pulping, bleaching, and effluent treatment technologies to reduce the environmental load of mill effluents (Pryke 1997, 2003). Biobleaching has attracted considerable attention worldwide (Paice et al. 1995). It relies on the ability of some microorganisms to depolymerize lignin directly or uses microorganisms or enzymes that attack hemi-celluloses, thus facilitating subsequent depolymerization (Beg et al. 2000). Many enzymes especially, xylanases and ligninases, have been put to trial for use in bio-bleaching processes, but xylanases have received significant results (Kantalinen et al. 1988; Viikari et al. 1991; Addleman and Archibald 1993). The application of xylanases in prebleaching of pulps is gaining importance as alternatives to toxic chlorine-containing chemicals (Singh and Roymoulik 1994; Ragauskas et al. 1994). Xylanases offer attractive and commercially viable option to eliminate chlorine in bleaching and reduce chlorinated organic compounds in bleach plant effluents, reduce residual lignin content in the pulp, and increase the brightness of the pulp (Techapun et al. 2003). Many research studies report the efficacy of xylanase for biobleaching for wood and non-wood raw materials (Roncero et al. 2000, 2003). The enzymatic treatment can be combined to different bleaching sequences (Suurnäkki et al. 1997). Xylanases can be applied in elementary chlorine and chlo-rine dioxide containing bleaching sequences, as well as in combination with oxy-gen, ozone, and hydrogen peroxide (Techapun et al. 2003). Xylanases are being used, primarily for the removal of the lignin carbohydrate complexes (LCC) gener-ated during the pulping process that act as physical barriers to the entry of bleaching chemicals (Paice et al. 1992). Besides bleaching through lignin removal, xylanases also helps to increase pulp fibrillation and reduce beating time in the original pulp (Jimenez et al. 1997), which results in better bonding and saving of energy.

Clark et al. (1991) suggested that xylanases loosen the hemicellulose structure to facilitate extraction of lignin. Xylanases hydrolyze some precipitated xylan onto the fibers, increasing their permeability and giving the bleaching chemicals an easier and smoother penetration and access to lignin. This results in an increase in pulp brightness and a decrease in the consumption of chemicals (Ragauskas et al. 1994). Apart from brightness gains, the xylanase-treated pulps also show improved vis-cosity, but a moderate loss in viscosity might also be observed (Tyagi et al. 2011; Medeiros et al. 2002). Similar observations have also been reported by Paice et al. (1989), while they worked on biological bleaching of hardwood kraft pulp with the fungus *C. versicolor*. Pulp drainage improvement has also been observed for xylanase-treated pulps (Singh et al. 2010, 2011; Singh and Dutt 2014), which might be due to removal of fiber materials with a high hydrophilic property, change of the fiber structure, and liberation of free water by enzymatic treatment (Zhao et al. 2006). Paper properties like tear index and tensile and burst indices are influenced by enzyme treatment (Zhao et al. 2002). Several studies indicated the reduction in AOX values of effluents from xylanase-bleached pulps (Jain et al. 2001; Leduc et al. 1995;

Senior and Hamilton 1992). Saleem et al. 2014 reported the production of xylanases and laccases for potential application in kraft pulp bleaching from *Phanerochaete sordid* MRL3, *Lentinus tigrinus* MRL6, and *Polyporus ciliatus* MRL7, where the highest xylanase production was recorded for *L. tigrinus*. When bleached kraft pine cellulosic pulp was treated with xylanases of *Thermomyces lanuginosus* (Buzala et al. 2016), the enzyme hydrolyzed xylan, loosened the fibers, decreased pulp refining energy, and improved paper strength. Prebleaching of eucalyptus kraft pulp and kraft bagasse pulps with xylanases and laccases of indigenously isolated white-rot fungi was done (Thakur et al. 2012) for potential implementation of xylanase prebleaching in an Indian paper mill. The study established that the enzyme treatment reduced the requirement of chlorine-based chemicals by more than 15% and decreasing AOX levels in the effluent by 20–25%. The optical properties of the pulp also improved, and the results were successfully validated at mill scale too. *P. chrysosporium, Pleurotus florida, and Pleurotus citrinopileatus were investigated for aerobic treatment of handmade paper industrial effluents* (Kulshreshtha et al. (2012), *where P. chrysosporium was found to be the most effective in reducing color and COD of effluent generated.* Different strategies were used by Yadav et al. (2012) to develop environmental-friendly pulping and bleaching approaches, using combinations of xylanase enzyme, conventional bleaching, and *Phanerochaete* sp. fungal pretreatment. The combinations were tested against chemical pulping and bleaching. Biobleaching using fungal strain and xylanase enzymes saved 37.3% and 20.3% of elemental chlorine (Cl_2) and 30.8% and 23.1% of chlorine dioxide (ClO_2), respectively, along with reduction of COD, BOD, color, and adsorbable organic halides (AOX).

3.5   Conclusion

Enzyme-based processes are the future of pulp and paper industry and need to be extensively applied in conventional chemical-based processes. Xylanases can play a big role in the same and paper mill trails using enzymes especially xylanases, have been in place (Jean et al. 1994; Scott et al. 1993; Turner et al. 1992) and studies confirm that xylanase treatment can be implemented with little capital investment (Burgt et al. 2002). Xylanase prebleaching technology is now in use at some mills worldwide, but Indian mills further need to increase its usage in pulp prebleaching. Also, other applications of xylanases in pulp and paper industry need to be extensively evaluated at mill scales, as such investigations are still limited. Further, for improving commercial applicability of xylanases in pulp and paper industries, extensive studies need to be carried out on regulatory mechanisms controlling the enzyme production, characteristics of the enzymes in detail, so as to determine effective utilization of this enzyme in various sections of the paper industry. Large-scale efficient integration of xylanases in pulp and paper processing requires these enzymes to exhibit high stability at high temperatures and pH, which make the process simple and economical too. Studies are also required for commercial production of these thermo-alkali-stable xylanases, which would further help in adoption of enzyme-based process in the pulp and paper industry. Purification strategies further need to be optimized and developed for xylanases to be used extensively in pulp and paper, as the recovery of the purified enzyme is to be increased for improving cost benefits as well as enzyme specificity. Having said that, crude xylanases are also attractive options for pulp and paper applications as the enzymatic cocktail present in these crude preparations might provide additional favorable properties to improve overall quality of the finished product. Genetic engineering can significantly contribute to tailor-make such xylanases for pulp and paper applications, but the cost issues might hinder investigations using such modified enzymes. Thus, a balance has to be stricken in exploring genetically engineered xylanases and finding newer strains of white-rot fungi (large number of white-rot fungi are still required to be explored for their xylanases) which can be potentially applied in pulp and paper industry.

References

Addleman K, Archibald FS (1993) Kraft pulp bleaching and delignification by dikaryons and monokaryons of *Trametes versicolor*. Appl Environ Microbiol 59:266–273

Agnihotri S, Dutt D, Tyagi CH, Kumar A, Upadhyaya JS (2010) Production and biochemical characterization of a novel cellulose-poor alkali-thermo-tolerant xylanase from *Coprinellus disseminatus* SW-1 NTCC1165. World J Microbiol Biotechnol 26:1349–1359

Aguilar CN, Augur C, Favela-Torres E, Viniegra-González G (2001) Production of tannase by *Aspergillus niger* Aa-20 in submerged and solid-state fermentation: influence of glucose and tannic acid. J Ind Microbiol Biotechnol 26:296–302

Ali M, Sreekrishnan TR (2001) Aquatic toxicity from pulp and paper mill effluents: a review. Adv Environ Res 5:175–196

Arbeloa M, De Leseleuc J, Goma G, Pommier JC (1992) An evaluation of the potential of lignin peroxidases to improve pulps. TAPPI J 75(3):215–221

Avalos OP, Noyola TP, Plaza IM, Torre M (1996) Induction of xylanase and β-xylosidase in *Cellulomonas flavigena* growing on different carbon sources. Appl Microbiol Biotechnol 46:405–409

Bajpai P (1999) Application of enzymes in the pulp and paper industry. Biotechnol Prog 15:147–157

Bajpai P, Bajpai PK (1992) Biobleaching of Kraft pulp. Process Biochem 27:319–325

Balakrishnan H, Srinivasan MC, Rele MV (1997) Extracellular protease activities in relation to xylanase secretion in an alkalophilic *Bacillus* sp. Biotechnol Lett 18:599–601

Beg QK, Bhushan B, Kapoor M, Hoondal GS (2000) Enhanced production of a thermostable xylanase from *Streptomyces* sp. QG-11-3 and its application in biobleaching of eucalyptus Kraft pulp. Enzyme Microbiol Technol 27:459–466

Blanchette R (1991) Delignification by wood-decay fungi. Ann RevPhytopath 29:381–398

Blanchette RA, Burnes TA (1988) Selection of white-rot Fungi for biopulping. Biomass 15:93–101

Blanchette RA, Farrel RL, Bunes TA, Wendler PA, Zimmerman W, Brush TS, Snyder RA (1992) Biological control of pitch in pulp and paper production by *Ophiostoma piliferum*. TAPPI J 75(12):102–106

Burgt Van der T, Tolan JS,Thibault LC (2002) US kraft mills lead in xylanase implementation In: 35th Annual Pulp and Paper Congress and Exhibition, 1–7

Buzała KP, Przybysz P, Kalinowska H, Derkowska M (2016) Effect of Cellulases and Xylanases on refining process and Kraft pulp properties. PLoS One 11(8):e0161575. https://doi.org/10.1371/journal.pone.0161575,1/14

Carabajal M, Levin L, Aberto E, Lechner B (2012) Effect of co-cultivation of two *Pleurotus* species on lignocellulolytic enzyme production and mushroom fructification. Int Biodeterior Biodegrad 66:71–76

Chanda SK, Hirst EL, Jones JKN, Percival EGV (1950) The constitution of xylan from esparto grass. J Chem Soc 0:1289–1297

Clark TA, Steward D, Bruce M, McDonald A, Singh A, Senior D (1991) Improved bleachability of Radiata pine Kraft pulps following treatment with hemicellulosic enzymes. Appita J 44:389–383

Dowarah P, Boruah P, Goswami T, Barkakati P (2015) Xylanase production from *Phanerochaete chrysosporium* using response surface methodology and its validation in a bioreactor. Int J Eng Tech Res (IJETR) 3(8):142–147

Easton MDL, Kruzynski GM, Solar II, Dye HM (1997) Genetic toxicity of pulp mill effluent on juvenile Chinook salmon (*Oncorhynchus tshawytscha*) using flow cytometry. Water Sci Technol 35(2–3):347–357

Eda S, Ohnishi A, Kato K (1976) Xylan isolated from the stalk of *Nicotiana tabacum*. Agric Biol Chem 40:359–364

Elisashvili V, Penninckx M, Kachlishvili E, Tsiklauri N, Metreveli E, Kharziani T, Kvesitadze G (2008) *Lentinusedodes* and *Pleurotus* species lignocellulolytic enzymes activity in submerged and solid-state fermentation of lignocellulosic *wastes of different composition*. Bioresour Technol 99(3):457–462

Flores ME, Perea M, Rodriguez O, Malváez A, Huitrón C (1996) Physiological studies on induction and catabolite repression of β-xylosidase and endoxylanase in *Streptomyces* sp. CH-M-1035. J Biotechnol 49:179–187

Flournoy DS, Paul JA, Kirk TK, Highley TL (1993) Changes in the size and volume of pores in sweetgum wood during simultaneous rot by *Phanerochaete chrysosporium* Burds. Holzforschung 47(4):297–301

Gallardo O, Fernández-Fernández M, Valls C, Valenzuela SV, Roncero MB, Vidal T, Díaz P, Pastor FIJ (2010) Characterization of a family GH5 Xylanase with activity on neutral oligosaccharides and evaluation as a pulp bleaching aid. Appl Environ Microbiol 76(18):6290–6294

Garg AP (1996) Biobleaching effect of *Streptomyces thermoviolaceous* xylanase preparation on birchwood Kraft pulp. Enzyme Microbiol Technol 18:263–267

Gessesse A, Mamo G (1999) High-level xylanase production by an alkalophilic *Bacillus* sp. by using solid-state fermentation. Enzyme Microbial Technol 25:68–72

Gierer J, Wännström S (1984) Formation of alkali-stable C-C bonds between lignin and carbohydrate fragments during Kraft pulping. Holzforschung 38:181–184

Govumoni SP, Gentela J, Koti S, Haragopal V, Venkateshwar S, VenkateswarRao L (2015) Extracellular Lignocellulolytic enzymes by *Phanerochaete chrysosporium* (MTCC 787) under solid-state fermentation of agro wastes. Int J Curr Microbiol App Sci 4(10):700–710

Gupta S, Bhushan B, Hoondal GS (1999) Enhanced production of xylanase from *Staphylococcus* sp. SG-13 using amino acids. World J Microbiol Biotechnol 15:511–512

Gupta S, Bhushan B, Hoondal GS, Kuhad RC (2001) Improved xylanase production from a haloalkalophilic *Staphylococcus* sp. SG-13 using inexpensive agricultural residues. World J Microbiol Biotechnol 17(1):5–8

Hasnul MB, Oorlidah NA, Ikineswary VS, Yusoff MDM (2015) Investigation of oil palm empty fruit bunches in Biosoda pulping by tropical white-rot fungi, Ganoderma austral (FR.) PAT. Malays Appl Biol 44(2):51–57

Hatakka A, Hammel K (2011) Fungal biodegradation of lignocelluloses. In: Hofrichter M (ed) The mycota, industrial applications, vol 10, 2nd edn. Springer, Berlin, pp 319–340

Hrmova M, Biely P, Vrsanka M, Petrakova E (1984) Induction of cellulose-and xylan-degrading enzyme complex in yeast *Trichosporon cutaneum*. Arch Microbiol 161:371–376

Hrmova M, Beily P, Vrsanka M (1989) Cellulose and xylan degrading enzymes of *Aspergillus terreus*. Enzym Microb Technol 11:610–616

Ikura Y, Horikoshi K (1987) Stimulatory effect of certain amino acids on xylanase production by alkalophilic *Bacillus* sp. Agric Biol Chem 51:3143–3145

Jaeger KE, Eggert T (2004) Enantioselective biocatalysis optimized by directed evolution. Curr Opin Biotechnol 15(4):305–313

Jain RK, Thakur VV, Manthan M, Mathur RM, Kulkarni AG (2001) Enzymatic prebleaching of pulps: challenges and opportunities in Indian paper industry. Ippta Convention Issue 57:1289–1297

James TY, Srivilai P, Kües U, Vilgalys R (2006) Evolution of the bipolar mating system of the mushroom *Coprinellus disseminatus* from its tetrapolar ancestors involves loss of mating-type-specific pheromone receptor function. Genetics 172:1877–1891

Jean P, Hamilton J, Senior DJ (1994) Mill trial experiences with xylanase: AOX and chemical reductions. Pulp and Paper Canada 25(12):126–128

Jeffries TW (1992) Enzymatic treatment of pulps: emerging technologies for materials and chemicals for biomass. ACS Symp Ser 476:313–329

Jiménez L, López F (1993) Characterization of paper sheets from agricultural residues. Wood Sci Technol 27(6):468–474

Jimenez L, Martinez C, Perez I, Lopez F (1997) Biobleaching procedures for pulp and agricultural residues using *Phanerochaete chrysosporium* and enzymes. Process Biochem 4:297–304

Johnsrud SC, Fenandez N, Lopez P, Gutierrez L, Saez A, Ericksson K-E (1987) Properties of fungal pretreated high yield bagasse pulps. Nordic Pulp Pap Res J 75:47–52

Kachlishvili E, Penninckx MJ, Tsiklauri N, Elisashvili V (2005) Effect of nitrogen source on lignocellulosic enzyme production by white-rot basidiomycetes under solid-state cultivation. World JMicrobiolBiotechnol. 22(4):391–397

Kantalinen A, Ratto M, Sundqvist J, Ranua M, Viikari L, Linko M (1988) Hemicellulases and their potential role in bleaching. In: 1988 International Pulp Bleaching Conference, Tappi Proceeding. Atlanta, pp 1–9

Kaur H, Dutt D, Tyagi CH (2011) Production of novel alkali-thermo-tolerant cellulase-poor xylanases from *Coprinopsis cinerea* HK-1 NFCCI-2032. Bioresources 6(2):1376–1391

Kibblewhite PR, Wong KKY (1999) Modification of a commercial radiate pine Kraft pulp using carbohydrate degrading enzymes. Appita J 52(4):300–304

Kirk TK, Cullen D (1998) Enzymology and molecular genetics of wood degradation by white-rot fungi. In: Young RA, Akhtar M (eds) Environmentally friendly technologies for the pulp and paper industry. Wiley, New York, pp 273–307

Kuhad RC, Singh A (1993) Lignocellulose biotechnology: current and future prospects. Crit Rev Biotechnol 13:151–172

Kuhad RC, Manchanda M, Singh A (1998) Optimization of xylanase production by a hyper-xylanolytic mutant strain of *Fusarium oxysporum*. Process Biochem 33:641–647

Kulkarni N, Shendye A, Rao M (1999) Molecular and biotechnological aspects of xylanases. FEMS Microbiol Rev 23:411–456

Kulshreshtha S, Mathur N, Bhatnagar P (2012) Aerobic treatment of handmade paper industrial effluents by white rot fungi. J Bioremed Biodeg 3:151. https://doi.org/10.4172/2155-6199.1000151

Lal M, Dutt D, Kumar A, Gautam A (2015) Optimization of submerged fermentation conditions for two xylanase producers *Coprinellus disseminatus* MLK-01NTCC-1180 and MLK-07NTCC-1181 and their biochemical characterization. Cellulose Chem Technol 49(5–6):471–483

Leduc C, Daneault C, Delaunois P, Jaspers C, Penninckx MJ (1995) Enzyme pretreatment of Kraft pulp to reduce consumption of bleach chemicals. Appita J 48(6):435–439

Levin L, Forschiassin F (1998) Influence of growth conditions on the production of xylanolytic enzymes by *Trametes trogii*. World J Microbiol Biotechnol 14:443–446

Levin L, Herrmann C, Papinutti VL (2008) Optimization of lignocellulolytic enzyme production by the white-rot fungus *Trametes trogii*. Biochem Eng J 39(1):207–214

Liu W, Zhu W, Lu Y, Kong Y, Ma G (1998) Production, partial purification and characterization of xylanase from *Trichosporon cutaneum* SL409. Process Biochem 33:331–326

Lopez C, Blanco A, Pastor FIJ (1998) Xylanase production by a new alkali-tolerant isolate of *Bacillus*. Biotechnol Lett 20:243–246

Mandal A (2015) Review on microbial xylanases and their applications. Int J Life Sci 4(3):178–187

McGeorge LJ, Lonis JB, Atherholt TB, McGarrity GJ (1985) Mutagenicity analyses of industrial effluents: results and considerations for integration into water pollution control program. In: Waters MD, Sandhu SS, Claxton J, Strauss G, Nesnow S (eds) Short-term bioassays in the analysis of complex environmental mixtures. Plenum Press, New York, pp 247–268

Medeiros RG, Silva FG Jr, Salles BC, Estelles RS, Filho EXF (2002) The performance of fungal xylan-degrading enzyme preparations in elemental chlorine-free bleaching for eucalyptus pulp. J Ind Microbiol Biotechnol 28:204–206

Mendonca RT, Jara JF, González V, Elissetche JP, Freer J (2008) Evaluation of the white-rot fungi *Ganoderma australe* and *Ceriporiopsis subvermispora* in biotechnological applications. J Ind Microbiol Biotechnol 35:1323

Milagres AMF, Magalhaes PO, Ferraz A (2005) Purification and properties of a xylanase from *Ceriporiopsis subvermispora* cultivated on *Pinus taeda*. FEMS Microbiol Lett 253:267–272

Montoya S, Sanchez OJ, Levin L (2015) Production of lignocellulolytic enzymes from three white-rot fungi by solid-state fermentation and mathematical modeling. Afr J Biotechnol 14(15):1304–1317

Motta FL, Andrade CCP, Santana MHA (2013) A review of xylanase production by the fermentation of xylan: classification, characterization and applications. In: Chandel AK, da Silva SS (eds) Sustainable degradation of lignocellulosic biomass- techniques, applications and commercialization. InTech, Croatia, pp 251–266

Nissan H (1981) The pulp and papermaking processes. In: Wangaard FF (ed) Paper, wood: its structure and properties. University Park, PA: Pennsylvania State University, pp 335

Paice M, Zhang X (2005) Enzymes find their niche. Pulp and Paper Canada 106(6):17–20

Paice MG, Jurasek L, Ho C, Bourbonnais R, Archibald F (1989) Direct biological bleaching of hardwood Kraft pulp with the fungus *Coriolus versicolor*. TAPPI J 72:217–221

Paice MG, Gurnagul N, Page DH, Jurasek L (1992) Mechanism of hemicellulose directed pre-bleaching of Kraft pulp. Enzym Microb Technol 14:272–276

Paice MG, Bourbonnais R, Reid ID, Archibald FS, Jurasek L (1995) Oxidative bleaching enzymes. J Pulp Pap Sci 21:J280–J284

Patrick K (2005) Enzyme technology improves efficiency, cost, safety of stickies. Pulp Pap Can 106(2):23–25

Pawar S, Mathur RM, Kulkarni AG (2002) Control of AOX discharges in pulp and paper industry-the role of new fiberlineIppta Convention Issue 57

Peck V, Daley R (1994) Toward a 'greener' pulp and paper industry. Environ Sci Technol 28(12):524A–527A

Prade RA (1995) Xylanases: from biology to biotechnology. Biotechnol Genet Eng Rev 13:100–131

Pryke DC (1997) Elemental chlorine-free (ECF): Pollution prevention for the pulp and paper industry http://www.ecfpaper.org/science/science.html

Pryke D (2003) ECF is on a roll. Pulp Pap Int 45(8):27

Puchart V, Katapodis P, Biely P, Kremnicky L, Christakopoulos P, Vrsanska M, Kekos D, Marcis BJ, Bhat MK (1999) Production of xylanases, mannanases, and pectinases by the thermophilic fungus *Thermomyces lanuginosus*. Enzym Microb Technol 24:355–361

Qinnghe C, Xiaoyu Y, Tiangui N, Cheng J, Qiugang M (2004) The screening of culture condition and properties of xylanase by white-rot fungus *Pleurotus ostreatus*. Process Biochem 39:1561–1566

Ragagnin de Menezes C, Silva IS, Pavarina EC, Fonseca de Faria A, Franciscon E, Durrant LR (2010) Production of xylooligosaccharides from enzymatic hydrolysis of xylan by white-rot fungi *Pleurotus*. Acta Scientiarum Technology Maringá 32(1):37–42

Ragauskas AJ, Poll KM, Cesternino AJ (1994) Effects of xylanasepretreatment procedures on non-chlorine bleaching. Enzyme Microbial Technol 16:492–495

Rättö M, Kantelinen A, Bailey M, Viikari L (1992) Potential of enzymes for wood debarking. TAPPI J 76(2):125–128

Reid ID (1989) Solid-state fermentations for biological delignification: review. Enzym Microb Technol 11:786–803

Roncero MB, Torres AL, Colom JF, Vidal T (2000) Effects of xylanase treatment on fiber morphology in total chlorine free bleaching (TCF) of eucalyptus pulp. Process Biochem 36:45–50

Roncero MB, Torres AL, Colom JF, Vidal T (2003) Effect of xylanase on ozone bleaching kinetics and properties of *Eucalyptus* Kraft pulp. J Chem Technol Biotechnol 78:1023–1031

Rytioja J, Hilden K, Yuzon J, Hatakka A, de Vries RP, Makela MR (2014) Plant-polysaccharide-degrading enzymes from Basidiomycetes. Microbiol Mol Biol Rev 78(4):614–649

Sachslehner A, Nidetzky B, Kulbe KD, Haltrich D (1998) Induction of mannanase, xylanase and endoglucanase activities in *Sclerotium rolfsii*. Appl Environ Microbiol 64:594–600

Saleem R, Khurshid M, Safia Ahmed S (2014) Xylanase, Laccase and manganese peroxidase production from white rot Fungi Iranica. J Energy Environ 5(1):59–66

Sanghi A, Garg N, Sharma J, Kuhar K, Kuhad RC, Gupta VK (2008) Optimization of xylanase production using inexpensive agro-residues by alkalophilic *Bacillus subtilis* ASH in solid state fermentation. World J Microbiol Biotechnol 24:633–640

Schumacher K, Sathaye J (1999) India's pulp and paper industry: productivity and energy efficiency. Environmental Energy Technologies Division, Ernest Orlando Lawrence Berkeley National Laboratory

Scott BP, Young F, Paice MG (1993) Mill-scale enzyme treatment of a softwood Kraft pulp prior to bleaching: brightness was maintained while reducing the active chlorine multiple. Pulp and Paper Canada 94(3):57–61

Senior DJ, Hamilton J (1992) Use of xylanases to decrease the formation of AOX in Kraft pulp bleaching. J Pulp and Paper Science 18(5):J165–J168

Shallom D, Shoham Y (2003) Microbial hemicellulases. Curr Opin Microbiol 6:219–228

Siedenberg D, Gerlach SR, Schugerl K, Giuseppin MLF, Hunik J (1998) Production of xylanase by *Aspergillus awamori* on synthetic medium in shake flask cultures. Process Biochem 33:429–433

Silva da R, Lago ES, Merheb CW, Macchione MM, Park YK, Gomes E (2005) Production of xylanase and CMCase on solid state fermentation in different residues by *Thermoascus aurantiacus* miehe. Braz J Microbiol 36:235–241

Singh S, Dutt D (2014) Mitigation of adsorbable organic halides in combined effluents of wheat straw soda-AQ pulp bleached with cellulase-poor crude xylanases of *Coprinellus disseminatus* in elemental chlorine free bleaching. Cellul Chem Technol 48(1–2):127–135

Singh SP, Roymoulik SK (1994) Role of biotechnology in the pulp and paper industry: a review part 2: biobleaching. IPPTA J 6(1):39–42

Singh S, Tyagi CH, Dutt D,Upadhyaya JS (2009)Production of high level of cellulase-poor xylanases by wild strains of white rot fungus *Coprinellus disseminatus* in solid state fermentation New Biotechnol. 26(¾): 165–170

Singh S, Dutt D, Tyagi CH, Upadhyaya JS (2010) Bio-conventional bleaching of wheat straw soda-AQ pulp with crude xylanases from SH-1 NTCC-1163 and SH-2 NTCC-1164 strains of *Coprinellus disseminatus* to mitigate AOX generation. New Biotechnol 28(1):47–57

Singh S, Dutt D, Tyagi CH (2011) Environmentally friendly total chlorine free bleaching of wheat straw pulp using novel cellulase poor xylanases of wild strains of *Coprinellus disseminatus*. Bioresources 6(4):3876–3882

Sood S (2007) India in the world paper markets and emerging Asia. 8th International Technical Conference on Pulp, Paper, Conversion and allied industry, New Delhi, 7–9 Dec 2007, pp 3–7

Srinivasan MC, Rele MV (1999) Microbial xylanases for paper industry. Curr Sci 77:137–142

Srivastava R, Srivastava AK (1993) Characterization of a bacterial xylanase resistant to repression by glucose and xylose. Biotechnol Lett 15:847–852

Subramaniyan S, Prema P (2002) Biotechnology of microbial xylanases: enzymology, molecular biology, and application. Crit Rev Biotechnol 22:33–64

Sun X, Zhang R, Zhang Y (2004) Production of lignocellulolytic enzymes by *Trametesgallica* and detection of polysaccharide hydrolase and laccase activities in polyacrylamide gels. J Basic Microbiol 44:220–231

Suurnäkki A, Tenkanen M, Buchert J, Viikari L (1997) Hemicellulases in the bleaching of chemical pulps. Adv Biochem Eng 57:261–287

Swanson S, Rappe C, Malmstorm J, Kringstad KP (1988) Emissions of PCDDs and PCDFs from the pulp industry. Chemosphere 17:681–691

Techapun C, Poosaran N, Watanabe M, Sasaki K (2003) Thermostable and alkaline-tolerant microbial cellulase-free xylanases produced from agricultural wastes and the properties required for use in pulp bleaching bioprocess: a review. Process Biochem 38:1327–1340

Tenkanen M, Siika-aho M, Hausalo T, Puls J, Viikari L (1996) Synergism of xylanolytic enzymes of *Trichoderma reesei* in the degradation of acetyl-4-O-methylglucuronoxylan. In: Srebotnik E, Messener K (eds) Biotechnology in the pulp and paper industry. Facultas-Universitatsverlag, Vienna, pp 503–508

Thakur VV, Jain RK, Mathur RM (2012) Studies on xylanase and laccase enzymatic prebleaching to reduce chlorine-based chemicals during CEH and ECF bleaching. Bioresources 7(2):2220–2235

Thompson G, Swain J, Kay M, Forster CF (2001) The treatment of pulp and paper mill effluent: a review. Biores Technol 77:275–286

Thompson V, Schaller K, Apel W (2003) Purification and characterization of a novel thermo-alkali-stable catalase from *Thermus brockianus*. Biotechnol Prog 19(4):1292–1299

Thorton JW, Eckerman C, Ekman R,Holbolm B(1992) Treatment of alkaline treated pulp for use in papermaking. European Patent Application, # 92304028. I

Turner JC, Skerker PS, Burns BJ, Howard JC, Alonso MA, Andres JL (1992) Bleaching with enzymes instead of chlorine-mill trials. Tappi J 75(12):83–89

Tyagi CH, Singh S, Dutt D (2011) Effect of two fungal strains of *Coprinellus disseminatus* SH-1 NTCC-1163 and SH-2 NTCC-1164 on pulp refining and mechanical strength properties of wheat straw soda-AQ pulp. Cellul Chem Technol 45(3–4):257–263

Viikari L, Kantelinen A, Rättö M, Sundquist J (1991) Enzymes in biomass conversion, pp 12–21. https://doi.org/10.1021/bk-1991-0460.ch002

Viikari L, Kantelinen A, Buchert J, Puls J (1994a) Enzymatic accessibility of xylans in lignocellulosic materials. Appl Microbiol Biotechnol 41:124–129

Viikari L, Kantelinen A, Sundquist J, Linko M (1994b) Xylanases in bleaching: from an idea to the industry. FEMS Microbiol Rev 13:335–350

Wesenberg D, Kyriakides I, Agathos SN (2003) White-rot fungi and their enzymes for the treatment of industrial dye effluents. Biotechnol Adv 22(1–2):161–187

William RK, Jeffries TW (2003) Enzyme processes for pulp and paper: a review of recent developments, chapter 12. In: Goodwell B, Darrel DN, Schultz TP (eds) Wood deterioration and preservation- advances in our changing world, ACS symposium series 845, pp 210–241

Xia Z, Beaudry A, Bourbonnais R (1996) Effects of cellulases on the surfactant assisted acidic deinking of ONG and OMG. Progress Paper Recycling 5(4):46–58

Xu J, Nogawa M, Okada H, Morikawa Y (1998) Xylanase induction by L-sorbose in a fungus *Trichoderma reesei* PC-3–7. Biosci Biotechnol Biochem 62:1555–1559

Yadav RD, Chaudhry S, Gupta S (2012) Novel application of fungal *Phanerochaete* sp. and xylanase for reduction in pollution load of paper mill effluent. J Environ Biol 33:223–226

Yamasaki T, Hosoya S, Chen C-L, Gratzl JS, Chang HM (1981) Characterization of residual lignin in Kraft pulp. In: The Ekman days international symposium on wood and pulping chemistry Stockholm, vol 2, pp 34–42

Zakariashvili NG, Elisashvili VI (1993) Regulation of *Cerrena unicolor*lignocellulolytic activity by a nitrogen source in culture medium. Microbiology (Eng Trans of Mikrobiologiya) 62:525–528

Zhao J, Li X, Qu Y, Gao P (2002) Xylanasepretreatment leads to enhanced soda pulping of wheat straw. Enzym Microb Technol 30:734–740

Zhao J, Li X, Qu Y (2006) Application of enzymes in producing bleached pulp from wheat straw. Bioresour Technol 97:1470–1476

Chapter 4
Fungal Enzymes Applied to Industrial Processes for Bioethanol Production

Cecilia Laluce, Longinus I. Igbojionu, and Kelly J. Dussán

Abstract The main biofuels are biodiesel obtained from natural oils and ethanol obtained by fermentation of agricultural residues. Residues of lignocellulosic materials are found worldwide and are one of the most promising renewable substrate for the production of bioenergy. However, the conversion of biomass into biofuel, using fast, cheap and efficient methodologies is the major challenge of biofuel production. The present work intends to cover the composition of lignocellulosic substrates, fungal enzymes, synergism between cellulases and industrial integrated processes for biomass utilization. Concerning applications, fungal enzymes are used in pulp and paper industry, textile, bioethanol, wine and brewery manufacturing, food processing, and animal feed.

4.1 General Aspects

The main and most abundant product in a biorefinery is ethanol produced from the lignocellulosic substrates, known as "biomass" ethanol, which are produced by fermentation of sugarcane molasses and other biomaterials. Plant biomass is found anywhere in the world, and it is considered the most promising substrate to renewable energy production. The main advantage of ethanol obtained from sugarcane consists of its use in flexible fuel vehicles (FFV). The economic constraints related to the conversion of biomass into ethanol are the main obstacle to be overcome. However, disadvantages of the biomass ethanol includes its lower energy density

C. Laluce · L. I. Igbojionu (✉)
IPBEN - Bioenergy Research Institute, Institute of Chemistry, São Paulo
State University - UNESP, Araraquara, SP, Brazil

K. J. Dussán
State University (Unesp), Institute of Chemistry, Department of Biochemistry
and Technological Chemistry, Araraquara, São Paulo, Brazil
e-mail: kelly.medina@iq.unesp.br

© Springer International Publishing AG, part of Springer Nature 2018
S. Kumar et al. (eds.), *Fungal Biorefineries*, Fungal Biology,
https://doi.org/10.1007/978-3-319-90379-8_4

compared to gasoline, corrosiveness, low flame luminosity, lower vapor pressures (hard ignition), miscibility with water, and toxicity to ecosystem (Ibeto et al. 2011). Based on advances in research and future prospects, the competitiveness of bioethanol in relation to costs is expected to be overcome in a few decades (Lynd et al. 1991; Bayer et al. 1998).

Currently, the United States is the world's leading producer of ethanol from maize, but its production depends on US government subsidies, and therefore it is considered less efficient compared to ethanol produced from sugarcane (Sanderson 2006). As sugarcane only grows efficiently in tropical climates, Brazil has one of the most successful biofuel programs in the world, producing ethanol from sugarcane without any subsidy from the government. Finally, the use of biomass residues as feedstock to produce energy and other useful by-products has been the subject of intense academic and industrial research (Mishra et al. 2004). In biorefineries, vinasse resulting from the processing of sugarcane can be converted into value-added chemicals such as proteins extracted from fungi for animal feed (Nitayavardhana and Khanal 2010).

4.2   Fungal Proteins as By-Products in Sugar-Based Industries

The integration of innovative fungal technology in sugar-based ethanol process could provide an opportunity for producing food-grade fungal protein for animal feed application (Nitayavardhana and Khanal 2010). The fungal biomass from *Rhizopus microsporus* (*var. oligosporus*) was found to have 45.55% crude protein with significantly high amount of essential amino acids and significantly reduced levels of organic matters such as organic acids (initial levels of 71% lactic acid and 100% acetic acid), which can be achieved within 5 days of fungal cultivation (Rasmussen et al. 2007). A significant reduction in organic matters and organic acids (71% lactic acid and 100% acetic acid) can be achieved within 5 days of fugal cultivation (Nitayavardhana and Khanal 2010). Thus, integration of fungal processing could provide protein for aquaculture feed at a cheaper price than the commercial protein sources for fish and shrimp feed applications.

4.3   Substrates for Fungal Enzymes

Vegetable biomass contains polysaccharides, which are polymers consisting of hundreds to several thousand units of glucose (Mathews et al. 1999). Such polymeric materials are constituted by three main biopolymers (Lee et al. 2014; Himmel et al. 2007): cellulose (~ 30–50% by weight), hemicellulose (~ 19–54%), and lignin (~ 15–35%). Cellulose is at the core of each crystalline unit, forming microfibrils in higher plants. The heterogeneous nature of the natural supramolecular structures of lignocellulosic matrices makes it difficult in understanding of the interactions between complex enzymes and substrates. However, it is evident that the efficacy of hydrolysis is linked to the innate structural characteristics of the substrates and/or the modifications occurring during saccharification (Mansfield et al. 1999).

The structure of the cellulose is made up of long spirals of glucose units connected to each other by β-acetal bonds (Cooper 2015). The enzymatic hydrolysis of lignocellulosic biomass is affected by several factors, including lignin and hemicellulose contents, cellulose crystallinity, degree of polymerization, surface areas accessible to degradation, and pore sizes. However, it is believed that lignin is the major obstacle for the enzymatic hydrolysis, since it forms a physical barrier that restricts the access of cellulases to the cellulose.

Actually, lignin shows a major hindrance to the enzymatic hydrolysis (Chang and Holtzapple 2000; Kim et al. 2003), because it is a physical barrier that restricts the access of cellulases to the cellulose. Thus, development of eco-friendly pretreatments and better lignin removal will increase cell wall porosity, rendering the fungal biomass more amenable to enzymatic hydrolysis (Barakat et al. 2014; Wi et al. 2015).

Several pretreatments (physical, chemical physicochemical, and biological) have been reported in the literature (Saritha et al. 2012; Jouzani and Taherzadeh 2015; Amin et al. 2017; Miranda et al. 2016) with the aim of improving hydrolysis yields, digestibility of the cellulose-enriched fractions, lower processing time, recalcitrance, and costs of a variety of chemical reagents. A successful alteration in the lignocellulose materials is supposed to cause rupture of the biomass structure leading to partial or total separation of its components, which allow greater accessibility of the enzymes to cellulosic fraction.

4.4 Effects of Pretreatments on Substrates Utilized by Fungal Enzymes

Decreases in cellulose crystallinity, degree of polymerization (DP), and particle sizes (Chundawat et al. 2007) decrease the physical barriers that hinder access of cellulase to cellulose structure, required for digestion (Yang et al. 2011). Cellulase enzymes must bind to the surface of the particles of lignin (Morrison et al. 2011, Nakagame et al. 2010) prior to the hydrolysis of the insoluble cellulose (Zhang and Lynd 2004).

Surface characterization, morphology, and analysis of the microstructure can be obtained by SEM (scanning electron microscopy) images, which also provide valuable information when untreated and pretreated samples of cellulosic materials are compared (Amiri and Karimi 2015).

Pretreatments to break up the interrelationships between lignocellulosic materials are necessary for the access of the hydrolytic enzymes before the hydrolysis step. This will enable saccharification of the pretreated material by enzymes such as cellulases, xylanases, and other carbohydrases. Recent approaches in this context (Khare et al. 2015) involve the use of enzymes produced by extremophilic organisms, as well as the development of pretreatments capable of reducing the levels of products toxic to yeast cells. Saccharification step remains one of the most critical bottlenecks to obtain the biomass ethanol.

A successful pretreatment has to largely remove the hemicellulose, leaving the cellulose available for hydrolysis (Olofsson et al. 2008). Since the most commonly used microorganisms for ethanol production solely utilize sugar monomers, the cellulose needs to be hydrolyzed.

A diagram illustrating the major steps in the production of ethanol is summarized in Fig. 4.1, where the pretreated lignocellulose materials are characterized (SEM, X-ray, FTIR, and composition analysis) prior to the hydrolysis, which is used to liberate sugars from cellulosic fractions for fermentation. Finally, hydrolysis (enzymatic and chemical) is necessary to validate the efficacy of partial degradation of lignin and hemicellulose present in lignocellulosic materials. In addition, cellulosic ethanol technologies have been researched in laboratories around the world, but it becomes an expensive process on an industrial biorefinery scale.

Cellulose can exist in several crystalline polymorphs. Cellulose I is a natural or native form, which shows polymorphic structures. However, cellulose I can be converted into other polymorphic forms (cellulose II, III, and IV) using various treatments (Agbor et al. 2011). The transformation of cellulose I into cellulose II using alkali is one of the most important reactions of cellulose processing. Alkaline pretreatment also breaks down the lignin structure by breaking the bonds between lignin and the other carbohydrate fractions, making carbohydrates more accessible to enzymatic degradation (Chandra et al. 2007; Chang and Holtzapple 2000; Galbe and Zacchi 2007; Mosier et al. 2005).

Accessibility of enzymes to crystal surfaces (ordered) and amorphous regions (disordered) vary in the following order (Ciolacu et al. 2011): cellulose II (38%) > cellulose I (24%) > cellulose III (10%). The binding of cellulosic enzymes to the surface of cellulose particles is a prerequisite for hydrolysis. Nevertheless, most crystalline cellulosic substrates exhibit a 10-fold smaller fraction of accessible bonds, a 10-fold smaller frequency of chain ends, and a much smaller fraction of bonds cleaved in the soluble phase during enzymatic hydrolysis as compared to starch (Zhang and Lynd 2004).

Crystalline index (CrI), X-ray diffraction, and solid-state C^{13} NMR are techniques used to evaluate relative amounts of crystalline materials in the structure of the celluloses. Starting from this, the behavior of both regions (amorphous and crystalline) was extensively studied to obtain responses at micro- and macroscale related to the structural changes of lignocellulose materials caused by thermal, hydrothermal, and chemical treatments (Ciolacu et al. 2011; Klemm et al. 1998). From a technical and commercial point of view, different industries require different cellulose polymorphs. As the paper industry requires high crystallinity, cellulose II is preferred, while the preparation of cellulose derivatives requires α-cellulose of low crystallinity and higher content of cellulose I (Gupta et al. 2013). Alteration

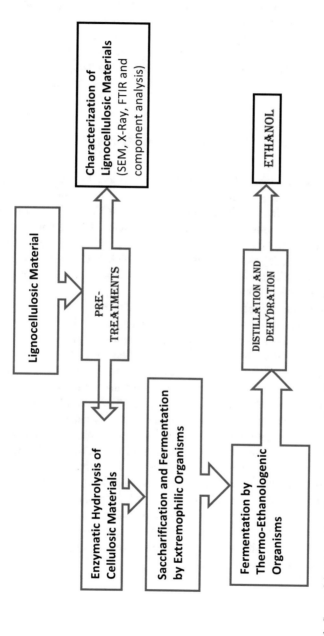

Fig. 4.1 Steps of the conversion of sugarcane bagasse into ethanol

of the crystalline organization of amorphous celluloses leads to simplification in the Fourier transform infrared (FTIR) spectral shape, due to the reduction in intensity or even disappearance of bands characteristic of the crystalline domains. Excellent correlations between crystallinity index and data supplied by XRD, DSC (differential scanning calorimetry), and FTIR were observed. Peaks detected in the DSC (differential scanning calorimetry) curves are in linear correlation with amounts of the amorphous material (%) present in the structure of the cellulose. The accessibility values obtained by DSC showed excellent agreement with crystallinity values determined by traditional techniques (Bertran and Dale 1986).

4.5   Recalcitrance of the Fungal Substrates

The lignocellulose materials are highly recalcitrant to microbial degradation (Himmel et al. 2007). As cellulose and hemicellulose are embedded in a recalcitrant and inaccessible material (Koppram et al. 2013), therefore, a pretreatment step is required to disrupt the structure and to make it accessible to subsequent steps. It is relevant to emphasize that recalcitrance is related to the presence of lignin (Grabber 2005), degree of crystallinity (Park et al. 2010), degree of polymerization (Merino and Cherry 2007), surface area, and moisture content of particles (Hendriks and Zeeman 2009).

The cellulose microfibrils have shown crystalline and amorphous regions, while the cellulose crystallinity has been considered as one of the important factors in determining hydrolysis rates of cellulosic substrates. A greater part of cellulose (around 2/3 of the total cellulose) is in crystalline form (Chang and Holtzapple 2000). However, cellulases readily hydrolyze the most accessible amorphous portion of crystalline cellulose, but are not effective in degrading the less accessible crystalline part of the celluloses. Thus, high-crystallinity cellulose is expected to be more resistant to enzymatic hydrolysis, and it is widely accepted that decreases in crystallinity also increase the digestibility of lignocellulose (Kumar and Wyman 2009). Heterogeneous nature of celluloses and the contribution of other components, such as lignin and hemicellulose, are not the only limiting factors for the efficient enzymatic hydrolysis of biomass (Kumar and Wyman 2009).

Because of the complexity of plant cell wall biosynthesis, the mechanisms by which lignin restrict degradation of the biomass fibers are poorly understood. Studies suggested that reductions in lignin contents, hydrophobicity, and cross-linking improve the enzymatic hydrolysis and utilization of structural polysaccharides for nutritional and industrial purposes.

4.6 Fungal Enzymes Involved in Biopolymer Degradations

The biocatalysts, known as extremozymes, are proteins produced by microorganisms that operate under extreme conditions. Due to extremely high stability, extremozymes offer opportunities for new types of biotransformation. Among examples of extreme enzymes are cellulases, amylases, xylanases, proteases, pectinases, keratinases, lipases, esterases, catalases, peroxidases, and phytases, which have potential applications in several biotechnological processes (Gomes and Steiner 2004).

The major cellulose-degrading enzymes belong to the glucohydrolase (GHs) group and are subgrouped as follows:

- Endoglucanases (EC 3.2.1.4), enzymes that randomly attack the β-1,4 linkages within the polymer chain releasing oligosaccharides
- Exoglucanases or cellobiohydrolases that cleave off cellobioses from the reducing end (EC 3.2.1.176) or nonreducing ends (EC 3.2.1.91) of the chains
- β-Glucosidases (EC 3.2.1.21), which degrade small-chain oligosaccharides to release nonreducing β-D-glucosyl residue (Bayer et al. 1998; Demain et al. 2005; Schwarz 2001; Wilson 2009)

Cellobiohydrolases attack the ends of cellulose chains, while endoglucanases cleave cellulose sites at the middle of cellulose chains reducing the degree of polymerization (Teeri 1997), whereas glucosidases are involved in the glycoprotein processing. The oligosaccharide region of these glycoconjugates plays key roles in biological processes, such as immune response, intercellular recognition, fertilization, cell differentiation, protein folding, and stability and solubility of proteins, which are part of the body's immune response, as well as pathological processes such as cancer and inflammation (Van Dyk and Pletschke 2012).

Currently, the best enzymatic cocktails proposed for saccharification of cellulosic materials are synergistic mixtures of well-defined enzymes to degrade celluloses (Himmel et al. 2007). The heterogeneous nature of the substrate and the complexity of the enzymatic cocktails suggested that traditional studies related to the enzymatic properties could not provide all the necessary information required by the industrial area (Van Dyk and Pletschke 2012).

β-Glucosidases from a new *Aspergillus* species can substitute commercial β-glucosidases in saccharification of lignocellulosic biomass, even with a better performance in some aspects compared to the Novozymes 188, whose production has been discontinued. The efficacy of commercial enzyme mixtures, as well as mixture of enzymes from the rumen, was improved through formulation with synergetic recombinant enzymes. This approach reliably identified supplemental enzymes that enhanced sugar release from alkaline pretreated alfalfa hay and barley straw (Badhan et al. 2014). Commercial enzymes are produced by Novozymes company, which operates in a number of countries, including China, India, Brazil, Argentina, the

United Kingdom, the United States, and Canada, to produce enzymes and chemicals for bioenergy and beverage industry, as well as for domestic uses, human hygiene, and agriculture.

4.7 Synergistic Interactions Between Enzymes Occurring During Hydrolyzes of Glucose Polymers

One of the major challenges facing cellulase researchers is the elucidation of the synergistic interactions between individual components. The degree of synergism is dependent upon the nature of the substrate and occurs when the activity exhibited by mixtures of components is greater than the sum of the activity of each enzymatic component evaluated separately (Walker and Wilson 1991; Woodward 1991). Synergy between cellulolytic enzymes is a term used for the observation that the overall degree of hydrolysis of a mixture of enzyme components is greater than the sum of the degrees of hydrolysis observed by the individual enzymes.

A common way to quantify the extent of synergy is to calculate a "degree of synergy" (DS), which is defined as the ratio of activity exhibited by a mixture of components divided by the sum of the activities of separate components (Andersen et al. 2008).

Synergism has been identified between the following enzymes (Zhang and Lynd 2004): (i) endoglucanases or 1,4-β-d-glucan-4-glucanohydrolases (EC 3.2.1.4), (ii) exoglucanases, including 1,4-β-d-glucan glucanohydrolases (also known as cellodextrinases, EC 3.2.1.74) and 1,4-β-d-glucan cellobiohydrolases (cellobiohydrolases, EC 3.2.1.91), and (iii) β-glucosidases or β-glucoside glucohydrolases (EC 3.2.1.21). Endoglucanases cut randomly at amorphous sites inside the cellulose polysaccharide chain, generating oligosaccharides of various lengths and consequently new chain ends. Exoglucanases act in a progressive manner on the reducing or nonreducing ends of cellulose polysaccharide chains, liberating either glucose (glucanohydrolases) or cellobiose (cellobiohydrolase) as major products. Exoglucanases can also act on microcrystalline cellulose, presumably peeling cellulose chains from the microcrystalline structure.

Cellulases are distinguished from other glycoside hydrolases by their ability to hydrolyze β-1,4-glucosidic bonds between glucosyl residues. The enzymatic breakage of the β-1,4-glucosidic bonds in cellulose proceeds through an acid hydrolysis mechanism, using a proton donor and nucleophile or a base. The hydrolysis products can result in either the inversion or retention (double replacement mechanism) of the anomeric configuration of carbon-1 at the reducing ends.

Fungus *Penicillium pinophilum* has provided some new information on the mechanism by which crystalline cellulose in the form of the cotton fiber is rendered soluble. It was observed that there was little or no synergistic activity either between purified cellobiohydrolases I and II or, contrary to previous findings, between the individual cellobiohydrolases and the endoglucanases. Cotton fiber was degraded to a significant degree only when three enzymes were present in the reconstituted enzyme mixture (Wood et al. 1989): these were cellobiohydrolases I and II and some specific endoglucanases.

Only a trace of endoglucanase activity was required to make the mixture of cellobiohydrolases I and II effectively active (Wood et al. 1989). The addition of cellobiohydrolases I and II individually to endoglucanases from other cellulolytic fungi resulted in little synergistic activity; however, a mixture of endoglucanases and both cellobiohydrolases was effective. It was suggested that the current mechanism of cellulase action may be the result of incompletely resolved complexes between cellobiohydrolase and endoglucanase activities. It was found that such complexes present in filtrates of *P. pinophilium* or *Trichoderma reesei* were easily resolved using affinity chromatography on a column of p-aminobenzyl-1-thio-beta-D-cellobioside (Beldam et al. 1988; Wood et al. 1989).

Proximity effect is a form of synergistic effect exhibited when cellulases work within a short distance from each other, and this effect can be a key factor in enhancing saccharification efficiency. Proximity effect was also observed for crystalline cellulose (Avicel). This effect is expected to increase the efficiency of the enzymes by improving cellulose degradation activity (Bae et al. 2015).

In a study by Andersen et al. (2008), a synergistic interaction opposite to that shown for cellulases was observed. Nevertheless, using Avicel (microcrystalline cellulose powder) as a substrate, synergistic interactions were not observed. Therefore, the degree of synergism appears to vary depending on the nature of the substrate, the specific nature of the enzymes, and the assay conditions as described by Woodward (1991).

Researchers have extensively studied the effects of pretreatments on synergistic interactions between branched fungal cellulases. However, the rate of hydrolysis of native bacterial cellulose increased dramatically with the combination of the two enzymes produced by *Trichoderma viride* (cellobiohydrolase I and endoglucanase II), while no synergism was observed during hydrolysis of cellulose pretreated with acid (Samejima et al. 1997).

4.8   Microorganisms Involved in Biomass Utilization

Fungi are heterogeneous group of heterotrophic eukaryotes, which depend on the availability of nutrients in the environment and can live as saprophytes, parasites, or symbionts, absorbing nutrients from dead substrates or living things (Osiewacz 2002). Concerning morphology, species of fungi show unicellular forms (yeast form) and complex and pluricellular forms (filamentous fungi). Fungi produce two types of enzymatic systems for degradation of lignocellulosic materials as free or complex enzymatic systems. At temperature range of 30–35 °C, free enzymes can be produced at industrial scale by bacteria and aerobic fungi, being *Trichoderma reesei* and *Aspergillus niger* the most studied (Zhang and Lynd 2004). *Trichoderma* cellulases are usually supplemented with extra p-glucosidases (Krishna et al. 2001,

Itoh et al. 2003), while most commercial cellulases are produced by *Trichoderma* spp. Cellulases are the most studied and best characterized enzymes at the molecular level (Aro et al. 2003). Concerning applications, fungal enzymes are used in pulp and paper industry, textile, bioethanol, wine and brewery manufacturing, food processing, and animal feed (Ali et al. 2016).

Despite their simple chemical composition, cellulose structures show a series of crystalline and amorphous topologies. Their insolubility and heterogeneity make native cellulose a recalcitrant substrate for enzymatic hydrolysis. Specific enzymes act in synergy to cause an efficient hydrolysis (Schwarz 2001). Efficient combinations of microbial communities and enzymes act in a sequential and synergistic manner to degrade plant cell walls (Wei et al. 2009)

Delignification processes are time-consuming and may last up to 8 weeks. Fungi and their corresponding substrates are the following (Saritha et al. 2012): *Phanerochaete chrysosporium* (dye decolorization), *Merulius tremellosus* (aspen wood), *Phanerochaete chrysosporium*, *Bjerkandera adusta*, *Pleurotus ostreatus*, *Phlebia tremellosa*, *Trametes versicolor* (barley straw, wood pulp), *Fusarium proliferatum* (industrial lignin removal), *Pleurotus* spp., *Lentinus edodes* (milled tree leaves, banana peel, apple peel, mandarin peel), *Aspergillus terreus*, *Cellulomonas uda*, *Trichoderma reesei*, *Zymomonas mobilis*, *Aspergillus awamori*, *Cellulomonas cartae*, *Bacillus macerans*, *Trichoderma viride* (sugarcane trash), Fungal isolate, RCK-1 (wheat straw), *Phanerochaete chrysosporium* (cotton stalks), *Echinodontium taxodii* 2538, *Trametes versicolor* G20 (bamboo culms), *Coriolus versicolor B1* (bamboo residues), *Phanerochaete chrysosporium* (oil palm empty fruit bunch), *Irpex lacteus* (cornstalks), *Ceriporiopsis subvermispora*, and *Ceriporiopsis subvermispora* (corn stover).

As lignin makes the access of cellulolytic enzymes to cellulose difficult, it is necessary to decompose the network of lignin prior to the enzymatic hydrolysis (Sun and Cheng 2002). To this end, biological treatments with lignin-degrading fungi have a great potential for application if the fungal treatment decomposes the network of lignin with minimum loss of polysaccharides. This suggests that biopulping fungi are the most promising for this purpose among a wide variety of wood-rot fungi with different wood decay patterns (Itoh et al. 2003).

Appropriate hydrolytic enzymes can be selected from a wide array of enzymes available in nature. For example, samples collected in the Brazilian "Cerrado" (hot climate, old and poor soil with rainy summers and dry winters) were assayed, and five strains showed the highest hydrolytic activity. *Acremonium strictum* (dimorphic fungi with a yeastlike form) showed the highest hydrolytic activity on carboxymethyl cellulose (CMC) and filter paper (Goldbeck et al. 2012). The enzymes from thermophiles, grown at temperatures ranging from 50 to 90 °C, are advantageous enzymes for hard industrial processes because they are less susceptible to contamination. The known thermophilic fungi are either *Ascomycetes* belonging to the orders Sordariales, Eurotiales, and Onygenales or *Zygomycetes* of the order *Mucorales* (Morgenstern et al. 2012). Heat-tolerant fungi were isolated from the environment and identified (morphological/molecular tools), while the production of enzymes by this type of fungi was evaluated by solid-state fermentation using

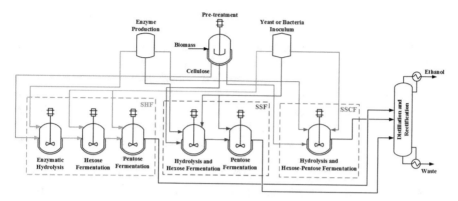

Fig. 4.2 Options for the SSF process: SSF (simultaneous saccharification and fermentation), SHF (separate hydrolysis and fermentation), and SSCF (simultaneous saccharification and co-fermentation)

lignocellulosic materials as substrates (de Cassia Pereira et al. 2015). The fungi *Myceliophthora thermophila* JCP 1-4 was the best producer of endoglucanase, β-glucosidase, xylanase, and avicelase. These enzymes were found to be most active at 55–70 °C and stable at 30–60 °C.

A remarkable synergism (Hoshino et al. 1997) was observed during saccharification of crystalline and amorphous celluloses using combinations of cellulases of *Trichoderma reesei* (endoenzyme) and isoenzyme *from Aspergillus niger* (isoenzymes). Improvements in the quantity of products and decrease in the costs of fermentable sugars from plant biomasses can be obtained using the A-xylosidase (Walton et al. 2016).

4.9 Integrating Fermentation Systems

From an energetic and environmental viewpoint, bioethanol produced by fermentation of lignocellulosic biomass (agricultural by-products, forest residues, or energy crops) shows several advantages compared to ethanol derived from starch materials. In addition, pretreatment, hydrolysis, and fermentation may have an impact on the enzymes involved with hydrolysis through processes that include enzymatic hydrolysis, production of saccharolytic enzymes (cellulases and hemicellulases), hydrolysis of the polysaccharides present in pretreated biomass, and finally fermentation of both the hexose and pentose. One possibility to reduce costs is using the consolidated bioprocessing (CBP) in one process step by integrating cellulase production, cellulose hydrolysis, and the fermentation of sugars (Antonov et al. 2017).

In Fig. 4.2, a flowchart illustrates the integration between SSF (simultaneous saccharification and fermentation) and two other process options (Hamelinck et al. 2005).

The three operational units or levels of process integration for the conversion of lignocellulose to ethanol (Fig. 4.2) are the following: separate (or sequential) hydrolysis and fermentation (SHF), simultaneous saccharification and fermentation (SSF), and SSCF (simultaneous saccharification and co-fermentation).

In the SHF (separated hydrolysis and fermentation) unit (Fig. 4.2), enzymatic hydrolysis of cellulose takes place followed by fermentation of both hexose and pentose in two different vessels, while the SSF (simultaneous saccharification and fermentation) involves the hydrolysis of cellulose and fermentation of both hexose and pentose in the same vessel. In SSF system, the enzymatic hydrolysis is performed simultaneously with glucose fermentation (Wingren et al. 2003; Canterella et al. 2004; Ohgren et al. 2007; Ask et al. 2012; Dahnum et al. 2015). Different biomass, microorganism, non-edible plants, and lignocellulosic materials can be used in SSF system (Alfani et al. 2000; Olofsson et al. 2008; Tomás-Pejó et al. 2008; Gao et al. 2014). Compared with separate enzymatic hydrolysis and fermentation, simultaneous saccharification and fermentation (SSF) systems require lower capital investment and lead to higher overall ethanol yields, which is highly beneficial for the overall process economy (Wingren et al. 2003; Söderström et al. 2005; Wyman et al. 1992). However, the main drawback of the SSF process is the result of the yeast cell recycling.

In the SHF process described in Fig. 4.2, the enzymatic hydrolysis was performed by cellulases (endoglucanase and exoglucanase) that break down cellulose into cellobiose, which is subsequently converted into two molecules of glucose by the β-glucosidase. However, the endoglucanase activity is inhibited by cellobiose, while β-glucosidase is inhibited by glucose. On the other hand, the co-fermentation of hexose and pentose (SSCF processes) is an alternative for optimizing ethanol production with less production costs (Ojeda et al. 2011; Moreno et al. 2013a, b).

Filamentous fungi are remarkable organisms that are naturally specialized in deconstructing plant biomass, and thus they have tremendous potential as components of CBP (Ali et al. 2016), which has been considered as an efficient and economical method of manufacturing bioethanol from lignocellulose. CBP integrates the hydrolysis and fermentation steps into a single process, thereby significantly reducing the amount of steps in the biorefining process.

In the SSCF process (Fig. 4.2), cellulose hydrolysis and fermentation of hexose and pentose occur. The advantages of SSCF process are lower capital investment and continuous removal of glucose and xylose (product of enzymatic hydrolysis). In simultaneous saccharification and co-fermentation processes, the use of a cellulose-controlled feed improves xylose conversion due to decreased inhibition of cellulases or β-glucosidases and increases in productivity (Olofsson et al. 2010a, b). This is a technology widely used for cellulosic ethanol production (e.g., Teixeira et al. 1999; Patel et al. 2005; Kang et al. 2010; Wang et al. 2016a, b), since it provides higher ethanol yields than those reported for hydrolysis and fermentation when performed separately (Koppram et al. 2013). Thus, SSCF process operated in a fed-batch mode offers the possibility to maintain glucose at low levels to allow efficient co-fermentation of glucose and xylose. However, the SSCF process has been a highly researched area but has not yet achieved commercial status (Gikonyo 2015). Poor mass transfer and inhibition of the yeast cells lead to decreased ethanol yield, titer (concentration of a solution as determined by titration), and productivity.

Pretreatment of wheat straw was carried out by multi-feed SSCF, a -fed-batch process with additions of substrate, enzymes, and cells to integrate yeast propagation and adaptation in the pretreated liquor (Wang et al. 2016a).

The main advantages of using thermotolerant organisms in integrated processes are economic, industrial and commercial aspects, such as reduced cooling costs, decreases in refrigeration units, better hydrolysis yields (Hasunuma and Kondo 2012).

4.10 Future Trends and Perspectives

As highlighted by the US Department of Energy (2011), future developments in biorefineries will stimulate new advances in the agriculture such as the emergence of new varieties of sugarcane, cultivation conditions, and increasing amounts of products.

Production of the second-generation ethanol at low cost depends on the availability of efficient and low-cost cellulolytic enzymes, use of new microbial strains, and improvements in the fermentation process (Sassner et al. 2008; Laluce et al. 2012). Most industrial fermentation processes are initiated with thermochemical hydrolysis of cellulosic materials, followed by enzymatic hydrolysis and fermentation of the sugars using microbial communities, where yeasts are dominant organisms (Lopes et al. 2016).

The increases in co-product formation (e.g., ethanol or enzymes) and lower production cost can be achieved by reducing energy consumption applying enzymatic hydrolysis in SSF systems, operating with higher substrate concentration, and continuous flux of recycling cells. As described above, SSCF process has several advantages, while the consolidated bioprocessing (CBP) seems very promising and deserves more attention. Multi-feed SSCF provides the possibilities to balance interdependent rates by systematic optimization of feeding strategies (Wang et al. 2016b).

The strategies applied to CBP systems include the use of native single strains with high cellulolytic activities or microorganisms genetically improved. However, microbial communities, containing dominant ethanologenic yeast, are preferred for industrial uses. Unfortunately, CBP technology is still in its early stages (Lynd et al. 2005; Jouzani and Taherzadeh 2015), while the productive sector is waiting for the development of microorganisms and/or ideal process conditions for industrial use. However, the production of some cellulolytic enzymes in the active form and sufficient amounts for use in CBP systems has been obtained. Cellulolytic fungi, such as *T. reesei,* naturally produce a wide variety of saccharolytic enzymes to digest lignocellulosic material to be used in the conversion of lignocellulose into bioethanol (Xu et al. 2009). Cost constraints can be overcome by designing and building up robust cellulolytic organisms for the production of bio-alcohol in a consolidated bioprocessing system (CBP).

Finally, advances in the improvement of plant depend on better social policies, new agricultural practices, as well as improved programs for domestication of plants and funding for research. However, the conversion of biomass into products using fast, inexpensive, and efficient methodologies to disintegrate and hydrolyze lignocellulosic biomass is the major challenge in the second-generation ethanol production.

Acknowledgment First, we would like to thank Dr. Sachin Kumar for encouraging us to prepare this book chapter. We are also grateful to FAPESP (Foundation for Research Support of the State of São Paulo) for financial aids granted to our research group over the years, as well as the scholarships awarded to the students by CNPq (National Council for Scientific and Technological Development, Brasília), and CAPES (Coordination of Improvement of Higher Education Personnel).

References

Agbor VB, Cicek N, Sparling R, Berlin A, Levin DB (2011) Biomass pretreatment: fundamentals toward application. Biotechnol Adv 29:675–685. https://doi.org/10.1016/j.biotechadv.2011.05.005

Alfani F, Gallifuoco A, Saporosi A, Spera A, Cantarella M (2000) Comparison of SHF and SSF processes for the bioconversion of steam-exploded wheat straw. J Ind Microbiol Biotechnol 25:184–192. https://doi.org/10.1038/sj.jim.7000054

Ali SS, Nugent B, Mullins E, Doohan FM (2016) Mini-review, fungal-mediated consolidated bioprocessing: the potential of Fusarium oxysporum for the lignocellulosic ethanol industry. AMB Express 6:1–13. https://doi.org/10.1186/s13568-016-0185-0

Amin FR, Khalid H, Zhang H, Rahman S, Zhang R, Liu G, Chen C (2017) Pretreatment methods of lignocellulosic biomass for anaerobic digestion. AMB Express 7:72. https://doi.org/10.1186/s13568-017-0375-4

Amiri H, Karimi K (2015) Improvement of acetone, butanol, and ethanol production from woody biomass by using organosolv pretreatment. Bioprocess Biosyst Eng 381:959–1972. https://doi.org/10.1007/s00449-015-1437-0

Andersen N, Johansen KS, Michelsen M, Stenby EH, Krogh KBRM, Olsson L (2008) Hydrolysis of cellulose using mono-component enzymes shows synergy during hydrolysis of phosphoric acid swollen cellulose (PASC), but competition on Avicel. Enzyme Microb Technol 42:362–370. https://doi.org/10.1016/j.enzmictec.2007.11.018

Antonov E, Ivan Schlembach I, Regestein L, Miriam A, Rosenbaum MA, Jochen Büchs J (2017) Process relevant screening of cellulolytic organisms for consolidated bioprocessing. Biotechnol Biofuels 10:106. https://doi.org/10.1186/s13068-017-0790-4

Aro N, Ilmén M, Saloheimo A, Penttilä M (2003) ACEI of *Trichoderma reesei* is a repressor of Cellulase and Xylanase expression. Appl Environ Microbiol 69:56–65. https://doi.org/10.1128/AEM.69.1.56-65.2003

Ask M, Olofsson K, Di Felice T, Ruohonen L, Penttilä M, Lidén G, Olsson L (2012) Challenges in enzymatic hydrolysis and fermentation of pretreated *Arundo donax* revealed by a comparison between SHF and SSF. Process Biochem 47:1452–1459. doi.org/10.1016/j.procbio.2012.05.016

Badhan A, Wang Y, Gruninger R, Patton D, Powlowski J, Tsang A, Mcallister T (2014) Formulation of enzyme blends to maximize the hydrolysis of alkaline peroxide pretreated alfalfa hay and barley straw by rumen enzymes and commercial cellulases. BMC Biotechnol 14:1–14. https://doi.org/10.1186/1472-6750-14-31

Bae J, Kouichi K, Ueda M (2015) Proximity effect among cellulose-degrading enzymes displayed on the *Saccharomyces cerevisiae* cell surface. Appl Environ Microbiol 81:58–56. https://doi.org/10.1128/AEM.02864-14

Barakat A, Chuetor S, Monlau F, Solhy A, Rouau X (2014) Eco-friendly dry chemo-mechanical pretreatments of lignocellulosic biomass: impact on energy and yield of the enzymatic hydrolysis. Appl Energy 113:97–105. https://doi.org/10.1016/j.apenergy.2013.07.015

Bayer EA, Chanzy H, Lamed R, Shoham Y (1998) Cellulose, cellulases and cellulosomes. Curr Opin Struct Biol 8:548–557. https://doi.org/10.1016/S0959-440X(98)80143-7

Beldam G, Voragem AGJ, Romboust FM, Pilnk W (1988) Synergisms in cellulose hydrolysis by exoglucanase and exoglucanase purified form Trichoderma viridie. Biotechnol Bioeng 31:173–178. https://doi.org/10.1002/bit.260310211

Bertran MS, Dale BC (1986) Determination of cellulose accessibility by differential scanning calorimetry. J Appl Polym Sci 32:4241–4253. https://doi.org/10.1002/app.1986.070320335

Cantarella M, Cantarella L, Gallifuoco A, Spera A, Alfani F (2004) Comparison of different detoxification methods for steam-exploded poplar wood as a substrate for the bioproduction of ethanol in SHF and SSF. Process Biochem 39:1533–1542. https://doi.org/10.1016/S0032-9592(03)00285-1

Chandra RP, R Bura P, Mabee WE, Berlin A, Pan X, Saddler JN (2007) Substrate pretreatment: the key to effective enzymatic hydrolysis of Lignocellulosics? Adv Biochem Engin/Biotechnol 108:67–93. https://doi.org/10.1007/10_2007_064

Chang VS, Holtzapple MT (2000) Fundamental factors affecting biomass enzymatic reactivity. Appl Biochem Biotechnol 84:5–37. https://doi.org/10.1385/ABAB:84-86:1-9:5

Chundawat SP, Venkatesh B, Dale BE (2007) Effect of particle size based separation of milled corn Stover on AFEX pretreatment and enzymatic digestibility. Biotechnol Bioeng 96:219–231. https://doi.org/10.1002/bit.21132

Ciolacu D, Ciolacu F, Popa VI (2011) Amorphous cellulose- structure and characterization cellulose. Cellul Chem Technol 45:13–12. https://doi.org/10.12691/nnr-4-3-1

Cooper GM (2015, preprinting) The cell: a molecular approach, 2nd edn. Sinauer Associates, Inc., Publishers. ISBN-13: 978-0878931064

Dahnum D, Tasum SO, Triwahyuni E, Nurdin M, Abimanyu H (2015) Comparison of SHF and SSF processes using enzyme and dry yeast for optimization of bioethanol production from empty fruit bunch. Energy Procedia 68:107–116. doi.org/10.1016/j.procbio.2012.05.016

de Cassia Pereira J, Paganini MN, Rodrigues A, Brito de Oliveira T, Boscolo M, da Silva R, Gomes E, Bocchini Martins DA (2015) Thermophilic fungi as new sources for production of cellulases and xylanases with potential use in sugarcane bagasse saccharification. J Appl Microbiol 118:928–939. https://doi.org/10.1111/jam.12757

Demain AL, Newcomb M, Wu JD (2005) Cellulase, clostridia, and ethanol. Microbiol Mol Biol Rev 69:124–154. https://doi.org/10.1128/MMBR.69.1.124-154.2005

Galbe M, Zacchi G (2007) Pretreatment of lignocellulosic materials for efficient bioethanol production. Adv Biochem Eng Biotechnol 108:41–65. https://doi.org/10.1007/10__070

Gao Y, Xu J, Yuan Z, Zhang Y, Liang C, Liu Y (2014) Ethanol production from high solids loading of alkali-pretreated sugarcane bagasse with an SSF process. Bioresources 9:3466–3479. https://doi.org/10.15376/biores.9.2.3466-3479

Gikonyo B (ed) (2015) Fuel production from non-food biomass corn Stover. Apple Academic Press Publishing, Oakville, p 157

Goldbeck R, Andrade CCP, Pereira GAG, Maugeri Filho F (2012) Screening and identification of cellulase producing yeast-like microorganisms from Brazilian biomes. Afr J Biotechnol 11:11595–11603. https://doi.org/10.5897//AJB12.422

Gomes J, Steiner W (2004) The biocatalytic potential of extremophiles and extremozymes. Food Technol Biotechnol 42:223–235. ISSN 1330-9862

Grabber JH (2005) How do lignin composition, structure, and cross-linking affect degradability? A review of Cell Wall model studies. Crop Sci 45:820–831. https://doi.org/10.2135/cropsci2004.0191

Gupta PK, Uniyal V, Naithan S (2013) Polymorphic transformation of cellulose I to cellulose II by alkali pretreatment and urea as an additive. Carbohydr Polym 94:843–849. https://doi.org/10.1016/j.carbpol.2013.02.012

Hasunuma T, Kondo A (2012) Consolidated bioprocessing and simultaneous saccharification and fermentation of lignocellulose to ethanol with thermotolerant yeast strains. Process Biochem 47:1287–1294. https://doi.org/10.1016/j.procbio.2012.05.004

Hamelinck CN, van Hooijdonk G, Faaij APC (2005) Ethanol from lignocellulosic biomass: techno-economic performance in short-, middle- and long-term. Biomass Bioenergy 28, 384–410. https://doi.org/10.1016/j.biombioe.2004.09.002.

Hendriks A, Zeeman G (2009) Pretreatment to enhance the digestibility of lignocellulosic biomass. Bioresour Technol Essex, 100:10–18. 10-8. https://doi.org/10.1016/j.biortech.2008.05.027

Himmel ME, Ding S, Johnson DK, Adney WS, Nimlos MR, Brady JW, Foust TD (2007) Biomass recalcitrance: engineering plants and enzymes for biofuels production. Science 315:804–807. https://doi.org/10.1126/science.1137016

Hoshino E, Shiroishi M, Amano Y, Nomura M, Kanda T (1997) Synergistic action of exo-type cellulases in the hydrolysis of cellulose with different crystallite. J Ferment Bioeng 84:300–306. https://doi.org/10.1016/S0922-338X(97)89248-3

Ibeto CN, Oloefule AU, Agbo KE (2011) Global overview of biomass potentials for bioethanol production; a renewable alternative fuel. Trends Appl Sci Res 6:410–425. https://doi.org/10.3923/tasr.2011.410.425

Itoh H, Wada M, Honda Y, Kuwahara M, Watanabe T (2003) Bio-organosolve retreatments for simultaneous saccharification and fermentation of beech wood by ethanolysis and white rot fungi. J Biotechnol 103:273–280. https://doi.org/10.1016/S0168-1656(03)00123-8

Jouzani GS, Taherzadeh MJ (2015) Advances in consolidated bioprocessing systems for bioethanol and butanol production from biomass: a comprehensive review. Biofuel Res J 5:152–195. https://doi.org/10.3389/fchem.2014.00066

Kang L, Wang W, Lee YY (2010) Bioconversion of Kraft paper mill sludge's to ethanol by SSF and SSCF. Appl Biochem Biotechnol 161:53–66. https://doi.org/10.1007/s12010-009-8893-4

Khare SK, Pandey A, Larroche C (2015) Current perspectives in enzymatic saccharification of lignocellulosic biomass. Biochem Eng J 102:38–44. https://doi.org/10.1016/j.bej.2015.02.033

Kim TH, Kim JS, Sunwoo C, Lee YY (2003) Pretreatment of corn Stover by aqueous ammonia. Bioresour Technol 90:39–47. https://doi.org/10.1385/ABAB:124:1-3:1119

Klemm D, Philipp B, Heinze T, Heinze U, Wagenknecht W (1998) Comprehensive cellulose chemistry. Wiley-VCH Verlag GmbH publishing, Weinheim, Germany.

Koppram R, Nielsen F, Alber E, Olsson L (2013) Simultaneous saccharification and co-fermentation for bioethanol production using corncobs at lab, PDU and demo scales. Biotechnol Biofuels 6:1–10. https://doi.org/10.1186/1754-6834-6-2

Krishna SH, Reddy TJ, Chowdary GV (2001) Simultaneous saccharification and fermentation of lignocellulosic wastes to ethanol using a thermotolerant yeast. Bioresour Technol 77:193–196. https://doi.org/10.1016/S0960-8524(00)00151-6

Kumar R, Wyman CE (2009) Effects of cellulose and xylanase enzymes on the destruction of solids from pretreated of poplar by leading technologies. Biotechnol Prog 25:301–314. https://doi.org/10.1002/btpr.102

Laluce C, Schenberg ACG, Gallardo JCM, Coradello LFC, Pombeiro-Sponchiado SR (2012) Advances and developments in strategies to improve strains of Saccharomyces cerevisiae and processes to obtain the lignocellulosic ethanol, a review. Appl Biochem Biotechnol 166:1908–1926. https://doi.org/10.1007/s12010-012-9619-6

Lee HV, Hamid SBA, Zain SK (2014) Review Article. Conversion of Lignocellulosic Biomass to Nanocellulose: Structure and Chemical Process. GFSCI World From lignin-derived aromatic compounds to novel Bio-based polymers. J, pp 20. https://doi.org/10.1155/2014/631013

Lopes ML, Paulillo SCL, Godoy A, Cherubin RA, Lorenzi MS, Giometti FHC, Bernardino CD, de Amorim NHB, de Amorim HV (2016) Ethanol production in Brazil: a bridge between science and industry. Braz J Microbiol 47S:64–76. https://doi.org/10.1016/j.bjm.2016.10.003

Lynd LR, Cushman JH, Nichols RJ, Wyman CE (1991) Fuel ethanol from cellulosic biomass. Science 251:1318–1323. https://doi.org/10.1126/science.251.4999.1318

Lynd LR, van Zyl WH, McBride JE, Laser M (2005) Consolidated bioprocessing of cellulosic biomass: and update. Curr Opin Biotechnol 16:577–583. https://doi.org/10.1016/j.
copbio.2005.08.009

Mansfield DS, Mooney C, Saddler JN (1999) Substrate and enzyme characteristics that limit
cellulose hydrolysis. Biotechnol Prog 15:804–816. https://doi.org/10.1021/bp9900864View/
savecitation

Mathews CK, van Holde KE, Kevin G, Ahern KG (1999) Biochemistry, 3rd edn. Prentice Hall
Publisher, New Jersey, USA. ISBN 10: 0805330666/ISBN 13: 9780805330663

Merino ST, Cherry J (2007) Progress and challenges in enzyme development for biomass utilization. Adv Biochem Eng Biotechnol 108:95–120. https://doi.org/10.1007/10_2007_066

Miranda I, Masiero MO, Zamai T, Capella M, Laluce C (2016) Improved pretreatments applied
to the sugarcane bagasse and release of lignin and hemicellulose from the cellulose-enriched
fractions by sulfuric acid hydrolysis. J Chem Technol Biotechnol 91:476–482. https://doi.
org/10.1002/jctb.4601View/savecitation

Mishra S, Mohanty AK, Drzal LT, Misra M, Hinrichse G (2004) A review on pineapple leaf
fibers, sisal fibers and their biocomposites. Macromol Mater Eng 289:955–974. https://doi.
org/10.1002/mame.200400132

Moreno AD, Tomás-Pejó E, Ibarra D, Ballesteros M, Olsson L (2013a) In situ laccase treatment
enhances the fermentability of steam-exploded wheat straw in SSCF processes at high dry matter
consistencies. Bioresour Technol 143:337–343. https://doi.org/10.1016/j.biortech.2013.06.011

Moreno AD, Tomás-Pejó E, Ibarra D, Ballesteros M, Olsson L (2013b) Fed-batch SSCF using
steam-exploded wheat straw at high dry matter consistencies and a xylose-fermenting
Saccharomyces cerevisiae strain: effect of laccase supplementation. Biotechnol Biofuels
6:160. https://doi.org/10.1186/1754-6834-6-160

Morgenstern I, Powlowski J, Ishmael N, Darmond C, Marqueteau S, Moisan MC, Quenneville
G, Tsang A (2012) A molecular phylogeny of thermophilic fungi. Fungal Biol 116:489–502.
https://doi.org/10.1016/j.funbio.2012.01.010

Morrison D, van Dik JS, Pletschke BI (2011) The effect of alcohols, lignin and phenolic compounds of the enzyme activity of Clostridium cellulovarans Xyna. Bioresources 6:3132–3141.
https://doi.org/10.1007/s13205-011-0011-y

Mosier N, Wyman CE, Dale BE, Elander R, Lee YY, Holtzapple MT, Ladisch M (2005) Features
of promising technologies for pretreatment of lignocellulosic biomass. Bioresour Technol
96:673–686. https://doi.org/10.1016/j.biortech.2004.06.025

Nakagame S, Chandra RP, Saddler JN (2010) The effect of isolated lignin obtained from a range
of pretreated lignocellulosic substrates, on enzymatic hydrolysis. Biotechnol Bioeng 105:871–
879. https://doi.org/10.1002/bit.22626

Nitayavardhana S, Khanal SK (2010) Innovative biorefinery concept for sugar-based ethanol
industries: production of protein-rich fungal biomass on vinasse as an aquaculture feed ingredient. Bioresour Technol 101:9078–9085. https://doi.org/10.1016/j.biortech.2010.07.048

Öhgren K, Bura R, Lesnicki G, Saddler J, Zacchi G (2007) A comparison between simultaneous saccharification and fermentation and separate hydrolysis and fermentation using steam-pretreated
corn Stover. Process Biochem 42:834–839. https://doi.org/10.1016/j.procbio.2007.02.003

Ojeda K, Sánchez E, El-Halwagi M, Kafarov V (2011) Exergy analysis and process integration
of bioethanol production from acid pre-treated biomass: comparison of SHF, SSF and SSCF
pathways. Chem Eng J 176/177:195–201. https://doi.org/10.1016/j.cej.2011.06.083

Olofsson K, Bertilsson M, Liden G (2008) A short review on SSF - an interesting process option
for ethanol production from lignocellulosic feedstocks. Biotechnol Biofuels 1:1–14. https://
doi.org/10.1186/1754-6834-1-7

Olofsson K, Palmqvist B, Lidén G (2010a) Improving simultaneous saccharification and co-
fermentation of pretreated wheat straw using both enzyme and substrate feeding. Biotechnol
Biofuels 3:1–17. https://doi.org/10.1186/1754-6834-3-17

Olofsson K, Wiman M, Lidén G (2010b) Controlled feeding of cellulases improves conversion of xylose in simultaneous saccharification and co-fermentation for bioethanol production. J Biotechnol 145:168–175. https://doi.org/10.1016/j.jbiotec.2009.11.001

Osiewacz HD (2002) Genes, mitochondria and aging in filamentous fungi. Ageing Res Rev 1:425–442. https://doi.org/10.1016/S1568-1637(02)00010-7

Park S, Baker JO, Himmel ME, Parilla PA, Johnson DK (2010) Cellulose crystallinity index: measurement techniques and their impact on interpreting cellulase performance. Biotechnol Biofuels 23:1–10. https://doi.org/10.1186/1754-6834-3-10

Patel MA, Ou MS, Ingram LO, Shanmugam KT (2005) Simultaneous saccharification and co-fermentation of crystalline cellulose and sugar cane bagasse hemicellulose hydrolysate to lactate by a thermotolerant acidophilic Bacillus sp. Biotechnol Prog 21:1453

Rasmussen M, Kambam Y, Khanal SK, Pometto AL, van Leeuwen J (2007) Thin stillage treatment from dry-grind ethanol plants with fungi. ASABE Annual International Meeting of American Society of Agricultural and Biological Engineers, June 17–20, Minneapolis, MN, USA

Samejima M, Sugiyama J, Igarashi K, Eriksson K-E L (1997) Enzymatic hydrolysis of bacterial cellulose. Carbohydr Res 305:281–288. https://doi.org/10.1016/S0008-6215(97)10034

Sanderson K (2006) US biofuels: a field in ferment. Nature 444:673–676. https://doi.org/10.1038/444673a

Saritha M, Arora A, Lata (2012) Biological pretreatment of Lignocellulosic substrates for enhanced delignification and enzymatic digestibility. Indian J Microbiol 52:122–130. https://doi.org/10.1007/s12088-011-0199-x

Sassner P, Galbe M, Zacchi G (2008) Techno-economic evaluation of bioethanol production from three different lignocellulosic materials. Biomass Bioenergy 32:422–430. https://doi.org/10.1016/j.biombioe.2007.10.014

Schwarz WH (2001) The cellulosome and cellulose degradation by anaerobic bacteria. Appl Microbiol Biotechnol 56:634–649. https://doi.org/10.1007/s002530100710

Söderström JM, Galbe M, Zacchi G (2005) Separate versus simultaneous saccharification and fermentation of two-step steam pretreated softwood for ethanol production. J Wood Chem Technol 25:187–202. https://doi.org/10.1080/02773810500191807

Sun Y, Cheng J (2002) Hydrolysis of lignocellulosic materials for ethanol production: a review. Bioresour Technol 83:1–11. https://doi.org/10.1016/S0960-8524(01)00212-7

Teeri TT (1997) Crystalline cellulose degradation: new insight into the function of cellobiohydrolases. Trends Biotechnol 15:160–167. https://doi.org/10.1016/S0167-7799(97)01032-9

Teixeira LC, Linden JC, Schroeder HA (1999) Optimizing per- acetic acid pretreatment conditions for improved simultaneous saccharification and co-fermentation (SSCF) of sugar cane bagasse to ethanol fuel. Renew Energy 16:1070–1073. https://doi.org/10.1016/S0960-1481(98)00373-5

Tomás-Pejó E, Oliva JM, Ballesteros M, Olsson L (2008) Comparison of SHF and SSF processes from steam-exploded wheat straw for ethanol production by xylose-fermenting and robust glucose-fermenting Saccharomyces cerevisiae strains. Biotechnol Bioeng 100:1122–1131. https://doi.org/10.1002/bit.21849

U.S. Billion-Ton Update: Biomass Supply for a Bioenergy and. Bioproducts Industry. R.D. Perlack and B.J. Stokes (Leads), ORNL/TM-2011/224. Oak Ridge National. Laboratory, Oak Ridge, TN. 227p. https://www1.eere.energy.gov/bioenergy/pdfs/billion_ton_update.pdf

Van Dyk JS, Pletschke BI (2012) A review of lignocellulose bioconversion using enzymatic hydrolysis and synergistic cooperation between enzymes-factors affecting enzymes, conversion and synergy. Biotechnol Adv 30:1458–1480. https://doi.org/10.1016/j.biotechadv.2012.03.002

Walker LP, Wilson DB (1991) Enzymatic hydrolysis of cellulose: an overview. Bioresour Technol 36:3–14. https://doi.org/10.1016/0960-8524(91)90095-2

Walton JD, Scott-craig JS, Melissa SB (2016) A-xylosidase enhanced conversion of plant biomass into fermentable sugars. United States Patent 9404136

Wang R, Unrean P, Franzen CJ (2016a) Model-based optimization and scale-up of multi-feed simultaneous saccharification and co-fermentation of steam pre-treated lignocellulose enables high gravity ethanol production. Biotechnol Biofuels 9:88. https://doi.org/10.1186/s13068-016-0500-7

Wang X, Taylor S, Wang Y (2016b) Improvement of radio frequency (RF) heating-assisted alkaline pretreatment on four categories of lignocellulosic biomass. Bioprocess Biosyst Eng 39:1539–1551. https://doi.org/10.1007/s00449-016-1629-2

Wei H, Xu Q, Taylor LE, Baker JO, Tucker MP, Ding S (2009) Natural paradigms of plant cell wall degradation. Curr Opin Biotechnol 20:330–333. https://doi.org/10.1016/j.copbio.2009.05.008

Wi SG, Cho EJ, Lee D, Lee SJ, Lee YJ, Bae H (2015) Lignocellulose conversion for biofuel: a new pretreatment greatly improves downstream biocatalytic hydrolysis of various lignocellulosic materials. Biotechnol Biofuels 8:228. https://doi.org/10.1186/s13068-015-0419-4

Wilson DB (2009) Cellulases and biofuels. Curr Opin Biotechnol 20:295–299. https://doi.org/10.1016/j.copbio.2009.05.007

Wingren A, Galbe M, Zacchi G (2003) Techno-economic evaluation of producing ethanol from softwood: comparison of SSF, SHF, and identification of bottlenecks. Biotechnol Prog 19:1109–1117. https://doi.org/10.1021/bp0340180

Wood MT, McCrae IS, Mahalingeshwara K (1989) The mechanism of fungal cellulase action synergism between enzyme components of Penicillium pinophilum cellulase in solubilizing hydrogen bond-ordered cellulose. Biochem J 260:37–43

Woodward J (1991) Synergism in cellulase systems. Bioresour Technol 36:67–75. https://doi.org/10.1016/0960-8524(91)90100-X. Get rights and content

Wyman CE, Spindler DD, Grohmann K (1992) Simultaneous saccharification and fermentation of several lignocellulosic feedstocks to fuel ethanol. Biomass Bioenergy 3:301–307. https://doi.org/10.1016/0961-9534(92)90001-7

Xu Q, Singh A, Himmel ME (2009) Perspectives and new directions for the production of bio-ethanol using consolidated bioprocessing of lignocellulose. Curr Opin Biotechnol 20:364–371. https://doi.org/10.1016/j.copbio.2009.05.006

Yang B, Dai Z, Ding S, Wyman CE (2011) Review: enzymatic hydrolysis of cellulosic biomass. Biofuels 2:421–450. https://doi.org/10.4155/bfs.11.116

Zhang YH, Lynd LR (2004) Toward an aggregated understanding of enzymatic hydrolysis of cellulose: noncomplex cellulase systems. Biotechnol Bioeng 30:797–824. https://doi.org/10.1002/bit.20282

Chapter 5
Economical Lactic Acid Production and Optimization Strategies

Sheelendra M. Bhatt

Abstract The best replacement of plastic is bioplastic. Lactic acid is a promising monomer of polylactide production. Therefore, lowering down lactic acid cost is a challenge. Biochemical production of lactic acid via starch or glucose is not economical. Therefore, the current chapter addresses strategies in economical lactic acid production using fungi, which is still looking for its place in the industry. Lactic acid bacteria (LAB) or fungi have their own merits and demerits. Fungi have several advantages against LAB, such as robust metabolic pathway, suitable for conversion of a diversity of substrates into lactic acid. Therefore, understanding of metabolic pathways and metabolic engineering strategies is crucial in developing a novel strain and increasing the yield of lactic acid. The central role of fungi especially *Rhizopus* has been discussed in detail. Simultaneous saccharification and fermentation is advantageous, and various factors affecting fermentation have also been discussed in detail. These factors are helpful in optimizing media conditions for reactors.

5.1 Introduction

Lactic acid is biochemically produced from lactic acid bacteria or fungi and is an important raw material for making thermostable biodegradable plastic and also for other biochemicals. Therefore, a tremendous rise in lactic acid demand has been observed in the recent past due to interest in the production of ballistic (Abdel-Rahman et al. 2013). The worldwide demand for lactic acid reported was around 130,000 to 150,000 (metric) tons per year (Mirasol 1999). The global PLA market may increase up to 500,000 (metric) tons per year by 2010 (Nexant Inc./Chem. System, Technical report, 2002). Although the demand for PLA is increasing, its current production is only 450,000 metric tons per year till date, which still lags behind 200 million metric tons per year of total plastics produced (Okano et al. 2010). Global demand of lactic acid was around 482 kilotons in 2016 which was

S. M. Bhatt (✉)
Sant Baba Bhag Singh University, Jalandhar, India

© Springer International Publishing AG, part of Springer Nature 2018
S. Kumar et al. (eds.), *Fungal Biorefineries*, Fungal Biology,
https://doi.org/10.1007/978-3-319-90379-8_5

Table 5.1 Application of lactic acid in food industry (Varadarajan and Miller 1999)

SN	Compound	Food application
1.	L (+) lactic acid	Antimicrobial agent, curing, and pickling flavoring
2.	D (−) lactic acid	Enhancer, adjuvant, pH control, solvent, and vehicle
3.	Calcium lactate	Flavoring enhancer, forming agent, leavening agent, supplement, and stabilizer thickener
4.	Ferrous lactate	Nutrient supplements and in infant formula
5.	Potassium lactate	Flavor enhancer, flavoring agent, humectant, and pH control
6.	Sodium lactate	Flavor enhancer, flavoring agent, humectant, pH control, and emulsifier
7.	Calcium stearoyl-2- lactate	Dough conditioner in bakery products, shipping agent in egg products, and conditioning agent in dehydrated potato
8.	Sodium stearoyl-2-lactate	Dough conditioner emulsifier, emulsifier processing aid in baked products, emulsifier, and stabilizing processing

14.2% more than was in 2010 (248 kilotons), and as per estimation of market research, current price of LA is around 0.8 USD/kg which should be less than half of the current price (Okano et al. 2010). As per estimation, current price of food-grade lactic acid (88%) is around $1400–1600/metric tons, while it can go up to 130,000–150,000 metric tons, and it can go over million tons up to 2020 (Upadhyaya et al. 2014).

Polylactic acid (PLA) polymer is also useful in making biomedical devices such as sutures, orthopedic implants, and other useful surgery items (Datta and Tsai 1997; Sarkar and Datta 2004; Abdel-Rahman et al. 2011). PLA biopolymers are biodegradable and tough, similar to thermoplastics and, therefore, can be a good substitute as compared to conventional plastics, which has been proved to be a serious threat to the environment because of its recalcitrant nature. Lowering down the price of lactic acid will certainly be going to make the overall price of PLA competitive and affordable in the public domain.

5.2 Historical Aspects

Scheele was the first to discover LA in 1780, and it was first produced commercially by Charles E. Avery at Littleton, Massachusetts, USA, in 1881. Pure cultures of *Lactobacillus delbrueckii* and *Lactobacillus bulgaricus* were first prepared for industrial use by Max Delbrück's Institute for the Fermentation Industries in Berlin (1896). Lactic acid exists in two optically active isomeric forms, D(−) and L(+) lactic acid, that differ in their optical properties but are identical in their physical and chemical characteristics with different applications (Table 5.1) (Gunatillake and Adhikari 2003; Ren 2011).

Lactic acid is a three-carbon organic acid (2-hydroxypropionic acid): one terminal carbon atom is part of an acid or carboxyl group; the other terminal carbon atom is part of a methyl or a hydrocarbon group and a central carbon atom having

Table 5.2 Properties of lactic acid

S.N.	Properties of lactic acid	Values
1.	Density at 20 °C (g/l)	1.249
2.	Melting point (°C)	52.8 (D) 53 (L) 16.8 (DL)
3.	Boiling point (°C)	82 (DL) at 0.55 mmHg 103 (D) at 15 mmHg
4.	Dissociation constant (pKa) at 25 (°C)	3.83 (D);3.79 (L)
5.	Heat capacity (kJ/mol °C) at 20 (°C)	190 (DL)
6.	Heat of fusion (kJ/mol)	16.86 (L);11.3 (DL)

Holten (1971), Perry et al. (1999)

an alcohol. Lactic acid is water-miscible organic solvent with low volatility. The presence of two functional groups, one hydroxyl (–OH) and other carboxyl (–COOH) groups, makes them very active. Thus, they play a very important role in synthesis of chemical feedstock for varied chemicals such as acrylic acid, propylene glycol, acetaldehyde, and 2,3-pentanedione and for making polylactide a biodegradable plastic (Varadarajan and Miller 1999). In addition, large-volume, low-cost lactic acid could give rise to commercial development of cost-competitive lactate esters (methyl, ethyl, butyl, etc.) as a substitute to those derived from petrochemical sources (Refer Table 5.2).

5.3 Lactic Acid Production

Two routes of lactic acid synthesis exist: (1) chemical route and (2) biochemical route. Nowadays 90% of lactic acid is produced commercially via biochemical route via fermentation of sugars by lactic acid bacteria or fungi.

Synthetic lactic acid produced chemically is not advantageous for making bioplastic since it is optically inactive due to racemic mixtures, whereas lactic acid produced by biochemical route is optically active and selectively produces either L (+) or D (−) form. L (+) LA is essentially needed for production of the quality biopolymer. Microbial lactic acid production occurs either via LAB or via fungi; both have its own merits and demerits.

Demerits of LA Production via LAB

1. LAB require costly synthetic media to grow.
2. Requirements of $CaCO_3$ as neutralizing agents, thus adds additional purification cost.
3. Limited metabolic enzyme, as a result widely available substrate lignocellulose can't be used in industrial scale for the production of lactic acid.
4. Substrate inhibition and end product inhibition lowers LA yield.
5. Formation of by-product during lactic acid.
6. Lactic acid production ceases at low pH.

5.4   Lactic Acid Production and Related Pathway

Lactic acid production requires specific metabolic enzymes in the cell which is used in the final bioconversion of complex carbohydrates into glucose and then to lactic acid. Cellulose is one such complex carbohydrate, largely available on earth round the year and difficult to metabolize by LAB. In addition, a few fungi such as *Rhizopus*, yeast, and *Candida* are known to produce specific metabolic enzymes for hydrolyzing a variety of agrowaste (Zhang et al. 2007a).

Lactic acid production using agrowaste has been reported by various workers using fungi *Rhizopus* and *Aspergillus*, which could metabolize cellulose and hemicellulose into xylose and other fermentable sugars. However, low lactic acid yield results without genetic modification (range of 0.7–2.5 g LA/l/h) as compared to LAB owing to its heterofermentative nature. Nowadays, at commercial scale, lactic acid is produced mostly by submerged fermentation of corn using LAB (Wang et al. 2016).

5.4.1   Lactic Acid Production from Fungi

Lactic acid production via fungi has *several advantages*: (1) they can easily hydrolyze components of agricultural waste such as cellulose and hemicellulose and (2) can grow on low-cost media for lactic acid production.

The *only disadvantage* is they use heterofermentative pathway, for lactic acid production. Different strategies have been used to check the by-product formation, such as mutation of gene, viz., pyruvate decarboxylase *(pdc1 and pdc2)* or alcohol dehydrogenase *(adh)* which had worked well and helped in diverting flux toward more lactic acid synthesis by reducing ethanol production (Ilmén et al. 2013; Yamada et al. 2017). After mutating these genes in *L. helveticus* strain, 92 g/l, LA was reported with a yield of 0.94 g/g glucose. Some workers effectively succeeded in increasing LA up to 38% (0.4gLA/g glucose) by adding alcohol dehydrogenase *(adh)* inhibitor (2,2,2-trifluoroethanol and 2-diazole) in the fermentation media (Thitiprasert et al. 2011). These were new strategies which are followed to get LA without any gene deletion (Zhang et al. 2016).

Diverse range of agricultural waste is highly useful for economical lactic acid production using robust microbes prepared by genetic modification of fungal cells to utilize more C5 and C6 sugars for LA productions and also to check by-product formation (Maas et al. 2006; Ghosh and Ray 2011). One such strain *R.* NRRL 395 was created by mutating *adh* and malate dehydrogenase *mdh gene* which resulted in dramatic improvement of lactic acid (up to 140 g/l LA) from starch hydrolysate. When encountering low LA production, due to high fungal cell density in the reactor, a pellet of fungal cell is suggested instead of cells, which produces long hyphae, which are supposed to block the agitator in fermenters (Liu et al. 2008). A pellet is a spherical or ellipsoidal mass of hyphae which are useful for the reactor because of compact bodies and don't create any mess with the agitator.

SSF mode is generally useful with solid substrate cellulose and fungal spores in the fermenter. Cellulose has lignin; therefore, only pretreated cellulosic mass is used in SSF; see Fig. 5.5. One obstacle observed is the production of inhibitors during pretreatment (such as furfural and HMF) which are known to cease cell growth and, thus, also inhibit LA production (Zhang et al. 2016).

Metabolic conversion of C5 sugars from hydrolyzed cellulosic mass is still a challenge. Some cells, such as modified yeast cells, directly are unable to hydrolyze pentoses, but *Pichia*, on the other hand, is able to hydrolyze pentose. The use of *R. oryzae* NBRC 5378 with wheat straw (after alkaline pretreatment) was able to ferment and convert cellulosic mass into lactic acid yield up to 2.2 g/g (Saito et al. 2012). *R. oryzae* has a benefit, that is, it is able to grow on mineral medium with glucose as a sole carbon source and can produce optically pure L(+)-lactic acid even from xylose with a yield of 0.7 g/g and also doesn't face a problem of product inhibition at low pH. To lower down the production of inhibitors during pretreatment, alkaline pretreatment was useful for wheat straw, which could result in high xylose production, and thus high concentration of glucose is produced (19.0 g/l from 10.0 g/l xylose), and final lactic acid was 6.8 g/l.

An encouraging high lactic acid production (355g lactic acid per kg corncob) is obtained when genetically engineered strain *R. oryzae* GY18 has been used with corn cob or sucrose (Guo et al. 2012). On the other hand, *R. oryzae* NRRL 395 strain could utilize empty fruit bunches fiber, and *R. oryzae* TS-61 can metabolize even chicken feather protein hydrolysate for lactic acid production in SSF (simultaneous saccharification and fermentation) mode (Hamzah et al. 2009; Taskin et al. 2012).

Sometime, consortia of fungi are highly useful in enhancing lactic acid production. In synergistic mode, a complete hydrolysis of lignocellulosic waste was observed for high lactic acid production. In consortia microbes experimented were *Mucor indicus*, *R. oryzae*, and *Saccharomyces cerevisiae* in SSF mode. *R. oryzae* has been used in lactic acid production (Karimi et al. 2006).

5.5 Metabolic Pathways Used for Lactic Acid Production

There are two metabolic categories of LAB, and thus microbial lactic acid production occurs either by (1) homofermentative or (2) heterofermentative pathway. Depending upon the presence of respective lactate dehydrogenases, two types of lactic acid produced either D-(−)- or L-(+)- lactic acid (Garvie 1980). Homofermentative strain

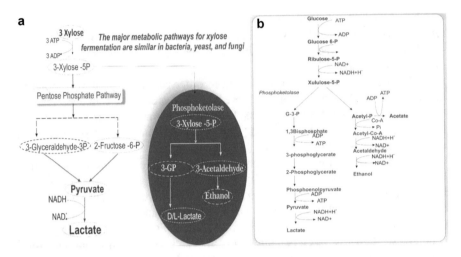

Fig. 5.1 Heterofermentative metabolism of glucose, (**a**) pentose phosphate pathway (PPP), (**b**) PPK phosphoketolase pathway. (Source: Schaechter (2009))

produces a single product, while heterofermentative strain produces lactic acid along with a by-product. The behavior of some homofermentative strain may change to heterofermentative under the growth-limiting levels of carbon source due to change expression of *ldh* which is central catalysts in pyruvate-mediated conversion of lactic acid (Kandler 1983).

Fungi mainly use heterofermentative pathway to produce lactic acid, and the by-product produced is either ethanol or acetic acid. In heterofermentative pathway, pyruvate is decarboxylated to acetaldehyde (which is the terminal electron acceptor) which is being reduced to ethanol via pyruvate using pentose phosphoketolase pathway (PPK) (refer Fig. 5.1).

Rhizopus is proved to be a versatile microbe naturally laden with several metabolic enzymes such as pectinase, cellulase, xylanase, amylase, lipase, and tannase. Therefore, xylose fermentation is common in fungi for glucose metabolism to produce lactic acid via EMP pathway or phosphoketolase pathway.

PPK pathway is common in many bacteria and fungi such as *Leuconostoc*, *Lactococcus*, *Pediococcus*, *Lactobacillus* (e.g., *L. brevis*, *L. lysopersici*, and *L. pentoaceticus*), *Bacillus*, *Acetobacter* (*A. aceti*), *Streptococcus*, *Microbacterium*, and *Rhizopus*. In this pathway, an equimolar amount of lactate, CO_2, and ethanol or acetate is produced using glucose (Fig. 5.1) (Schaechter 2009).

5.5.1 Ldh-Mediated Metabolic Engineering

In the recent past, most metabolic engineering has been focused on *ldh*-mediated pathway to engineer overproducer strain (Liaud et al. 2015). An overproducing strain of *R. oryzae* was developed (R1021) by mutation (Bai et al. 2004). *Aspergillus* and *Rhizopus* become more robust to digest the cellulose, hemicellulose, and other complex carbohydrates such as starch for lactic acid production at low pH.

Low-pH condition in media prompts a shift in metabolic pathway in most of the LAB, thus lowering down the yield of lactic acid production (Bhatt 2013). Liaud et al. 2015 modified *Aspergillus* cells for production of lactic acid from agricultural waste (Liaud et al. 2015). In this experiment *A. brasiliensis* BRFM1877 was integrated with 6 *ldhA* gene copies, to enhance intracellular LDH activity, and up to 9.2×10^{-2} U/mg of enzyme activity was observed; as a result lactic acid concentration observed was 13.1 g/l at pH 1.6 (conversion yield, 26%, w/w) in 138 h using synthetic media (glucose ammonium).

Production of L (+) or D (−) lactic acid depends on specific isomeric *ldh* enzyme which is present in cell in both the forms L or D (Kandler 1983). According to Garvie (1980), L-lactic acid is the major form produced during the early growth phase, while D-lactic acid is produced during the late growth phase (Garvie 1980). Activity of LDH depends on pH and internal pyruvate concentration and accordingly L or D lactic acid is produced. NADH holds the key position in serving as an e-/H+ acceptor during the lactic acid formation to continue the process. It's important to mention that during NADH limitation, other electron acceptor (ethanol) is produced under low-pH condition. Recent finding suggests that regulation of lactococcal dehydrogenase (LDH) at substrate level occurs by at least two mechanisms: the fructose-1,6- bisphosphate/phosphate ratio and NADH/NAD+ ratio. It was reported that the key steps in parent strain where most of the carbon lost during lactic acid formation are reactions from pyruvate to acetyl Co-A, oxaloacetate, and ethanol (Skory 2004).

Aspergillus is naturally adapted for production of diverse range of acids such as citric acid (from *A. niger*) and itaconic acid (from *A. itaconicus* and *A. terreus*) at industrial scales (Chambergo and Valencia 2016; Yang 2017). Further, many experiment had proved that *Aspergillus* can grow well at very low pH, while *Rhizopus* still needs pH to be controlled (de Vries 2003), and thus, *Aspergillus* can metabolize pentose even at low pH. But still strain modification is done to abolish the by-product of ethanol production. Many strains of yeast which is a natural ethanol producer, when modified with the insertion of *ldh*, were able to produce L-lactic acid without any loss. However, this strategy works well and is helpful in almost doubling the production of lactic acid by using recombinant strains using metabolic tools (Fig. 5.2).

A very good review has been published by (Skory 2000) regarding role of two ldh genes in lactic acid production, viz., *ldh a* and *ldh b*. Activation of *ldh a/ldh b* depends on the presence of specific metabolite which results in more lactic acid production. On gene transfer of *ldh-a* in other microbes, it was reported that glucose > xylose >trehalose is the best substrate for activation of this gene (Skory 2000; Okino et al. 2008). During L-lactate production mostly *ldh-b* is activated. Therefore, when *ldh-a* gene transferred to yeast, it results in more L (+) lactic acid production. The resulting construct, pLdhA68X, is produced around 38 g D (−) lactic acid/l (40% more lactic acid) with a yield of 0.44 g lactic acid/g glucose in 30 h at pH 5.0 (Skory 2003).

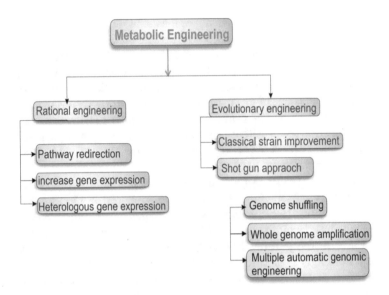

Fig. 5.2 Lactic acid production strategies via metabolic engineering. (Upadhyaya et al. 2014)

5.6 Factors Affecting Lactic Acid Production

Several media conditions affect lactic acid fermentation such as carbohydrate source, temperature, pH, sugar concentration, oxygen, metals, and some growth factor. Therefore, we will discuss about the role of (1) pH, (2) temperature, (3) metal ions, (4) $CaCO_3$, (5) carbon substrate, (6) culture condition, and (7) Tween 80, in overall lactic acid production.

1. *Effect of pH:* pH is an important media condition which lowers down continuously as lactic acid is produced; as a result product inhibition occurs (Itoh et al. 2003; Wang et al. 2014; Juturu and Wu 2016). One alternative way is to remove lactic acid regularly from the fermentation broth. But this could be stressful because of additional requirement of labor and time. Other suggested method is to use fungal cells that could grow well in acidic conditions, viz., *Rhizopus* and *Aspergillus*. Advantages of using LAB for lactic acid production are they can grow under anaerobic condition, at neutral or slightly acidic pH, and temperature range between 30–42 C, depending on species, while fungal growth requires different pH and temperature conditions for growth and production (low pH and Temp.). With the use of genetic engineering tools, they can be made more robust (Meussen et al. 2012).

   Optimization of media condition allows understanding of the actual impact of pH. There are several optimization tools which could be helpful in optimizing the pH and other physiological conditions such as Minitab, Design Expert, and Qualitek. Yields and rates of lactic acid production depend on the pH.

2. *Temperature:* Temperature is one of the most crucial factors affecting various production parameters both at lab scale or in reactors and, thus, microbial

growth and enzyme activation (Yun et al. 2003). Different strains have different temperature requirements. Thus, the growth of microbes lies in the range of 25–45 °C (Abdel-Rahman et al. 2013). LAB have low temperature requirements (15–25 °C), while fungi have high temperature requirements (27–40 °C) (Abdel-Rahman et al. 2013).

Moreover, hydrolysis of many substrates is dependent at high temperature (55 °C) which could increase yield up to 0.22 g/g at 37 °C (Liang et al. 2014). *Rhizopus* can grow up to 40 °C after metabolic modification (Meussen et al. 2012). More thermophilic strains have been reported for lactic acid production.

Lactic acid production is reported to improve at optimum temperature (22–30 °C) using *R. arrhizus* 36017 and *R. oryzae* 2062 from wastewater in SSF condition (Huang et al. 2005). *Rhizopus* in a reactor condition is shown to grow well even at 40 °C (Meussen et al. 2012). Temperature control is maintained with heat transfer coils using circulating water, and cells are retained in a mixed suspension by mechanical agitators or pump circulation.

3. *Metal ions:* Metal ions are important part of many metabolic enzymes and thus in many reports found to improve lactic acid. They are essentially required in minute amount to make an enzyme active. However, divalant metal ions have more important role in lactic acid production than monovalant metal ions. Zn^{++}, Mn^{++}, and Mg^{++} are highly beneficial in improving lactic acid since *ldh* is activated by synergistic interaction of metals and Ca^{++} is more effective in separation of lactic acid as a salt of lactate. Sometime monovalent cations such as Na or NH_3 consequently are used in industrial separation of lactate acid and also zinc lactate or zinc carbonate (Komesu et al. 2017).

4. *Effect of $CaCO_3$:* $CaCO_3$ is used to maintain pH during lactic acid production using LAB. In large commercial-scale fermentations, downstream processing is one of the most important cost-effective parameters (Bigelis and Tsai 1994). CO_2 is shown to stimulate growth of many lactic acid bacteria, while some fungi require it in least amount. Homolactic fermentation in LAB makes devoid CO_2 produced. Therefore, supplementation of calcium carbonate is essential in order to neutralize lactic acid for pH control during fermentation (Bigelis and Tsai 1994). Calcium carbonate is usually added around (12–15%) during the batch fermentation to control the pH, and the highest concentration of lactic acid is limited by the precipitation of calcium lactate from the broth.

Precipitation of lactic acid by $CaCO_3$ is done to obtain technical grade LA (22–44%); however, highly pure lactic acid (90%) can be obtained after esterification of lactic acid with methanol or ethanol (Datta and Henry 2006; Komesu et al. 2017). However, $CaCO_3$ addition increases the final cost of production, as large amount of wastewater is incurred during reagent separation, filtration, and separation. This could be avoided by the use of acid-tolerant species *Aspergillus brasiliensis* which could grow at pH 3.5 (Abdel-Rahman and Sonomoto 2016)).

5. *Feedstock:* Feedstock used for lactic acid production must meet the following norms: they should be of (1) low cost, (2) should be contaminant free, (3) easily could be hydrolyzed, (4) have less by-product formation, (5) and should be available year-round (Vickroy 1985). Various substrates which have been used in recent past are cane molasses, corncob starch, and lignocellulosic waste such as

Table 5.3 Substrate being used for lactic acid production by Rhizopus sp.

Microbes	Substrate
R. arrhizus	Glucose, xylose, starch (Zhang et al. 2007a, b)
R. oryzae	Glucose, xylose(Saito et al. 2012; Turner et al. 2015),cornstarch (Yin et al. 1997)
R. oryzae MTCC 8784	Uses food waste (fruit and vegetable peel/waste) like sapota, banana, papaya, potato, corncob, and carboxymethyl cellulose (Kumar and Shivakumar 2014)
R. oryzae GY18	Corncob, sucrose
R. oryzae NRRL 395	Biodiesel glycerol into lactic acid (Vodnar et al. 2013)
Mucor indicus, R. oryzae, and *Saccharomyces cerevisiae*	Wheat straw (Karimi et al. 2006)
R. arrhizus, strain DAR 36017	Starch waste (Jin et al. 2003)
Yeasts (genetically engineered)	
Kluyveromyces lactis	Glucose (Porro et al. 1999)
Saccharomyces cerevisiae	Glucose, xylose with PDC/adh deletion (Turner et al. 2015)
Pichia stipitis	Glucose, xylose (Ilmén et al. 2007)

wheat straw and paddy straw (Zhang et al. 2012); see Table 5.3. A list of such substrate being used is given in Table 5.2. Since substrates cost around 30–60% of the total production cost (Ruengruglikit and Hang 2003), its choice is critical in deciding the quality and purity of lactic acid produced. Long-distance transport of feedstocks also is not suggested since additional cost in transportation will be involved and, thus, should be avoided. Complex carbohydrates as feedstocks require pretreatment which consumes electricity. Purified sucrose (from sugar beets and sugarcane), glucose from hydrolyzed starch sources such as corn, and whey containing lactose are the major raw materials used today since they insure lower cost of the final product recovery.

However, some work has been done in which integrated lignocellulosic hydrolysis for lactic acid production process has been developed where carboxymethyl cellulose and xylose were used as substrates (Zhang et al. 2007a, b; John et al. 2009). Woody biomass with hemicellulose content can be used for LA production, but lignin removal is required prior to fermentation. In recent past, biological pretreatment method is an attractive option, where high level of economy can be achieved along with an environmental protection. Xylose present in hemicellulose fraction can be fermented to lactic acid by the use of *R. arrhizus* strain in synthetic media, but the productivity observed was very low (Maas et al. 2006) Table 5.3.

Corncob (rich in cellulose and hemicellulose) can be hydrolyzed and transformed to lactic acid by using engineered fungi *R. arrhizus* NRRL 395 and *Rhizopus* species MK-19 after alkali pretreatment. Corncob could be effectively converted into lactic acid. More encouraging result has been obtained from the mixture of microbes such as *Rhizopus* along with other cellulase producer strain;

Acremonium thermophilus ATCC 24622 in SSF mode proved to be efficient for LA production. This strategy is shown to have better hydrolysis of cellulose. Low-oxygen condition employed with the mixture of nitrogen source could favor high lactic acid production. Miura et al. (2004) reported that the ratio of C/N also affects lactic acid production. Now it has proved that low ratio of C/N is better for high lactic acid production, while low phosphate is better for cell growth and high lactic acid production (0.2–1.6 gram lactic acid) which can be produced using prescribed media conditions (Zhang et al. 2007a, b).

Research shows that *Rhizopus* mainly uses glucose or starch for lactic acid productions. Works of some researchers prove that (Oda et al. 2002) various food and agriculture wastes can be used for lactic acid productions such as pulp and starch waste (Kumar and Shivakumar 2014) by the use of *R. oryzae or R. arrhizus*. Addition of lytic enzymes such as xylolytic enzyme (Maas et al. 2006; Saito et al. 2012) could enhance lactic acid production further. Further inclusion of pectinases enhances hydrolysis of pectin that could result in enhanced lactic acid production (Oda et al. 2002) in SSF made with *R. oryzae* with potato pulp. SSF mode is a form of fermentation wherein one-step complete hydrolysis and saccharification can be performed.

6. *Tween 80:* Tween 80 is a surfactant, and its addition increases lactic acid production. Some lactic acid bacteria require added fatty acids or fatty acid esters (Tweens 80), while fungi need only basic media. Therefore, supplementation of Tween 80 is essential sometime. Often fatty acids can be synthesized by these organisms when the vitamin, biotin, is added to the fermentation broth (Bigelis and Tsai 1994).

7. *Effect of nitrogen content:* Nitrogen is one of the most important factors in deciding the yield of lactic acid since it is a part of nucleic acid and protein content also. *Aspergillus brasiliensis* BRFM1877 was reported to produce lactic acid in acidic condition. The use of yeast extract and ammonium sulfate resulted in enhanced lactic acid production with *R. oryzae* TS-61. The benefit of using excessive nitrogen source is that mycelia floc did not form during lactic acid production (Yu et al. 2007). Using Na nitrate as a nitrogen source has a benefit that can help in the maintenance of pH and thus results in high (1.8-fold) lactic acid production. The productivity of large batch commercial reactors was low, at 1–3 g/l.hr. (Liaud et al. 2015).

8. *Reactor condition:* Batch, fed-batch, repeated batch, and continuous fermentations are the most frequently used methods for lactic acid production. However, higher lactic acid concentrations have been reported in batch and fed batch than in continuous cultures, whereas higher productivity may be achieved by the use of continuous cultures. Another advantage of the continuous culture compared to the batch culture is the possibility to continue the process for a longer period of time. The factors affecting the batch process have been reviewed (Litchfield 1996). Other factors affecting LA production are corrosive properties of lactic acid, construction materials used by the fermenter, and downstream processing equipment which are a major cost item. Therefore, reactor equipments made up

of copper, copper alloys, steel, chrome steel, and high-nickel steels are all unsatisfactory. High-molybdenum stainless steel like SS316 is satisfactory. Plastic linings of fermentation tanks have been used successfully, and new developments in ceramics and plastics may provide future choices (Litchfield 1996). Semicontinuous fermentation mode using palette of Rhizopus cells resulted in high lactic acid concentration (103.7 g/l), and the volumetric productivity was 2.16 g/(l h) for the first cycle; and after 19 repeated cycles, the final LAC reached 81–95 g/l, and the volumetric productivities were 3.40–3.85 g/(l h) (Wu et al. 2011).

9. *Economics of lactic acid production:* The complete economics of lactic acid fermentation is associated with substrates used, inexpensive nitrogen sources used, and other media conditions required for the cell growth and production. In addition, some more factors are there which affect the economics of lactic acid production such as the cost of downstream processing and overall purification steps. High lactic acid must have adaquate pH control (5–7) and other conditions but they may not be cost effective. For economical production of lactic acid, the use of *Rhizopus* seems more beneficial as compared to LAB owing to need of costly media composition and costly recovery and purification steps. Genetically modified strain produced by mutation in many fungi resulted in high production, thus, making the process economical with reduced fermentation time (Juturu and Wu 2016).

The major economics of lactic acid depends on various factors as discussed previously (Juturu and Wu 2016). Various economical raw materials have been used for lactic acid production via improvement of lactic acid production in recent past which are rich in lignocellulosic biomass.

5.7   Simultaneous Saccharification and Fermentation (SSF)

SSF mode has been explored extensively in recent past, for lactic acid production. Lignocellulosic biomass or other agricultural waste has been used in the past for ethanol production (Juturu and Wu 2016) and single-cell protein production but less for lactic acid production (Jin and Zhang 2008). In SSF mode, *R. arrhizus* DAR 36017 and *R. arrhizus* NRRL 2062 have been used for lactic acid production in shake-flask condition. SSF kinetics is reported to be affected by pH, temperature, substrate hydrolysis, and lactic acid concentration (Takano and Hoshino 2016), and optimized conditions for cell growth were temperature, 30 °C; pH, 6; and starch concentration, 20 g/l, resulting in lactic acid yield of 2.92 g/gram of substrate, in 36–48 h. High starch concentration (100 g/l) is possible to be used at industrial scale for lactic acid production after proper modeling of media condition (de J. C. Munanga et al. 2016).

In SSF, there is no need of complete hydrolysis, and the process starts even prior to fermentation such as cell growth, and microbial production starts simultaneously.

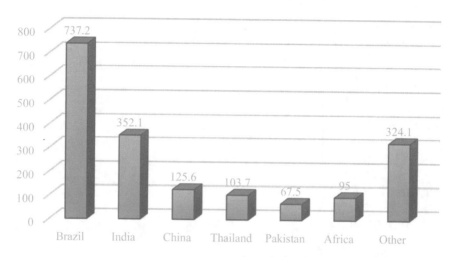

Fig. 5.3 Availability of sugarcane in 2014 million t/y source. (Mandegari et al. 2017)

This mechanism could be helpful in checking substrate inhibition, and thus enhanced saccharification could result in enhanced lactic acid productivity. In addition to reduced reactor volume, less capital cost is utilized. In solid state fermentation, many metabolic processes start even prior to fermentation such as cell growth, and microbial based lactic acid production starts simultaneously (Romaní et al. 2008).

Techno-Economical Aspects
Bio-based biochemical industry has a sustainability challenge which needs continuous scientific innovations, for example, lactic acid production using first- and second-generation biomass (Rocha et al. 2014; Mandegari et al. 2017). It needs a lot of extensive technical evaluation for economical calculation based on many feasible models. The model is evaluated on various agrowastes, biomass used for lactic acid production or for production of other biochemicals.

Biomass, such as lignocellulose, sugarcane bagasse, and other food waste, calculation shows that sustainability issue lies in raw material used, transportation, electricity, purification, and the final product as shown in Fig. 5.3.

In Fig. 5.3, it can be observed that Brazil has the highest production of sugarcane, million tons per year, while India lies at the second place. The systematic tool where economic viability can be analyzed is via SWOT analysis where both positive and negative aspects during manufacturing of product can be calculated. It was reported that total capital cost (TCI) was high using normal strain, while with acid-tolerant strain, TCI was low (Gezae Daful and Görgens 2017). Further experimental work based on simulation shows that lignocellulose agricultural waste has high TCI and operating cost (OPEX). In addition, gypsum-based processing of lactic acid has higher TCI and OPEX as compared

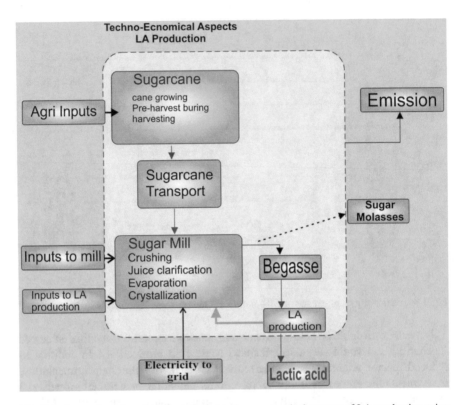

Fig. 5.4 System boundary for calculation of techno-economical aspects of LA production using sugarcane. (Gezae Daful and Görgens 2017)

to Mg(OH)2-based processing for lactic acid extraction for economical consideration. For economical consideration, it is mandatory to have low TCI and OPEX, for example, low TCI (7.08%) has been observed using gypsum-free process (using acid-resistant strain) as compared to using gypsum or Mg (OH)2 process. Internal rate of return IRR was higher in case of gypsum-free process using cellulose or hemicellulose process (Daful and Görgens 2017).

In Haylock 2016, various simulation tools have been used in the recent past, for example, correlation matrix has been well applicable for various bio-based products. In Fig. 5.4, it can be observed that during techno-economical aspect calculation, various system boundaries as input and output must be included: first agri-input, second input tools, third input tool activation production and for electricity consumption of lactic acid production let it is a production; see Fig. 5.4 (Haylock 2016) (Table 5.4).

Table 5.4 Techno-economical aspects of LA production

S.N.	Chemicals	Price	Unit	Ref.
1.	Mg (OH)₂	300	$/Mt	ChemFine international co., ltd. http://chemfine.en.forbuyers.com
2.	Trimethylamine	1850–1870	$/Mt	ChemFine international co., ltd. http://chemfine.en.forbuyers.com
3.	Lactic acid	1300–5000	$/Mt	Cellulac
4.	Bagasse	9.94	$/Mt	Gezae Daful and Görgens (2017)
5.	Ethanol	200	$/Mt	Gezae Daful and Görgens (2017)
6.	Water	0.00359	$/L	http://www.sawater.com.au/SAWater/YourHome/YourAccountBillPaymentCharges/Pricing+Information.htm
7.	Calcium hydroxide	300	$/Kg	ChemFine international co., ltd. http://chemfine.en.forbuyers.com
8.	H₂ SO₄ 93%	0.08798	$/Mt	Humbird et al. (2011)
9.	Electricity	0.0714	$/Kw.h	Mashoko et al. (2010)
10.	Enzyme nutrients	0.821	$/Kg	Humbird et al. (2011)
11.	Molasses	320	$/ton	Dias et al. (2011) and ChemFine international co., ltd.
12.	Sugar	760	$/ton	Dias et al. (2011)

Gezae Daful and Görgens (2017)

Table 5.5 Production of lactic acid by R. oryzae

SN	Substrate	Process	Yield (%)	Productivity (g/l/h)	Concentration (g/l)	References
1	Cotton flocs	Batch	87			
2	Cellulosic biosludges	Batch fed batch	37.8	0.87	39.4 g/L	Romaní et al. (2008)
3	Potato starch wastewater	Batch	85–92	na	na	Huang et al. (2005)
4	Xylose (below 40%)	Batch	41–71	na	na	Maas et al. (2006)
5	Wheat straw powder	Stirred tank	23	na	na	Saito et al. (2012)
6	Corncob	Airlift	na	na	24 g/l	Miura et al. (2004)
7	Cassava pulp	Batch	20.4	na	na	Phrueksawan et al. (2012)

5.8 Process Optimization

As shown in Table 5.5, lactic acid production varies in different modes such as batch, stirred tank, and airlift bioreactor conditions. Many authors show good productivity in airlift bioreactor (Maneeboon et al. 2010), and optimized condition lactic acid concentration was around 85% (102 g/l) using *R. oryzae* NRRL395 from cornstarch partially hydrolyzed with amylase. Process optimization shows ammonium sulfate as one of the best nitrogen sources out of yeast extract, CSL, and peptone. As compared to batch mode, airlift model has a benefit, that is, mycelia do not have an opportunity to make flocs, and thus production occurs in continuous mode. In 5 L stirred tank bioreactor, experiment was done, and it was reported that lactic acid concentration was 60 g/l in SSF mode with a yield of 0.34 g/g dry waste (Zhang et al. 2015). As compared to free cells, *Rhizopus oryzae, an immobilized cell has better performance in static bed bioreactor*, and observed yield was 2.9 g/l/h from 70 g/l glucose (Fig. 5.5).

Process optimization is recommended for enhanced lactic acid production at industrial scale. Conventional methods are not only cumbersome but also laborious, where many experiments are required to conduct where one factor is constant while other factors are variable.

There are advanced mechanisms to optimize process parameters affecting lactic acid production such as central composite design (CCD), response surface methodology (RSM) (Maneeboon et al. 2010), Box-Behnken design (Gupta et al. 2010), and Taguchi (Bhatt and Srivastava 2008). Taguchi methodology has advantages that individual factor can be accounted for lactic acid production and beside knowledge of interaction between factors and their overall impact on lactic acid production.

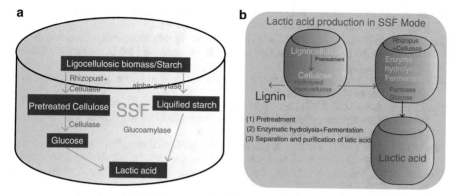

Fig. 5.5 SSF (**a**) Starch versus cellulose (**b**) Cellulosic SSF for LA production which involves [(1) pretreatment (2) enzymatic hydrolysis + fermentation (3) separation and purification of lactic acid]. (Zhang et al. 2007a, b; da Silva et al. 2013)

5.9 Future Prospectus and Conclusion

PLA finds major applications in tissue engineering (Gunatillake and Adhikari 2003) owing to its nature of being biodegradable and of having thermoplastic quality for sustainable use (Gupta et al. 2007). Implants prepared from PLA (Gad et al. 2008) had proved better in treatments of periodontitis, a tooth disease where teeth are damaged. In this context, more technological innovation is required, for large-scale use of agricultural waste for economical lactic acid production. In this respect, certain fungi may prove boon and can be made robust after metabolic engineering, for example, *R. oryzae* and *Aspergillus oryzae* (Ruengruglikit and Hang 2003). In conclusion, economical lactic acid production can be achieved which will be helpful in reducing prices of lactic acid further.

Acknowledgment The author is expressing a mammoth of thanks to our Vice Chancellor Prof. J. S. Bal for extending his encouragement and support. The author is also thankful to Prof. (Dr.) Vijay Dhir whose extended support was felt everywhere. This work can't be completed without the support of Dr. Sachin, a leading scientist in SSS-NIRE Kapurthala. No conflict of interest lies in preparation of this manuscript.

References

Abdel-Rahman MA, Sonomoto K (2016) Opportunities to overcome the current limitations and challenges for efficient microbial production of optically pure lactic acid. J Biotechnol 236:176–192

Abdel-Rahman MA, Tashiro Y, Zendo T, Shibata K, Sonomoto K (2011) Isolation and characterisation of lactic acid bacterium for effective fermentation of cellobiose into optically pure homo l-(+)-lactic acid. Appl Microbiol Biotechnol 89:1039–1049

Abdel-Rahman MA, Tashiro Y, Sonomoto K (2013) Recent advances in lactic acid production by microbial fermentation processes. Biotechnol Adv 31:877–902

Bai D-M, Zhao X-M, Li X-G, Shi-Min X (2004) Strain improvement and metabolic flux analysis in the wild-type and a mutant lactobacillus lactis strain for l (+)-lactic acid production. Biotechnol Bioeng 88(6):681–689

Bhatt SM (2013) Developments in Cellulase activity improvements intended towards biofuel production. J Bacteriol Parasitol 4:e120

Bhatt SM, Srivastava SK (2008) Lactic acid production from cane molasses by lactobacillus delbrueckii NCIM 2025 in submerged condition: optimization of medium component by Taguchi DOE methodology. Food Biotechnol 22(2):115–139

Bigelis R, Tsai SP (1994) Microorganisms for organic acid productions. Chapter 6. In: Hui YH, Khachatourians GG (eds) Food biotechnology: microorganisms. VCH Publishers, New York

Chambergo FS, Valencia EY (2016) Fungal biodiversity to biotechnology. Appl Microbiol Biotechnol 100(6):2567–2577

da Silva, Sant'Ana A, Teixeira RSS, Moutta R d O, Ferreira-Leitão VS, de Barros R d RO, Ferrara MA, Bon EP d S (2013) Sugarcane and woody biomass pretreatments for ethanol production. In: Sustainable degradation of Lignocellulosic Biomass-techniques, applications and commercialization. InTech

Datta R, Henry M (2006) Lactic acid: recent advances in products, processes and technologies – a review. J Chem Technol Biotechnol 81(7):1119–1129

Datta R, Tsai SP (1997) Lactic acid production and potential uses: a technology and economics assessment. In: Fuels and chemicals from biomass, vol 666, pp 224–236

de Munanga JC, Bettencourt GL, Grabulos J, Mestres C (2016) Modeling lactic fermentation of Gowé using lactobacillus starter culture. Microorganisms 4(4):pii: E44

de Vries R (2003) Regulation of Aspergillus genes encoding plant cell wall polysaccharide-degrading enzymes; relevance for industrial production. Appl Microbiol Biotechnol 61(1):10–20

Dias MOS, da Cunha MP, Filho RM, Bonomi A, Jesus CDF, Rossell CEV (2011) Simulation of integrated first and second generation bioethanol production from sugarcane: comparison between different biomass pretreatment methods. J Ind Microbiol Biotechnol 38(8):955–966

Gad HA, El-Nabarawi MA, El-Hady SSA (2008) Formulation and evaluation of PLA and PLGA in situ implants containing secnidazole and/or doxycycline for treatment of periodontitis. AAPS PharmSciTech 9(3):878

Garvie EI (1980) Bacterial lactate dehydrogenases. Microbiol Rev 44(1):106–139

Gezae Daful A, Görgens JF (2017) Techno-economic analysis and environmental impact assessment of lignocellulosic lactic acid production. Chem Eng Sci 162:53–65

Ghosh B, Ray RR (2011) Current commercial perspective of Rhizopus oryzae: a review. J Appl Sci 11(14):2470–2486

Gunatillake PA, Adhikari R (2003) Biodegradable synthetic polymers for tissue engineering. Eur Cell Mater 5(1):1–16

Guo F, Fang Z, Zhou TJ (2012) Conversion of fructose and glucose into 5-hydroxymethylfurfural with lignin-derived carbonaceous catalyst under microwave irradiation in dimethyl sulfoxide-ionic liquid mixtures. Bioresour Technol 112:313–318

Gupta B, Revagade N, Hilborn J (2007) Poly(lactic acid) fiber: an overview. Prog Polym Sci 32(4):455–482. http://linkinghub.elsevier.com/retrieve/pii/S007967000700007X

Gupta S, Cox S, Abu-Ghannam N (2010) Process optimization for the development of a functional beverage based on lactic acid fermentation of oats. Biochem Eng J 52(2):199–204

Hamzah F, Idris A, Rashid R, Ming SJ (2009) Lactic acid production from microwave-alkali pretreated empty fruit bunches fibre using Rhizopus oryzae pellet. J Appl Sci 9(17):3086–3091

Haylock RA (2016) Life-cycle assessment, techno-economic analysis, and statistical modeling of bio-based materials and processes. Iowa State University

Holten CH (1971) Lactic acid. Properties and chemistry of lactic acid and derivatives. Verlag Chemie GmbH, Weinheim, German Federal Republic

Huang LP, Bo J, Lant P, Zhou J (2005) Simultaneous saccharification and fermentation of potato starch wastewater to lactic acid by Rhizopus oryzae and Rhizopus arrhizus. Biochem Eng J 23(3):265–276

Humbird D, Davis R, Tao L, Kinchin C, Hsu D, Aden A, Schoen P, Lukas J, Olthof B, Worley M, et al (2011) Process design and economics for biochemical conversion of lignocellulosic biomass to ethanol: dilute-acid pretreatment and enzymatic hydrolysis of corn stover

Ilmén M, Koivuranta K, Ruohonen L, Suominen P, Penttilä M (2007) Efficient production of L-lactic acid from xylose by *Pichia stipitis*. Appl Environ Microbiol 73:117–123. http://www.pubmedcentral.nih.gov/articlerender.fcgi?artid=1797125&tool=pmcentrez&rendertype=abstract

Ilmén M, Koivuranta K, Ruohonen L, Rajgarhia V, Suominen P, Penttilä M (2013) Production of L-lactic acid by the yeast Candida sonorensis expressing heterologous bacterial and fungal lactate dehydrogenases. Microb Cell Factories 12(1):53

Itoh H, Wada M, Honda Y, Kuwahara M, Watanabe T (2003) Bioorganosolve pretreatments for simultaneous saccharification and fermentation of beech wood by ethanolysis and white rot fungi. J Biotechnol 103:273–280

Jin T, Zhang H (2008) Biodegradable polylactic acid polymer with nisin for use in antimicrobial food packaging. J Food Sci 73(3):M127

Jin B, Li PH, Lant P (2003) Rhizopus arrhizus--a producer for simultaneous saccharification and fermentation of starch waste materials to L(+)-lactic acid. Biotechnol Lett 25(23):1983–1987

John RP, Anisha GS, Madhavan Nampoothiri K, Pandey A (2009) Direct lactic acid fermentation: focus on simultaneous saccharification and lactic acid production. Biotechnol Adv 27(2):145–152

Juturu V, Wu JC (2016) Microbial production of lactic acid: the latest development. Crit Rev Biotechnol 36(6):967–977

Kandler O (1983) Carbohydrate metabolism in lactic acid bacteria. Antonie Van Leeuwenhoek 49(3):209–224

Karimi K, Emtiazi G, Taherzadeh MJ (2006) Ethanol production from dilute-acid pretreated rice straw by simultaneous saccharification and fermentation with Mucor indicus, Rhizopus oryzae, and Saccharomyces cerevisiae. Enzym Microb Technol 40(1):138–144

Komesu A, Maciel MRW, Filho RM (2017) Separation and purification Technologies for Lactic Acid–a Brief Review. Bioresources 12(3):6885–6901

Kumar R, Shivakumar S (2014) Production of L-lactic acid from starch and food waste by amylolytic Rhizopus oryzae MTCC 8784. Int J ChemTech Res 6(1):527–537

Liang S, McDonald AG, Coats ER (2014) Lactic acid production with undefined mixed culture fermentation of potato peel waste. Waste Manag 34(11):2022–2027

Liaud N, Rosso M-N, Fabre N, Crapart S, Herpoël-Gimbert I, Sigoillot J-C, Raouche S, Levasseur A (2015) L-lactic acid production by Aspergillus brasiliensis overexpressing the heterologous ldha gene from Rhizopus oryzae. Microb Cell Factories 14(1):66. http://www.pubmedcentral.nih.gov/articlerender.fcgi?artid=4425913&tool=pmcentrez&rendertype=abstract

Litchfield JH (1996) Microbiological production of lactic acid. Adv Appl Microbiol 42:45–95

Liu Y, Liao W, Chen S-l (2008) Co-production of lactic acid and chitin using a pelletized filamentous fungus Rhizopus oryzae cultured on cull potatoes and glucose. J Appl Microbiol 105(5):1521–1528

Maas RHW, Bakker RR, Eggink G, Weusthuis RA (2006) Lactic acid production from xylose by the fungus Rhizopus oryzae. Appl Microbiol Biotechnol 72(5):861–868

Mandegari MA, Farzad S, Görgens JF (2017) Recent trends on techno-economic assessment (TEA) of sugarcane biorefineries. Biofuel Res J 4(3):704–712

Maneeboon T, Vanichsriratana W, Pomchaitaward C, Kitpreechavanich V (2010) Optimization of lactic acid production by pellet-form Rhizopus oryzae in 3-L airlift bioreactor using response surface methodology. Appl Biochem Biotechnol 161(1–8):137–146

Mashoko L, Mbohwa C, Thomas VM (2010) LCA of the south African sugar industry. J Environ Plan Manag 53(6):793–807

Meussen BJ, de Graaff LH, Sanders JPM, Weusthuis RA (2012) Metabolic engineering of Rhizopus oryzae for the production of platform chemicals. Appl Microbiol Biotechnol 94(4):875–886

Mirasol FELIZA (1999) Lactic acid prices falter as competition toughen. Chem Mark Report 255:16

Miura S, Arimura T, Itoda N, Dwiarti L, Feng JB, Bin CH, Okabe M (2004) Production of l-lactic acid from corncob. J Biosci Bioeng 97(3):153–157

Oda Y, Saito K, Yamauchi H, Mori M (2002) Lactic acid fermentation of potato pulp by the fungus Rhizopus oryzae. Curr Microbiol 45(1):1–4

Okano K, Tanaka T, Ogino C, Fukuda H, Kondo A (2010) Biotechnological production of enantiomeric pure lactic acid from renewable resources: recent achievements, perspectives, and limits. Appl Microbiol Biotechnol 85(3):413–423

Okino S, Suda M, Fujikura K, Inui M, Yukawa H (2008) Production of D-lactic acid by Corynebacterium glutamicum under oxygen deprivation. Appl Microbiol Biotechnol 78(3):449–454

Perry's RH, Chilton CH, Kirkpatrick SD (1999) Chemical engineers handbook

Phrueksawan P, Kulpreecha S, Sooksai S, Thongchul N (2012) Direct fermentation of L(+)-lactic acid from cassava pulp by solid state culture of Rhizopus oryzae. Bioprocess Biosyst Eng 35(8):1429–1436

Ren J (2011) Lactic acid. In: Ren J (ed) Biodegradable poly (lactic acid): synthesis, modification, processing and applications. Tsinghua University Press, Beijing and Springer-Verlag Berlin Heidelberg, pp 4–14

Rocha MH, Capaz RS, Lora EES, Nogueira LAH, Leme MMV, Renó MLG, del Olmo OA (2014) Life cycle assessment (LCA) for biofuels in Brazilian conditions: a meta-analysis. Renew Sust Energ Rev 37:435–459

Romaní A, Yáñez R, Garrote G, Luis Alonso J (2008) SSF production of lactic acid from cellulosic biosludges. Bioresour Technol 99(10):4247–4254

Ruengruglikit C, Hang YD (2003) L(+)-lactic acid production from corncobs by Rhizopus oryzae NRRL-395. LWT Food Sci Technol 36(6):573–575

Saito K, Hasa Y, Abe H (2012) Production of lactic acid from xylose and wheat straw by Rhizopus oryzae. J Biosci Bioeng 114(2):166–169

Sarkar D, Datta R (2004) Arsenic fate and bioavailability in two soils contaminated with sodium arsenate pesticide: an incubation study. Bull Environ Contam Toxicol 72(2):240–247

Schaechter M (2009) Encyclopedia of microbiology. Academic Press

Skory CD (2000) Isolation and expression of lactate dehydrogenase genes from Rhizopus oryzae. Appl Environ Microbiol 66(6):2343–2348. http://www.ncbi.nlm.nih.gov/pmc/articles/PMC110528/pdf/am002343.pdf

Skory CD (2003) Lactic acid production by Saccharomyces cerevisiae expressing a Rhizopus oryzae lactate dehydrogenase gene. J Ind Microbiol Biotechnol 30(1):22–27. http://www.ncbi.nlm.nih.gov/pubmed/12545382

Skory CD (2004) Lactic acid production by Rhizopus oryzae transformants with modified lactate dehydrogenase activity. Appl Microbiol Biotechnol 64(2):237–242

Takano M, Hoshino K (2016) Lactic acid production from paper sludge by SSF with thermotolerant Rhizopus sp. Bioresour Bioproces 3(1):29

Taskin M, Esim N, Ortucu S (2012) Efficient production of l-lactic acid from chicken feather protein hydrolysate and sugar beet molasses by the newly isolated Rhizopus oryzae TS-61. Food Bioprod Process 90(4):773–779

Thitiprasert S, Sooksai S, Thongchul N (2011) In vivo regulation of alcohol dehydrogenase and lactate dehydrogenase in Rhizopus oryzae to improve L-lactic acid fermentation. Appl Biochem Biotechnol 164(8):1305–1322

Turner TL, Zhang GC, Kim SR, Subramaniam V, Steffen D, Skory CD, Ji YJ, Byung Jo Y, Jin YS (2015) Lactic acid production from xylose by engineered Saccharomyces cerevisiae without PDC or ADH deletion. Appl Microbiol Biotechnol 99(19):8023–8033

Upadhyaya BP, DeVeaux LC, Christopher LP (2014) Metabolic engineering as a tool for enhanced lactic acid production. Trends Biotechnol 32:637–644

Varadarajan S, Miller DJ (1999) Catalytic upgrading of fermentation-derived organic acids. Biotechnol Prog 15(5):845–854

Vickroy TB (1985) Lactic acid

Vodnar DC, Dulf FV, Pop OL, Socaciu C (2013) L (+)-lactic acid production by pellet-form Rhizopus oryzae NRRL 395 on biodiesel crude glycerol. Microb Cell Factories 12(1):92. http://www.microbialcellfactories.com/content/12/1/92

Wang Y, Abdel-Rahman MA, Tashiro Y, Xiao Y, Zendo T, Sakai K, Sonomoto K (2014) L-(+)-lactic acid production by co-fermentation of cellobiose and xylose without carbon catabolite repression using enterococcus mundtii QU 25. RSC Adv 4(42):22013–22021. https://doi.org/10.1039/C4RA02764G

Wang Y, Chen C, Di C, Wang Z, Qin P, Tan T (2016) The optimization of l-lactic acid production from sweet sorghum juice by mixed fermentation of Bacillus coagulans and lactobacillus rhamnosus under unsterile conditions. Bioresour Technol 218:1098–1105

Wu X, Jiang S, Mo L, Pan L, Zheng Z, Luo S (2011) Production of L-lactic acid by Rhizopus oryzae using semicontinuous fermentation in bioreactor. J Ind Microbiol Biotechnol 38(4):565–571

Yamada R, Wakita K, Mitsui R, Ogino H (2017) Enhanced d-lactic acid production by recombinant Saccharomyces cerevisiae following optimization of the global metabolic pathway. Biotechnol Bioeng 114(9):2075–2084

Yang R (2017) Production of ethanol from Sudanese sugar cane molasses and evaluation of its quality. J Food Process Technol 3(7):3–5. http://www.omicsonline.org/2157-7110/2157-7110-3-163.digital/2157-7110-3-163.html

Yin P, Nishina N, Kosakai Y, Yahiro K, Park Y, Okabe M (1997) Enhanced production of L(+)-lactic acid from corn starch in a culture of Rhizopus oryzae using an air-lift bioreactor. J Ferment Bioeng 84(3):249–253

Yu MC, Wang RC, Wang CY, Duan KJ, Sheu DC (2007) Enhanced production of L(+)-lactic acid by floc-form culture of Rhizopus oryzae. J Chin Inst Chem Eng 38(3–4):223–228

Yun JS, Wee YJ, Ryu HW (2003) Production of optically pure L(+)-lactic acid from various carbohydrates by batch fermentation of enterococcus faecalis RKY1. Enzym Microb Technol 33:416–423

Zhang ZY, Jin B, Kelly JM (2007a) Production of lactic acid from renewable materials by Rhizopus Fungi. Biochem Eng J 35:251–263

Zhang ZY, Jin B, Kelly JM (2007b) Production of lactic acid and byproducts from waste potato starch by Rhizopus arrhizus: role of nitrogen sources. World J Microbiol Biotechnol 23(2):229–236

Zhang Y, Han B, Ezeji TC (2012) Biotransformation of furfural and 5-hydroxymethyl furfural (HMF) by Clostridium acetobutylicum ATCC 824 during butanol fermentation. New Biotechnol 29:345–351

Zhang L, Li X, Yong Q, Yang ST, Ouyang J, Yu S (2015) Simultaneous saccharification and fermentation of xylo-oligosaccharides manufacturing waste residue for l-lactic acid production by Rhizopus oryzae. Biochem Eng J 94:92–99

Zhang L, Li X, Yong Q, Yang S-T, Ouyang J, Yu S (2016) Impacts of lignocellulose-derived inhibitors on l-lactic acid fermentation by Rhizopus oryzae. Bioresour Technol 203(March):173–180

Chapter 6
Exploiting Innate and Imported Fungal Capacity for Xylitol Production

Shaik Jakeer

Abstract Xylitol, a pentahydroxy chiral polyol, is a natural noncaloric sweetener with a wide spectrum of applications in food, confectionary, and pharmaceutical industries because of its advantageous properties. Industrial-scale production of xylitol from D-xylose derived from hemicellulosic hydrolysates is usually done by a chemical process by catalytic hydrogenation under high pressure and temperature. However, the sustainability issue boosted the biotechnological process. Much of the research is being focused on engineering metabolic pathways to improve the biological production of xylitol in both native xylitol-producing and nonproducing-fungal strains. This chapter provides a limelight on native fungal strains and the advances made in fungal metabolic engineering to increase the production of xylitol.

6.1 Introduction

Xylitol is a five-carbon sugar alcohol (molecular formula $C_5H_{12}O_5$), also known as a polyol or polyhydroxy alcohol acyclic. As an alternative sweetener, xylitol has attracted the attention of food and pharmaceutical industries as it has same sweetness as sucrose, with an energy value of only 2.4 cal/g when compared to sucrose that gives 4 cal/g. D-xylose is a major C5 sugar derived from hydrolysis of lignocellulosic biomass after glucose (Fig. 6.1a). Like any other carbohydrate, D-xylose is also susceptible to transformation, including reduction. For xylitol production, both chemical and biotechnological processes have been developed using D-xylose as the carbon source (Jeon et al. 2012; Ko et al. 2006; Moyses et al. 2016; Mussatto and Roberto 2003). In the chemical process, D-xylose is hydrogenated to xylitol

S. Jakeer (✉)
International Centre for Genetic Engineering and Biotechnology, Aruna Asaf Ali Marg, New Delhi, India

Shraga Segal Department of Microbiology, Immunology, and Genetics, Ben-Gurion University of the Negev, Beer-Sheva, Israel
e-mail: jakeer@post.bgu.ac.il

© Springer International Publishing AG, part of Springer Nature 2018
S. Kumar et al. (eds.), *Fungal Biorefineries*, Fungal Biology,
https://doi.org/10.1007/978-3-319-90379-8_6

107

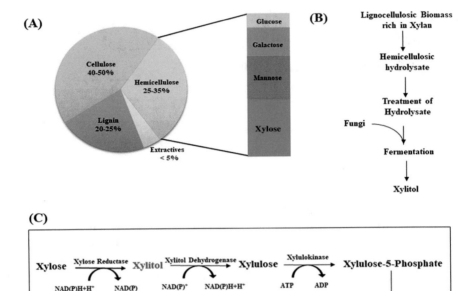

Fig. 6.1 Process involved in xylose extraction and xylitol production from plant biomass. (**a**) Typical composition of lignocellulose and hemicellulose. (**b**) Schematic presentation of xylose extraction and xylitol production using lignocellulosic biomass as substrate. (**c**) Metabolic pathway of xylitol production in fungi (native fungal pathway)

using nickel, Raney nickel, ruthenium, rhodium, and palladium catalysts (Yadav et al. 2012) at high temperature (80–140 °C) and high pressure (up to 50 atm) (Mussatto and Roberto 2003). This process demands the need for high purity D-xylose to reduce purification cost and a series of other purification steps that requires various operation units such as ion exchange, chromatographic separation, and discoloration (Sampaio et al. 2003). Concerning the sustainability issue, the biotechnological process for xylitol production using various alternative feedstocks has been evaluated using various microorganisms. This process of xylitol production is conventionally employed as it is less expensive and advantageous over the chemical process. Xylitol, being an intermediary product formed during carbohydrate metabolism, can be produced through microbiological routes, where a microorganism converts D-xylose to xylitol. The genera *Candida* (Li et al. 2012; Zhang et al. 2015), *Debaryomyces* (Carvalheiro et al. 2007), *Kluyveromyces* (Rocha et al. 2011; Zhang et al. 2014), and *Pichia* (Min-Soo, 2000; Rodrigues et al. 2011) have been explored and developed to be used in the microbial conversion of D-xylose to xylitol (Meinander and Hahn-Hägerdal 1997; Min-Soo 2000; Oh et al. 2012; Rao et al. 2006; Walfridsson et al. 1997; Zhang et al. 2015). Apart from using native organisms, studies have focused on the genus *Saccharomyces* (Moyses et al. 2016), to improve the enzymatic process, to achieve a theoretical yield of 100% D-xylose to xylitol conversion as these fungal species cannot not use xylitol for metabolic diversions or cell maintenance (Chung et al. 2002; Hallborn et al. 1994, 1991; Kwon et al. 2006a; Lee et al. 2000).

6.2 Microbial Production of Xylitol

Xylose is the second most abundant sugar component of lignocellulosic biomass after glucose (Fig. 6.1a). Many fungal and yeast species have been evaluated to produce xylitol using biotechnological methods. Moreover, it is less expensive and has the potential to be an advantageous alternative to the conventionally employed chemical process. During xylose metabolism, the enzyme activity of XR (xylose reductase) and NADPH supply play a significant role in xylitol production (Fig. 6.1b). High activity of XR is a major factor in wild-type strains and XR over-expressing recombinant strains for efficient xylitol production. Among the several yeast strains that contain native xylose reductase gene, the genus *Candida* has been marked as the highest xylitol-producing organism. These yeast strains use xylose as a carbon source for both xylitol production and energy metabolism. *C. tropicalis, C. guilliermondii, C. parapsilosis, C. mogii, C. shehatae*, and others can be included in the category (Winkelhausen and Kuzmanova 1998). In Table 6.1, we have listed both filamentous fungi and yeast strains that have been employed for xylitol production (Park et al. 2014).

In natural xylose-utilizing yeast, the conversion of xylose to xylulose occurs in two essential steps. First, xylose gets reduced to xylitol by NADH- or NADPH-dependent xylose reductase (XR). Later, xylitol is either secreted out of the cell or further gets oxidized to xylulose by NAD- or NADP-dependent xylitol dehydrogenase (XDH) (Alexander et al. 1988; Bruinenberg et al. 1984; Smiley and Bolen 1982). These two reactions are the rate-limiting steps in both xylose fermentation and xylitol production. Subsequently, xylulose is phosphorylated to xylulose-5-phosphate and is metabolized via pentose phosphate pathway (Fig. 6.1c).

Among the native xylitol-producing microorganisms, yeasts have proven to be the best xylitol producers. The genera of *Candida* have been extensively studied for their ability to produce xylitol (Granstrom et al. 2007). For example, *Candida tropicalis* have been extensively studied for xylitol production as they have advantages over the metabolically engineered *S. cerevisiae* for being natural D-xylose consumers and maintaining the reduction-oxidation balance during xylitol accumulation (Prakasham et al. 2009; García martín et al. 2011; Ko et al. 2008; Ling et al. 2011; Ping et al. 2013). With xylose as carbon source, higher xylitol productivities and yields have been reported in these yeasts. In a fed-batch fermentation with D-xylose as substrate and D-glucose as co-substrate, Kwon et al. (2006b) achieved 12 g/l/h of xylitol productivity. Zhang et al. (2012) reported xylitol production from D-xylose and horticultural waste hydrolysate with *Candida athensensis* SB18. The strain SB18 completely metabolized 250 g/l of xylose within 252 h and reached a maximum level of 207.7 g/l of xylitol concentration, which corresponds to 0.83 g/g of

Table 6.1 Xylitol production by native fungal and yeast strains

Microorganism	Substrate	Culture conditions pH	Culture conditions T°C	Culture conditions Time (h)	Fermentation mode	Maximum xylitol production (g/l)	Yield (g/g) / [productivity (g/l/h)	Reference
Candida athensensis SB18	Vegetable waste	7	30	102	Batch mode (bioreactor)	100.1	0.81/0.98	Zhang et al. (2012)
Candida guilliermondii FTI20037	Sugarcane bagasse	5.5	30	120	Batch mode (Erlenmeyer flasks)	50.5	0.81/0.6	de Arruda et al. (2011)
Candida guilliermondii FTI20037	Rice straw	–	30	116	Batch mode (bioreactor)	66.1	0.84/0.17	Mussatto and Roberto (2003)
Candida magnolia	Bamboo culm	–	30	30	Batch mode (bioreactor)	10.5	0.59/0.42	Miura et al. (2012)
Candida tropicalis as 2.1776	Corncob	5.5	30	120	Fed-batch mode (bioreactor)	96.5	0.83/1.01	Li et al. (2012)
Candida tropicalis BCRC 20520	Wood sawdust	5	30	96	Batch mode (Erlenmeyer flasks)	41.4	0.7/0.43	Ko et al. (2008)
Candida tropicalis CCTCC M2012462	Corncob	6	35	14	Fed-batch mode (bioreactor)	38.8	0.7/0.46	Ping et al. (2013)
Candida tropicalis HDY-02	Corncob	7	35	78	Fed-batch mode (bioreactor)	58	0.73/0.74	Ling et al. (2011)
Candida tropicalis JH030	Rice straw	6	30	80	Batch mode (Erlenmeyer flasks)	31.1	0.71/0.44	Huang et al. (2011)
Candida tropicalis NBRC 0618	Olive pruning waste	5	30	25	Batch mode (bioreactor)	53	0.49	García martín et al. (2011)
Candida tropicalis W103	Corncobs	6	35	70	Fed-batch mode (bioreactor)	68.4	0.7/0.95	Cheng et al. (2009)
Debaryomyces hansenii	Sugarcane bagasse	6	40	156	Batch mode (immobilized cells)	71.2	0.82/0.46	Prakash et al. (2011)

Debaryomyces hansenii CCMI 941	Grain brewery	5.5	30	72	Batch mode (Erlenmeyer flasks)	24	0.57/0.51	Carvalheiro et al. (2007)
Debaryomyces hansenii NRRL Y-7426	Vine waste	6	30	116	Batch mode (Erlenmeyer flasks)	27.5	0.54	Garcia-dieguez et al. (2011)
Hansenula polymorpha ATCC 34438	Sunflower stalks	5.5	30	169	Batch mode (bioreactor)	0.31	0.0023	Martínez et al. (2012)
Kluyveromyces marxianus CCA 510	Cashew apple bagasse	6	30	96	Batch mode (Erlenmeyer flasks)	6.76	–	Rocha et al. (2014)
Kluyveromyces marxianus CE 25		4.5	40	72	Batch mode (Erlenmeyer flasks)	4.8	–	Rocha et al. (2011)
Pichia stipitis YS-30	Corn Stover	5.6	30	72	Batch mode (Erlenmeyer flasks)	12.5	0.61/0.18	Rodrigues et al. (2011)
Thamnidium elegans	Xylose and glucose	6	28	–	Batch mode (Erlenmeyer flasks)	31	–	Zikou et al. (2013)
Debaryomyces hansenii UFV-170	Xylose			24	–	76.6	0.73	Sampaio et al. (2004)

xylitol yield. And around 100.1 g/l of xylitol was obtained from the bioconversion of detoxified horticultural waste hemicellulosic hydrolysate using the strain SB18 (Zhang et al. 2012). *Candida tropicalis* W103 was used to produce xylitol from dilute acid treated corncobs hemicellulosic hydrolysate. Under the optimum conditions, a maximum of 68.4 g/l of xylitol concentration (yield 0.7 g/g and productivity 0.95 g/l/h) was achieved after 72 h of fermentation (Cheng et al. 2009). Mussatto and Roberto (2003) investigated the production of xylitol with *Candida guilliermondii* FTI 20037 from rice straw hemicellulosic hydrolysate containing nearly 85 g/l of xylose and achieved a yield of 0.84 g/g.

Under semi-aerobic conditions, a maximum xylitol production of 76.6 g/l and yield of 0.73 g/g was obtained with *Debaryomyces hansenii* UFV-170 (Sampaio et al. 2004). D-xylose-fermenting yeast *P. stipitis* is among the few organisms that use both D-xylose and D-glucose and can produce ethanol or xylitol from D-xylose (Hahn-Hagerdal et al. 2007). (Rodrigues et al. 2011), tested *Pichia stipitis* YS-30 for xylitol production from corn stover DEO hydrolysate. The highest xylitol yield (0.61 g/g) and volumetric productivity (0.18 g/l) were obtained in hydrolysate neutralized with phosphoric acid.

6.3 Genetic Engineering Fungi for Xylitol Production

Native xylitol-producing yeast metabolizes D-xylose for both xylitol production and cell growth and maintenance; hence, they cannot reach maximum xylitol yield. Many recombinant fungal strains have been developed to achieve higher yield and productivity using metabolic engineering approaches. Table 6.2 shows a list of engineered fungal strains (Pal et al. 2016).

With a long history in recombinant protein production and fermentation, *S. cerevisiae*, a natural ethanologenic yeast has been serving the food and drug industries. Even though it can utilize a broad spectrum of sugars and has strong tolerance to various inhibitors present in lignocellulosic hydrolysates (Dubey et al. 2016), it cannot metabolize D-xylose as it lacks the essential enzymes D-xylose reductase (XR) and xylitol dehydrogenase (XDH). As xylose is not a preferred carbon source for *S. cerevisiae*, it does not provide sufficient energy for growth and metabolism (Sonderegger et al. 2004). Because of this, *S. cerevisiae* cannot rapidly regenerate its NADPH from its pentose phosphate pathway during xylose metabolism, and this is what causes the metabolic bottleneck (Kotter and Ciriacy 1993; Hallborn et al. 1994). Many *S. cerevisiae* strains were engineered for xylose utilization in the early 1990s. As for D-xylose to xylitol, Hallborn et al. (1991) reported a highly efficient conversion (95% of theoretical) of xylose to xylitol.

Genetic engineering of *S. cerevisiae* was done (Lee et al. 2000) by introducing the XR coding gene from *P. stipitis* for xylitol production. *XYL1* gene that encodes XR was amplified from the genome of *P. stipitis* using polymerase chain reaction and was overexpressed in *S.cereveciae*. The *XYL1* gene from *P. stipitis* does not have introns (Hallborn et al. 1991), and the XR protein encoded by *XYL1* gene can

Table 6.2 Molecular techniques and recombinant fungal strains for xylitol production

Host strain	Gene and its source	Carbon source	Xylitol yield (g/g)	Xylitol productivity (g/l/h)	Reference
Aspergillus oryzae xdhA2–1	*xdhA* disruption	Xylose	–	0.1	Mahmud et al. (2013)
	ladA disruption		–	0.03	
Candida glycerinogenes WL2002–5	Over expression of *XYL1* from *S. stipitis*	Xylose	0.98	0.83	Zhang et al. (2015)
Candida tropicalis ARSdR-16	*XYL2* (SDM)	Xylose	–	0.62	Ko et al. (2011)
Candida tropicalis BN-1	Over expression of *XYL1* from *C. parapsilosis*	Xylose	0.91	5.09	Lee et al. (2003)
Candida tropicalis BSXDH-3	Disruption *of XYL2*	Xylose	–	3.23	Ko et al. (2006)
Candida tropicalis CT-OVC5M1	*XDH* (SDM)	Xylose	–	0.74	Kumar et al. (2010)
Candida tropicalis LNG2	Over expression of XR from *N. crassa*	Xylose	NR	1.44	Jeon et al. (2012)
Debaryomyces hansenii DBX11	*XDH* (disruption)	Xylose	–	0.31	Pal et al. (2013)
Kluyveromyces marxianus YZB014	Over expression of XR and N272D from P. *stipitis*	Xylose	0.6	0.35	Zhang et al. (2013)
Kluyveromyces marxianus YZJ015	Over expression of XR from *N. crassa*	Xylose	0.83	1.49	Zhang et al. (2014)
Pichia pastoris GS225	Over expression of DalD from *K. pneumoniae, XDH* from *G. oxydans*	Glucose	0.078	0.29	Cheng et al. (2014)
Saccharomyces cerevisiae BJ3505	Chromosomal δ-integration of *XYL1*	Xylose + glucose	~100	2.34	Bae et al. (2004)
Saccharomyces cerevisiae BJ3505	Episomal expression of *GRE3*	Xylose + glucose	~100	0.5	Skim et al. (2002)
Saccharomyces cerevisiae BJ3505	Episomal expression of *XYL1*	Xylose + glucose	60	0.7	Chung et al. (2002)
Saccharomyces cerevisiae BJ3505	Chromosomal δ-integration of *XYL1*	Xylose + glucose	90	1.1	Chung et al. (2002)

(continued)

Table 6.2 (continued)

Host strain	Gene and its source	Carbon source	Xylitol yield (g/g)	Xylitol productivity (g/l/h)	Reference
Saccharomyces cerevisiae BJ3505	Chromosomal δ-integration of *XYL1*	Xylose + glucose	~100	1.23	Bae et al. (2004)
Saccharomyces cerevisiae EH13.15	Episomal expression of *XYL1*	Xylose + glucose	90	0.7	Lee et al. (2000)
Saccharomyces cerevisiae EH13.15	Episomal expression of *XYL1*	Xylose + glucose	95	1.69	Lee et al. (2000)
Saccharomyces cerevisiae S3-TAL-TKL	Over expression of *XYL1*, *XYL2*, *TKL1*, and *TAL1* from *P. stipitis*	Xylose	0.82	0.04	Walfridsson et al. (1997)
Saccharomyces cerevisiae δ XR	*XR* from *P. stipitis*	Xylose	0.9	1.1	Chung et al. (2002)
Trichoderma reesei	*Disruption of xdh1*	Organosolv-pretreated barley straw	–	0.06	Dashtban et al. (2013)
	Disruption of xdh1 + 1ad1	Xylose	–	0.08	
Trichoderma reesei ZY15	Antisense inhibition of *xdh1*	Xylose	–	0.02	Wang et al. (2005)

be actively expressed in *S. cerevisiae* (Chung et al. 2002). As the recombinant *S. cerevisiae* expressing *XYL1* gene lacks the key metabolic enzymes that fully oxidize the xylitol, the engineered strain is provided with a cheap carbon source such as glucose, for its cell growth and viability. *S. cerevisiae* prefers glucose to D-xylose. Hence, Lee et al. (2000) designed a fed-batch fermentation strategy to feed a highly concentrated glucose solution and to add D-xylose at a set point of 30 g/l, occasionally to elevate D-xylose concentration. When glucose concentration of about 0.35 g/l was upheld in the culture medium, the recombinant *S. cerevisiae* produced 105.2 g/l of xylitol concentration with 1.69 g/l/h xylitol productivity and 0.95 g/g xylitol yield. Bae et al. (2004) opted batch fermentation supplemented with 20 g/l xylose and 18 g/l glucose to achieve 20.1 g/l of xylitol yield and nearly 100% conversion yield with recombinant *Saccharomyces cerevisiae* BJ3505/δXR harboring the xylose reductase gene from *Pichia stipitis*.

C. tropicalis* is the best xylitol producer to date, among the native xylitol-producing strains. To increase the conversion rate of xylose to xylitol, Lee et al. (2003) expressed *C. parapsilosis XYL1* gene in *C. tropicalis* and achieved 5.09 g/l/h of xylitol productivity and 91.6% of xylitol yield with xylose as carbon source. Ko et al. (2006) deleted two copies of *XYL2* gene coding for xylitol dehydrogenase (XDH) that were destroyed in the diploid yeast *C. tropicalis* using the URA-blasting method. The resulting mutant was not able to grow in the minimal medium with xylose as sole carbon source, because of the absence of the *XYL2* gene. When glycerol was supplied as co-substrate for cell growth and NADPH regeneration, the *XYL2* mutant *C. tropicalis* strain could produce xylitol with 3.23 g/l/h of productivity and 98% of xylitol yield, volumetrically (Ko et al. 2006). To release the repression state, Ko et al. (2006) expressed the codon-optimized XR gene from *Neurospora crassa* under the control of the consecutive promoter glycerol 3-phosphate dehydrogenase from *C. tropicalis* in the *XYL2*-deficient *C. tropicalis* mutant. The resulting recombinant *Candida* strain produced 70% higher xylitol than the control strain without *N. crassa* XR gene, using xylose and glucose in the batch fermentation (Ko et al. 2006).

The other yeast strain studied for xylitol production is *Pichia stipitis*, a budding yeast that can metabolize xylose. But, under aerobic conditions, the mutant strain *P. stipitis* FPL-YS30 also deficient in the XYL3 gene consumed 100 g/l pure xylose completely to produce 26 g/l xylitol (Jin et al. 2005) and was able to produce 13 g/l xylitol from 21 g/l xylose in corn stover hydrolysates (Rodrigues et al. 2011). *P. stipitis* was screened by chemical mutagenesis using ethyl methanesulfonate (Min-Soo 2000). *When gluconate* was used as co-substrate (Ko et al. 2011). Xylitol concentration reached to 44.8 g/l with 100% xylitol yield, and 0.42 g/l/h of xylitol productivity was achieved in 4 days. More recently, Cheng et al. (2014) used recombinant *P. pastoris expressing DalD and XDH genes of Klebsiella pneumoniae and Gluconobacter oxydans, respectively, to achieve the xylitol productivity of* 0.29 g/l/h (yield of 0.078 g/g) with glucose as the carbon source.

Most of the studies on yeast have shown that XR has marked preference for NADPH, the main exception being *P. stipitis* that shows specificity for NAD+ and *P. tannophilus* whose XDH shows a higher activity with NADP+ than NAD+ (Fuente-hernandez et al. 2013). Barbosa et al. (1988) proposed a theoretical maximum xylitol yield in yeasts, 0.905 mol/mol of xylose when NADH was efficiently used as a cofactor by XR or under the aerobic condition where the NADH can be oxidized regenerated to NAD+ in the respiratory chain. Under anaerobic conditions, the theoretical yield drops to 0.875 mol/mol of xylose. This yield follows the Eqs. 6.1 and 6.2, respectively:

$$126 \text{ xylose} + 3O_2 + 6ADP + 6Pi$$
$$+ 48 \text{ } H_2O \rightarrow 114 \text{ xylitol} + 6 \text{ ATP} + 60 \text{ } CO_2 \tag{6.1}$$

$$48 \text{ xylose} + 15 \text{ } H_2O \rightarrow 42 \text{ Xylitol} + 2 \text{ ethanol} + 24 \text{ } CO_2 \tag{6.2}$$

Aeration is a crucial parameter under oxygen-limited xylitol production when aiming better yield in both xylitol and ATP. In general, xylitol production increases when a certain threshold of oxygen is allowed in the culture medium. However again, this preference is yeast specific (Ligthelm et al. 1988; Roseiro et al. 1991; Sampaio et al. 2004). For example, Debus et al. (1983) reported that under anoxic conditions *P. tannophilus* could achieve maximum xylitol yields (22.9 g/l). Most XRs prefer NADPH to NADH. As a proton donor, XR requires 1 mole of NADPH to convert 1 mole of D-xylose to xylitol (Rizzi et al. 1988). To supply sufficient

NADPH, genes encoding for metabolic enzymes that are engaged directly or indirectly in NADPH regeneration were overexpressed in the recombinant *S. cerevisiae* strain, harboring the *XYL1* gene.

Oxidative part of the pentose phosphate pathway is mainly responsible for the regeneration of NADPH, where glucose-6-phosphate dehydrogenase (G6PDH) and 6-phosphogluconate dehydrogenase generates 2 moles of NADPH by utilizing 1 mole of glucose via glucose-6-phosphate. Kwon et al. (2006a) overexpressed ZWF1 gene that encoded G6PDH in a xylitol-producing recombinant *S. cerevisiae*. Higher activity of G6PDH increased final cell mass, xylitol productivity, and concentration by about 24% in comparison to the control strain. Similarly, G6PDH and 6-PGDH were overexpressed in the recombinant *C. tropicalis* strain to increase the NADPH levels (Ahmad et al. 2012; Ko et al. 2006).

Pentose phosphate pathway (PPP) is the primary NADPH biosynthesis pathway, and efforts have been made to increase the flux. Glucose-6-phosphate is an important metabolite in the sugar metabolism that connects glycolysis and the pentose phosphate pathway. To prove the hypothesis that reduction of the carbon flux to the glycolysis increases the flux toward the pentose phosphate pathway, a strategy was applied to produce xylitol. Using promoter replacement technology, the activity of phosphoglucose isomerase (PGI) present in the glycolysis was reduced (Oh et al. 2007). The promoter region located at the upstream of PGI1 gene in *S. cerevisiae* was replaced with the low transcription ability ADH1 promoter. Contradictory to the expectation, tempering the PGI activity did not affect xylitol production in both batch and fed-batch fermentations. However, the simultaneous modulation of the ZWF1 overexpression and the reduced activity of PGI1 not only improved specific xylitol productivity by a 1.9-fold, it also lowered the amount of glucose consumed in a glucose-limited fed-batch cultivation (Oh et al. 2007).

Bacterial transhydrogenase from *Azotobacter vinelandii* that transfers the reducing equivalent between NADP(H) and NAD(H) and controls the redox balance in bacterial cells was investigated in *S. cerevisiae*. This recombinant *S. cerevisiae* strain with Bacterial transhydrogenase was studied for the effects of redox potential perturbation on xylitol production (Jeun et al. 2003). Although the prokaryotic transhydrogenase was actively expressed in the recombinant strain, its expression did not affect xylose consumption and xylitol production. Jeun et al. (2003) suggested that the bacterial transhydrogenase expressed in the recombinant *S. cerevisiae* was favorable for transfer of a proton in NADPH to NAD+, reducing the availability of NADPH for xylose conversion to xylitol.

Acetaldehyde dehydrogenase 6 and acetyl-CoA synthetase have also been transformed into yeast to challenge the cofactor imbalance. Recombinant *S. cerevisiae* strains, expressing acetaldehyde dehydrogenase 6 and acetyl-CoA synthetase, were also evaluated to increase the NAD(P)H pools. Acetaldehyde dehydrogenase 6 converts acetaldehyde to acetate and produces NADPH parallel, while acetyl-CoA synthetase makes acetyl-CoA with coenzyme A, ATP, and acetate produced by acetaldehyde dehydrogenase 6. A recombinant strain of *S. cerevisiae*, overexpressing acetyl-CoA synthetase, produced an 11% higher concentration of xylitol when compared to the recombinant strain overexpressing acetaldehyde dehydrogenase 6.

Even though acetyl-CoA synthetase does not produce NADPH directly, overexpressing it reduces the by-product (ethanol and acetate) formation more than acetaldehyde dehydrogenase 6, which might be advantageous to achieve higher xylitol yield (Oh et al. 2012).

Studies with filamentous fungi for xylitol production are rare, and a maximum of 13.22 g/l of xylitol production has been achieved using xylose as the carbon source with mutant *Trichoderma reesei* (Dashtban et al. 2013). Genetically modified strains of *T. reesei* has been considered to produce xylitol (6.1 g/l) from barley straw pretreated with organic solvents; fermentation was done for 168 h at 30 °C. Among fungi, yeast is still the primary focus of intense research that aims to produce xylitol, and most studies have been restricted to certain species of *Candida*. *Saccharomyces* and *Pichia* have been genetically engineered to express the heterologous XR gene or to disrupt the endogenous XDH gene.

6.4 Conclusion

Even though factors affecting xylitol production by microbial conversion is extensively cited in the literature, most of the limitations of xylose to xylitol are due to factors affecting xylitol production at cellular or molecular level. This chapter focused on the production of xylitol using various molecular strategies such as metabolic pathway engineering, altering cofactor dependency, and using enzyme technology for an increased level of xylitol production. Although the abovementioned strategies reported improved xylitol production, there are still challenges such as screening of novel strains with high tolerance for lignocellulosic hydrolysate inhibitors, development of robust and stable metabolically engineered strain, process scale-up, and development of downstream processing technologies for efficient product recovery with minimum loss and highest purity that are needed to be further addressed to make economical production of xylitol a reality.

References

Ahmad I, Shim WY, Jeon WY et al (2012) Enhancement of xylitol production in Candida tropicalis by co-expression of two genes involved in pentose phosphate pathway. Bioprocess Biosyst Eng 35:199–204

Alexander MA, Chapman TW, Jeffries TW (1988) Xylose metabolism by Candida shehatae in continuous culture. Appl Microbiol Biotechnol 28:478–486

de Arruda PV, Rodrigues Rde C, da Silva DD et al (2011) Evaluation of hexose and pentose in pre-cultivation of Candida guilliermondii on the key enzymes for xylitol production in sugarcane hemicellulosic hydrolysate. Biodegradation 22:815–822

Bae SM, Park YC, Lee TH et al (2004) Production of xylitol by recombinant Saccharomyces cerevisiae containing xylose reductase gene in repeated fed-batch and cell-recycle fermentations. Enzym Microb Technol 35:545–549

Barbosa MFS, Medeiros MB, Mancilha IM et al (1988) Screening of yeasts for production of xylitol from d-xylose and some factors which affect xylitol yield in Candida guilliermondii. J Ind Microbiol 3:241–251

Bruinenberg PM, de Bot PHM, Van dijken JP et al (1984) NADH-linked aldose reductase: the key to anaerobic alcoholic fermentation of xylose by yeasts. Appl Microbiol Biotechnol 19:256–260

Carvalheiro F, Duarte LC, Medeiros R et al (2007) Xylitol production by Debaryomyces hansenii in brewery spent grain dilute-acid hydrolysate: effect of supplementation. Biotechnol Lett 29:1887–1891

Cheng KK, Zhang JA, Ling HZ et al (2009) Optimization of pH and acetic acid concentration for bioconversion of hemicellulose from corncobs to xylitol by Candida tropicalis. Biochem Eng J 43:203–207

Cheng H, Lv J, Wang H et al (2014) Genetically engineered Pichia pastoris yeast for conversion of glucose to xylitol by a single-fermentation process. Appl Microbiol Biotechnol 98:3539–3552

Chung YS, Kim MD, Lee WJ et al (2002) Stable expression of xylose reductase gene enhances xylitol production in recombinant Saccharomyces cerevisiae. Enzym Microb Technol 30:809–816

Dashtban M, Kepka G, Seiboth B et al (2013) Xylitol production by genetically engineered Trichoderma reesei strains using barley straw as feedstock. Appl Biochem Biotechnol 169:554–569

Debus D, Methner H, Schulze D et al (1983) Fermentation of xylose with the yeast Pachysolen tannophilus. Eur J Appl Microbiol Biotechnol 17:287–291

Dubey R, Jakeer S, Gaur NA (2016) Screening of natural yeast isolates under the effects of stresses associated with second-generation biofuel production. J Biosci Bioeng 121:509–516

Fuente-hernandez A, Corcos PO, Beauchet R et al (2013) Biofuels and co-products out of hemicelluloses. Intech publishing, Croatia

García martín JF, Sánchez S, Bravo V et al (2011) Xylitol production from olive-pruning debris by sulphuric acid hydrolysis and fermentation with Candida tropicalis. Holzforschung 65:59–65

Garcia-dieguez C, Salgado JM, Roca E et al (2011) Kinetic modelling of the sequential production of lactic acid and xylitol from vine trimming wastes. Bioprocess Biosyst Eng 34:869–878

Granstrom TB, Izumori K, Leisola M (2007) A rare sugar xylitol. Part II: biotechnological production and future applications of xylitol. Appl Microbiol Biotechnol 74:273–276

Hahn-Hagerdal B, Karhumaa K, Fonseca C et al (2007) Towards industrial pentose-fermenting yeast strains. Appl Microbiol Biotechnol 74:937–953

Hallborn J, Walfridsson M, Airaksinen U et al (1991) Xylitol production by recombinant Saccharomyces cerevisiae. Biotechnology (N Y) 9:1090–1095

Hallborn J, Gorwa MF, Meinander N et al (1994) The influence of cosubstrate and aeration on xylitol formation by recombinant Saccharomyces cerevisiae expressing the XYL1 gene. Appl Microbiol Biotechnol 42:326–333

Huang CF, Jiang YF, Guo GL et al (2011) Development of a yeast strain for xylitol production without hydrolysate detoxification as part of the integration of co-product generation within the lignocellulosic ethanol process. Bioresour Technol 102:3322–3329

Jeon WY, Yoon BH, Ko BS et al (2012) Xylitol production is increased by expression of codon-optimized Neurospora crassa xylose reductase gene in Candida tropicalis. Bioprocess Biosyst Eng 35:191–198

Jeun YS, Kim MD, Park YC et al (2003) Expression of Azotobacter vinelandii soluble transhydrogenase perturbs xylose reductase-mediated conversion of xylose to xylitol by recombinant Saccharomyces cerevisiae. J Mol Catal B Enzym 26:251–256

Jin YS, Cruz J, Jeffries TW (2005) Xylitol production by a Pichia stipitis D-xylulokinase mutant. Appl Microbiol Biotechnol 68:42–45

Ko BS, Kim J, Kim JH (2006) Production of xylitol from D-xylose by a xylitol dehydrogenase gene-disrupted mutant of Candida tropicalis. Appl Environ Microbiol 72:4207–4213

Ko CH, Chiang PN, Chiu PC et al (2008) Integrated xylitol production by fermentation of hardwood wastes. J Chem Technol Biotechnol 83:534–540

Ko BS, Kim DM, Yoon BH et al (2011) Enhancement of xylitol production by attenuation of intra-
    cellular xylitol dehydrogenase activity in Candida tropicalis. Biotechnol Lett 33:1209–1213
Kotter P, Ciriacy M (1993) Xylose fermentation by Saccharomyces cerevisiae. Appl Microbiol
    Biotechnol 38:776–783
Kumar J, Reddy MS, Rao LV (2010) Strain improvement of Candida tropicalis OVC5 for xylitol
    production by random mutagenesis. IIOABJ 1:24–28
Kwon DH, Kim MD, Lee TH et al (2006a) Elevation of glucose 6-phosphate dehydrogenase
    activity increases xylitol production in recombinant Saccharomyces cerevisiae. J Mol Catal B
    Enzym 43:86–89
Kwon SG, Park SW, Oh DK (2006b) Increase of xylitol productivity by cell-recycle fermentation
    of Candida tropicalis using submerged membrane bioreactor. J Biosci Bioeng 101:13–18
Lee WJ, Ryu YW, Seo JH (2000) Characterization of two-substrate fermentation processes for
    xylitol production using recombinant Saccharomyces cerevisiae containing xylose reductase
    gene. Process Biochem 35:1199–1203
Lee JK, Koo BS, Kim SY (2003) Cloning and characterization of the xyl1 gene, encoding an
    NADH-preferring xylose reductase from Candida parapsilosis, and its functional expression in
    Candida tropicalis. Appl Environ Microbiol 69:6179–6188
Li M, Meng X, Diao E et al (2012) Xylitol production by Candida tropicalis from corn cob
    hemicellulose hydrolysate in a two-stage fed-batch fermentation process. J Chem Technol
    Biotechnol 87:387–392
Ligthelm ME, Prior BA, du Preez JC (1988) The oxygen requirements of yeasts for the fermenta-
    tion of d-xylose and d-glucose to ethanol. Appl Microbiol Biotechnol 28:63–68
Ling H, Cheng K, Ge J et al (2011) Statistical optimization of xylitol production from corncob
    hemicellulose hydrolysate by Candida tropicalis HDY-02. New Biotechnol 28:673–678
Mahmud A, Hattori K, Hongwen C et al (2013) Xylitol production by NAD(+)-dependent xylitol
    dehydrogenase (xdhA)- and l-arabitol-4-dehydrogenase (ladA)-disrupted mutants of aspergil-
    lus oryzae. J Biosci Bioeng 115:353–359
Martínez ML, Sánchez S, Bravo V (2012) Production of xylitol and ethanol by Hansenula poly-
    morpha from hydrolysates of sunflower stalks with phosphoric acid. Ind Crop Prod 40:160–166
Meinander NQ, Hahn-Hägerdal B (1997) Fed-batch xylitol production with two recombinant
    Saccharomyces cerevisiae strains expressing XYL1 at different levels, using glucose as a
    cosubstrate: a comparison of production parameters and strain stability. Biotechnol Bioeng
    54:391–399
Min-Soo K (2000) Enhancement of xylitol yield by xylitol dehydrogenase defective mutant of
    Pichia stipitis. Korean J Biotechnol Bioeng 15:113–119
Miura M, Watanabe I, Shimotori Y et al (2012) Microbial conversion of bamboo hemicellulose
    hydrolysate to xylitol. Wood Sci Technol 47:515–522
Moyses DN, Reis VC, de Almeida JR et al (2016) Xylose fermentation by Saccharomyces cerevi-
    siae: challenges and prospects. Int J Mol Sci 17:207
Mussatto SI, Roberto IC (2003) Xylitol production from high xylose concentration: evaluation
    of the fermentation in bioreactor under different stirring rates. J Appl Microbiol 95:331–337
Oh YJ, Lee TH, Lee SH et al (2007) Dual modulation of glucose 6-phosphate metabolism to
    increase NADPH-dependent xylitol production in recombinant Saccharomyces cerevisiae.
    J Mol Catal B Enzym 47:37–42
Oh EJ, Bae YH, Kim KH et al (2012) Effects of overexpression of acetaldehyde dehydrogenase 6
    and acetyl-CoA synthetase 1 on xylitol production in recombinant Saccharomyces cerevisiae.
    Biocatal Agric Biotechnol 1:15–19
Pal S, Choudhary V, Kumar A et al (2013) Studies on xylitol production by metabolic pathway
    engineered Debaryomyces hansenii. Bioresour Technol 147:449–455
Pal S, Mondal AK, Sahoo DK (2016) Molecular strategies for enhancing microbial production of
    xylitol. Process Biochem 51:809–819
Park YC, Kim SK, Seo JH (2014) Recent advances for microbial production of xylitol. Wiley,
    Chichester, pp 497–518
Ping Y, Ling HZ, Song G et al (2013) Xylitol production from non-detoxified corncob hemicel-
    lulose acid hydrolysate by Candida tropicalis. Biochem Eng J 75:86–91

Prakash G, Varma AJ, Prabhune A et al (2011) Microbial production of xylitol from D-xylose and sugarcane bagasse hemicellulose using newly isolated thermotolerant yeast Debaryomyces hansenii. Bioresour Technol 102:3304–3308

Prakasham RS, RS R, PJ H (2009) Current trends in biotechnological production of xylitol and future prospects. Curr Trends Biotechnol 3:8–36

Rao RS, Jyothi CP, Prakasham RS et al (2006) Strain improvement of Candida tropicalis for the production of xylitol: biochemical and physiological characterization of wild-type and mutant strain CT-OMV5. J Microbiol 44:113–120

Rizzi M, Erlemann P, Bui-thanh NA et al (1988) Xylose fermentation by yeasts. Appl Microbiol Biotechnol 29:148–154

Rocha MV, Rodrigues TH, Melo VM et al (2011) Cashew apple bagasse as a source of sugars for ethanol production by Kluyveromyces marxianus CE025. J Ind Microbiol Biotechnol 38:1099–1107

Rocha MVP, Rodrigues THS, Albuquerque d et al (2014) Evaluation of dilute acid pretreatment on cashew apple bagasse for ethanol and xylitol production. Chem Eng J 243:234–243

Rodrigues RC, Kenealy WR, Jeffries TW (2011) Xylitol production from DEO hydrolysate of corn Stover by Pichia stipitis YS-30. J Ind Microbiol Biotechnol 38:1649–1655

Roseiro JC, Peito MA, Gírio FM et al (1991) The effects of the oxygen transfer coefficient and substrate concentration on the xylose fermentation by Debaryomyces hansenii. Arch Microbiol 156:484–490

Sampaio FC, Silveira WBD, Chaves-alves VM et al (2003) Screening of filamentous fungi for production of xylitol from D-xylose. Braz J Microbiol 34:321–324

Sampaio FC, Torre P, Passos FM et al (2004) Xylose metabolism in Debaryomyces hansenii UFV-170. Effect of the specific oxygen uptake rate. Biotechnol Prog 20:1641–1650

Skim MD, Jeun YS, Kim SG et al (2002) Comparison of xylitol production in recombinant Saccharomyces cerevisiae strains harboring XYL1 gene of Pichia stipitis and GRE3 gene of S. cerevisiae. Enzym Microb Technol 31:862–866

Smiley KL, Bolen PL (1982) Demonstration of D-xylose reductase and D-xylitol dehydrogenase in Pachysolen tannophilus. Biotechnol Lett 4:607–610

Sonderegger M, Jeppsson M, Hahn-Hagerdal B et al (2004) Molecular basis for anaerobic growth of Saccharomyces cerevisiae on xylose, investigated by global gene expression and metabolic flux analysis. Appl Environ Microbiol 70:2307–2317

Walfridsson M, Anderlund M, Bao X et al (1997) Expression of different levels of enzymes from the Pichia stipitis XYL1 and XYL2 genes in Saccharomyces cerevisiae and its effects on product formation during xylose utilisation. Appl Microbiol Biotechnol 48:218–224

Wang TH, Zhong YH, Huang W et al (2005) Antisense inhibition of xylitol dehydrogenase gene, xdh1 from Trichoderma reesei. Lett Appl Microbiol 40:424–429

Winkelhausen E, Kuzmanova S (1998) Microbial conversion of d-xylose to xylitol. J Ferment Bioeng 86:1–14

Yadav M, Mishra DK, Hwang JS (2012) Catalytic hydrogenation of xylose to xylitol using ruthenium catalyst on NiO modified TiO2 support. Appl Catal A Gen 425–426:110–116

Zhang J, Geng A, Yao C et al (2012) Xylitol production from D-xylose and horticultural waste hemicellulosic hydrolysate by a new isolate of Candida athensensis SB18. Bioresour Technol 105:134–141

Zhang B, Li L, Zhang J et al (2013) Improving ethanol and xylitol fermentation at elevated temperature through substitution of xylose reductase in Kluyveromyces marxianus. J Ind Microbiol Biotechnol 40:305–316

Zhang J, Zhang B, Wang D et al (2014) Xylitol production at high temperature by engineered Kluyveromyces marxianus. Bioresour Technol 152:192–201

Zhang C, Zong H, Zhuge B et al (2015) Production of xylitol from D-xylose by overexpression of xylose reductase in osmotolerant yeast Candida glycerinogenes WL2002-5. Appl Biochem Biotechnol 176:1511–1527

Zikou E, Chatzifragkou A, Koutinas AA et al (2013) Evaluating glucose and xylose as cosubstrates for lipid accumulation and gamma-linolenic acid biosynthesis of Thamnidium elegans. J Appl Microbiol 114:1020–1032

Chapter 7
Insights into Fungal Xylose Reductases and Its Application in Xylitol Production

Yogita Lugani and Balwinder Singh Sooch

Abstract Xylose reductase (EC 1.1.1.21), an aldo-keto reductase enzyme, catalyzes the conversion of xylose into xylitol. It is present in animals, plants, and many microorganisms. In microorganisms, in addition to its production by many fungal (yeasts and molds) cultures, a few members of bacteria such as *Corynebacterium* sp. and *Enterobacter* sp. have also been reported to produce NADPH-dependent xylose reductase (XR). In fungi, XR directly converts xylose into xylitol during the metabolism of xylose by using NADH and/or NADPH as coenzyme. The tetrad of amino acids (Tyr, His, Asp, and Lys) at catalytic site is responsible for XR activity. Several attempts have been made to improve XR production using recombinant DNA technology by introducing xylose reductase gene (*xyl1*) into different fungal strains from other microorganisms for efficient conversion of xylose to xylitol. Site-directed mutagenesis at the catalytic site is another approach to increase the turnover number and catalytic efficiency of XRs. Xylitol is a rare pentol sugar whose global market is increasing at a very fast pace due to its applications in food, cosmetic, odontological, pharmaceutical, and medical sector. The microbial production of xylitol is emerging as a good alternative due to abundance of agriculture waste material. The present chapter will describe the different aspects of fungal XRs including their structural characteristics, sources, production, purification and characterization, immobilization, patent status, and xylitol applications.

7.1 Introduction

Xylose reductase (XR) (EC 1.1.1.21), an aldo-keto reductase (*AKR*) enzyme, catalyzes the conversion of xylose into xylitol. *AKRs* are exclusively involved in reversible reduction of aldehydes and/or ketones to their corresponding alcohols by utilizing NADPH and/or NADH as coenzyme. *AKRs* are present in animals, plants,

Y. Lugani · B. S. Sooch (✉)
Enzyme Biotechnology Laboratory, Department of Biotechnology, Punjabi University, Patiala, Punjab, India
e-mail: soochb@pbi.ac.in

© Springer International Publishing AG, part of Springer Nature 2018
S. Kumar et al. (eds.), *Fungal Biorefineries*, Fungal Biology,
https://doi.org/10.1007/978-3-319-90379-8_7

and many microorganisms; however, *AKRs* from lower organisms (*AKR* families 2–5 and 8–11) differ widely from the higher organisms with respect to structure and function. Different members of *AKR* family differ in their amino acid sequence, stereo structure, and substrate specificity. The amino acid position is mainly involved in their substrate discrimination (Costanzo et al. 2009). Most of the members of *AKRs* are active as monomers, whereas *AKR*2, *AKR*6, and *AKR*7 family members are active in multimeric form (Hyatt et al. 2013). In higher mammals, birds, amphibians, and fishes, some tissues like kidney and liver showed highest aldehyde reductase activity. *AKR* genes catalyze a variety of metabolic oxidation-reduction reactions involving the reduction of glucose, glucocorticoids, and small carbonyl metabolites to glutathione conjugates and phospholipid aldehydes in higher animals (Barski et al. 2008). XR is found almost in all the microorganisms like bacteria (*Corynebacterium* sp. and *Enterobacter* sp.), fungi including yeasts and molds (*Candida mogii, C. peltata, C. pelliculosa, C. boidinii, C. guilliermondii, C. intermedia, C. tropicalis, Saccharomyces* sp., *Debaryomyces* sp., *Pichia* sp., *Hansenula* sp., *Torulopsis* sp., *Kloeckera* sp., *Trichosporon* sp., *Cryptococcus* sp., *Rhodotorula* sp., *Monilia* sp., *Kluyveromyces* sp., *Pachysolen* sp., *Neurospora crassa, Penicillium* sp., *Aspergillus* sp., *Fusarium* sp., *Mucor* sp., *Rhizopus* sp., *Gliocladium* sp., *Byssochlamys* sp., *Myrothecium* sp.), and algae (*Galdieria sulphuraria*). The affinity of XR from different species of yeast generally varies for the cofactor, i.e., NADH or NADPH, but NADPH is usually preferred for the catalytic reduction of xylose (Kim et al. 2002; Kavanagh et al. 2003). The major metabolic pathway for the conversion of xylose into xylitol by NADH or NADPH utilizing XRs is found exclusively in fungi, and very few reports have been published with bacterial and algal XRs. The tetrad of amino acids (Tyr, His, Asp, and Lys) at catalytic site is responsible for acid base catalysis of XR with pK of 6.5–7.0. Both Tyr and His can form hydrogen bond with substrate carbonyl group, but Tyr was found to be most active for substrate binding by removing water before hydride transfer reactions (Ye et al. 2001). The maximum previous reports on production, purification, and immobilization of XR have been published with fungal (yeast and mold) strains, and very little data has been available on bacterial and algal XRs. However, in a previous few years, the new strains of bacteria and fungus have been developed by cloning and expressing XR genes by site-directed mutagenesis and recombination DNA technology to enhance the enzyme activity. The major industrial application of XR is the production of xylitol from xylose or xylan-containing agricultural waste materials.

Xylitol ($C_5H_{10}O_5$) is a rare sugar with sweetening power similar to sucrose and found naturally in fruits and vegetables in small amount. It has received wide global demand mainly due to its insulin-independent metabolism, anticariogenicity, laxative nature, sweetening power similar to that of sucrose but with low calories, absence of Maillard reaction for retaining the nutritional value of food and various other food, therapeutic and pharmaceutical applications (Russo 1977; Uhari et al. 2000; Granstrom et al. 2004; Kauko and Makinen 2010; Rafiqul and Sakinah 2012; Lakshmi et al. 2014; Yin et al. 2014; Lugani et al. 2017). Xylitol is classified as "GRAS" (generally recognized as safe) by the Food and Drug Administration, USA (Aguiar et al. 1999), and is approved for usage in foods, pharmaceuticals, and oral

Table 7.1 Significant properties of xylitol

Properties of xylitol
White, crystalline, optically inactive
Density, 1520 kg/m³; melting point, 92–96 °C; boiling point, 216 °C; molecular weight, 152.15; caloric value, 4.06 cal/g; heat of solution, 36.61 cal/g
Low glycemic index
Safe profile equivalent to sorbitol and mannitol
Slow utilization from intestine with little change in blood glucose
Appropriate xylitol dose enhances carbohydrate tolerance
Possess anticariogenic and antiketogenic properties
Provide protection to proteins against denaturation and other damages
Ability to fight against infections
Stabilizes salivary calcium and phosphate ions
Absence of Maillard reaction
Xylitol fatty acid esters possess emulsion stability
Enhances flavor in beverages
Cool and fresh sensation
Stimulation of remineralization of teeth
Maintain moisture content in cleanser, lotions, and beauty creams
Humectant and emollient properties

health products (Parajo et al. 1998). A list of significant properties of xylitol is shown in Table 7.1.

The global market of xylitol is increasing at a very fast pace and it is estimated to be US$6.30 billion by 2022 (Markets and Markets 2016). Currently, most of the xylitol produced is manufactured at the industrial level by a chemical hydrogenation of the five-carbon sugar D-xylose in the presence of nickel catalyst at elevated temperature and pressure (Melaja and Hamalainen 1977; Granstrom et al. 2007). This chemical process is laborious and cost- and energy-intensive. This process is also expensive due to the extensive separation process which involves number of purification stages (Parajo et al. 1998). These facts about xylitol production have encouraged the development of alternative technologies to lower the production cost with improved yield. The best eco-friendly alternative is bioconversion of xylose to xylitol by employing this specific enzyme, i.e., xylose reductase (XR). D-xylose is one of the most abundant pentose sugars found in nature and is the predominate hemicellulosic sugar of hardwoods and agricultural residues, accounting for up to 25% of the dry weight biomass of some plant species (Ladisch et al. 1983). In plant tissues, it exists primarily in the anhydride form (xylan) and can easily be separated and saccharified into monomeric units by either mild chemical or enzymatic treatment. Xylose is a major part of the hemicellulose fraction of hardwood and agro-based lignocellulosic material. Agricultural residues containing different amount of xylose are given in Table 7.2.

Upon acid hydrolysis of these cheap and abundant natural lignocellulosic materials, high yield of xylose-rich hydrolysates can be recovered in good yield (Nigam and Singh 1995). The abundance and ease of isolation of D-xylose from natural sources make it a potential feedstock for the production of xylitol through enzymatic route.

Table 7.2 Percent of xylose in different agriculture wastes on dry weight basis

S.No.	Agriculture waste hydrolysate	Xylose (%)	Reference
1.	Bagasse	23.2	Lee (1997)
2.	Rice straw	14.8–20.2	Karimi et al. (2006)
3.	Wheat straw	22.2 ± 0.3	Erdei et al. (2010)
4.	Corncob	28.1	Boonmee (2012)
5.	Peanut shell	6.7	Boonmee (2012)
6.	Rice hull	21.7	Boonmee (2012)
7.	Sugarcane leaf and stalk	27.4	Boonmee (2012)
8.	Sweet sorghum leaf and stalk	24.0	Boonmee (2012)
9.	Cottonseed hulls	13.6	Sharma (2014)
10.	Pea peel	18.85 ± 1.18	Rehman et al. (2015)

7.2   Structure of Xylose Reductase

Xylose reductase, a pentose reductase enzyme, is a member of the aldo-keto reductase family 2 (*AKR2*), which catalyzes the first step of five carbon metabolism by reducing xylose and arabinose to xylitol and arabitol. *AKRs* belong to a superfamily of enzymes comprising 15 separate families with approximately 120 identified members (Jez and Penning 2001). *AKRs* catalyze the reversible reduction of aldehydes and/or ketones to their corresponding alcohols, almost exclusively utilizing NADPH as coenzyme. A few are capable of dual NADPH/NADH specificity, but to date, only one *E. coli* enzyme has been shown to be specific for only NADH (Luccio et al. 2006). The most common motifs, which are observed in the crystal structure of aldo-keto reductase (*AKR*) protein, fold into $(\beta/\alpha)_8$ barrel (Branden 1991) with apo- and $NADP^+$-bound form in the xylose reductase (Kavanagh et al. 2002). An unusual hydrophilic patch with residues from loop 4 causes the diamerization of this protein. The loop 4 is involved in connecting strand 4 with helix 4, interactions from helix 5 and helix 6, as well as C-terminal loop. The antiparallel conformation of the protein dimer is provided by all of the above orientations (Kavanagh et al. 2003). The importance of C-terminal region in determining the substrate specificity was reported by Jez and Penning (2001) using site-directed mutagenesis. The detailed mechanism of catalytic reaction of xylose reductase was detected by the structural information, and it involves a tetrad of amino acids (Tyr, His, Asp, and Lys) at the catalytic site (Ellis 2002). XR catalyzes the reduction of xylose into xylitol by using NADH (k_m = 25.4 µM; k_{cat} = 18.1 s^{-1}) or NADPH (k_m = 4.8 µM; k_{cat} = 21.9 s^{-1}) as cofactor (Lunzer et al. 1998). The affinity of XRs from different species of yeast generally varies for the cofactor, but NADPH is usually preferred (>10 fold) for the catalytic reduction of xylose (Kavanagh et al. 2003). However, the catalytic efficiency of 2,5-diketo-D-gluconic acid reductase A (a member of *AKR* family 5) has been improved seven fold by using the co-substrate NADH (Banta et al. 2002).

The *xyl1* gene is involved in encoding the NADPH-dependent XR (M.wt-35.8 KDa) in *Pichia stipitis*, and it does not contain introns. The 954 bp open reading frame of this gene is involved in encoding the polypeptide with 318 amino acid residues (Hallborn et al. 1991; Takuma et al. 1991). Four other yeast xylose reductase genes have also been cloned and sequenced in the past from *C. tropicalis* (Yokoyama et al. 1995), *Kluyveromyces lactis* (Billard et al. 1995), and *Pachysolen tannophilus* (Bolen et al. 1985). The DNA sequence coding for the aldose reductase is available in *Saccharomyces cerevisiae* through complete genome study. The GenBank study of aldose reductase gene product from *S. cerevisiae* has shown similarity with other aldose reductase gene products (Kuhn et al. 1995). The comparison of amino acid sequence of *P. stipitis* xylose reductase with human, rat, and bovine *AKR* by using BLASTp software showed the sequence similarity of 42.4%, 40.9%, and 40.1%, respectively (Hallborn et al. 1991). The two dimers (four molecules) of xylose reductase were observed in the crystallographical asymmetric unit. From the total residues present in XR, 90.9% of the residues were observed in the core areas of Ramachandran plot generated by PROCHECK, and the rest of the 9.1% residues were observed in the allowed regions (Laskowski et al. 1993).

Despite the availability of several purified enzymes and cloned genes, much information is not available about the catalytic mechanism, substrate binding sites or molecular determinants of substrate, and coenzyme specificity in yeast xylose reductases.

7.3   Sources of Xylose Reductase and Their Metabolism

XR has been produced in various organisms like bacteria, fungi (yeasts and molds), and algae. The capability of bioconversion of D-xylose to xylitol by microorganisms using different metabolic pathways is important for industrial production of xylitol and has been studied in various microorganisms (Milessi et al. 2011; Rafiqul and Sakinah 2012; Kim et al. 2013; Zhang et al. 2015a, b; Moyses et al. 2016; Kogje and Ghosalkar 2016). The metabolic pathway involved in bioconversion of xylose to xylitol in fungal strains is illustrated in Fig. 7.1.

The NADPH-dependent xylose reductase in *Zymomonas mobilis* was found to be responsible for xylitol production (Feldmann et al. 1992). It has been postulated that yeasts which overproduce xylitol must experience some metabolic bottleneck in steps subsequent to xylose reductase. This bottleneck may be in the form of a redox imbalance as a result of the existence of NADPH-specific xylose reductase and NAD-specific xylitol dehydrogenase activities of the producing strain (Hagerdal et al. 1994). Some bacterial species like *Enterobacter liquefaciens* and *Corynebacterium* sp. are reported for the production of xylitol using XR enzyme system (Yoshitake et al. 1971, 1973; Horitsu et al. 1992; Parajo et al. 1998). The major metabolic pathway for conversion of xylose into xylitol by involving XR has been found in fungal (yeast and molds) species. In yeasts, XR reduces D-xylose into xylitol in the presence of NADPH or NADH coenzymes (Parajo et al. 1998). XR is present in the cytoplasm of many xylose-fermenting or xylose-utilizing yeasts, where they catalyze the first step of xylose metabolism, which is a rate-limiting step

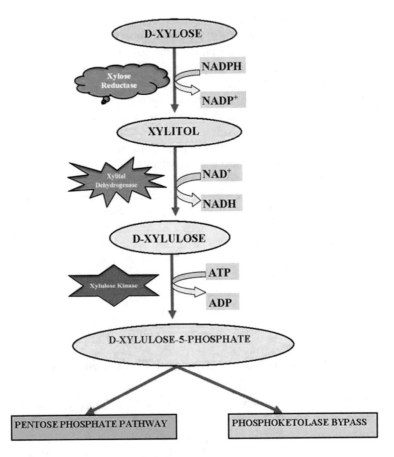

Fig. 7.1 Xylose metabolic pathway in fungus

of xylose metabolism in these microorganisms. This enzyme is essential for growth and utilization of xylose by yeasts (Lee 1998). Some yeast XRs exhibit a strict requirement for NADPH while others may function with either NADPH or NADH (Schneider 1989; Rafiqul and Sakinah 2012). In most yeasts, XR shows higher preference for NADPH over NADH. The catalytic activity of XR for the production of xylitol is favored under oxygen-limited conditions due to accumulation of NADH or NADPH, which results in inhibition of NAD-linked xylitol dehydrogenase (XDH). This XDH is responsible for conversion of xylitol to xylulose. The inability of yeast to compensate for accumulation of NADH or NADPH due to absence of transhydrogenase activity is known as the Custer effect (Dijken and Scheffers 1986). In some yeast species like *C. boidinii*, xylose is reduced to xylitol by two metabolic pathways involving two reduction steps utilizing NADPH with low efficiency. The first pathway includes direct reduction of xylose into xylitol, and the second pathway involves the initial isomerization of xylose into xylulose by xylose isomerase, which is subsequently reduced to xylitol (Vongsuvanlert and Tani 1988). The xylose

Table 7.3 Fungal sources of xylose reductase

Source	Reference
Yeast	
Pachysolen tannophilus	Verduyn et al. (1985a), Converti et al. (1999)
Pichia stipitis	Verduyn et al. (1985b), Bicho et al. (1988)
Cryptococcus flavus	Mayr et al. (2003)
C. shehatae	Yablochkova et al. (2003)
C. mogii	Mayerhoff et al. (2004)
Saccharomyces cerevisiae	Attfield and Bell (2006), Kogje and Ghosalkar (2016)
C. boidinii	Khoury et al. (2009)
Talaromyces emersonii	Fernandes et al. (2009)
Ogataea siamensis N22	Kokaew et al. (2009)
Kluyveromyces marxianus IMB	Mueller et al. (2011)
Hansenula polymorpha	Ahmed et al. (2011)
Debaryomyces nepalensis NCYC3413	Kumar and Gummadi (2011)
C. guilliermondii	Milessi et al. (2011), Branco et al. (2011)
C. tenuis	Vogl et al. (2011)
Debaryomyces hansenii	Biswas et al. (2012)
C. tropicalis	Ronzon et al. (2012), Rafiqul and Sakinah (2015)
Ogataea siamensis	Boontham et al. (2014)
Molds	
Penicillium sp.	Parajo et al. (1998)
Aspergillus sp.	Parajo et al. (1998)
Fusarium sp.	Parajo et al. (1998)
Mucor sp.	Parajo et al. (1998)
Myrothecium sp.	Parajo et al. (1998)
Gliocladium sp.	Parajo et al. (1998)
Byssochlamys sp.	Parajo et al. (1998)
Neurospora crassa	Zhao et al. (1998); Woodyer et al. (2005)
Talaromyces emersonii	Fernendes et al. (2009)
Trichoderma reesei	Hong et al. (2014)

reductase is mainly produced by fungal cultures, and major XR-producing yeasts and molds are shown in Table 7.3. The widely reported yeast species for the production of XR belongs to genus *Candida* including *C. mogii* ATCC18364 (Sirisansaneeyakul et al. 1995), *C. peltata* (Saha and Bothast 1999), *C. pelliculosa*, *C. boidinii*, *C. guilliermondii* (Lee et al. 1996; Cortez et al. 2006; Milessi et al. 2011), *C. intermedia* (Mayr et al. 2000), and *C. tropicalis* (Su et al. 2010; Rafiqul and Sakinah 2012; Ronzon et al. 2012). Other genera of yeast investigated for xylose reductase production include *Saccharomyces* sp., *Debaryomyces* sp., *Hansenula* sp., *Torulopsis* sp., *Kloeckera* sp., *Trichosporon* sp., *Cryptococcus* sp., *Rhodotorula* sp., *Monilia* sp., *Kluyveromyces* sp., *Pachysolen* sp., *Pichia* sp. (Verduyn et al. 1985a, b; Yablochkova et al. 2003; Kumar and Gummadi 2011), *Ambrosiozyma* sp., and *Torula* sp. (Saha and Bothast 1997).

Some molds are also reported to produce the enzyme xylose reductase like *Aspergillus* sp., *Byssochlamys* sp., *Fusarium* sp., *Gliocladium* sp., *Mucor* sp., *Myrothecium* sp., *Neurospora crassa* (Zhao et al. 1998), *Penicillium* sp., *Rhizopus* sp., and *Talaromyces* sp. (Parajo et al. 1998; Fernandes et al. 2009; Zhang et al. 2015a, b). Xylose reductase was also found in some algae like *Galdieria sulphuraria* (Gross et al. 1997) and *Chlorella sorokiniana* (Zheng et al. 2014).

7.4 Production of Xylose Reductase

The production of xylose reductase has been investigated from different microorganisms by various workers by altering their nutritional and process parameters. The xylose reductase production varies with the type of source and physicochemical conditions. Xylose reductase was produced by *C. tenuis*, and its activity was found to be 1.50 IU/mg of protein after 24 h (Neuhauser et al. 1997). *N. crassa* X1 cultivated on liquid medium containing 1% (w/v) xylose showed highest level of NADPH-dependent xylose reductase activity (220 U/ml) and NADH-dependent xylose reductase activity (100 U/ml) after 96 h (Zhao et al. 1998). Xylose reductase was produced from *C. guilliermondii*, and the enzymatic activity was found to be 129–2190 IU/mg of protein after 24 h of fermentation (Rosa et al. 1998). The affinity of xylose reductase obtained from *P. tannophilus* for both cofactors, i.e., NADH and NADPH, has been reported by Yablochkova et al. (2003). The xylose reductase enzyme activity with NADPH cofactor was found to be much higher than NADH. Xylose reductase, active toward benzaldehyde, was discovered and characterized from *Z. mobilis* ZM4 by Agrawal and Chen (2011). Thermotolerant *K. marxianus* cells grown on a synthetic media containing yeast extract (0.50 g/L), KH_2PO_4 (2.0 g/L), $(NH_4)_2SO_4$ (1.0 g/L), $MgSO_4.7H_2O$ (1.0 g/L), and xylose (20.0 g/L) were able to exhibit xylose reductase activity of 2.91 IU/mL after 18 h of fermentation (Mueller et al. 2011). The variable enzyme activity of NADPH-linked xylose reductase from *P. tannophilus* was obtained when growing on different sources of carbon, i.e., L-arabinose (534 nmol/min/mg), D-galactose (50 nmol/min/mg), and D-xylose (226 nmol/min/mg). Other carbon sources, such as D-glucose, D-fructose, glycerol, D-mannitol, D-mannose, and D-sorbitol, did not increase the enzymatic activity to a significant extent (Bolen and Detroy 1985). Similarly the inhibition of glucose on xylose reductase activity was reported in another study on *C. tropicalis* given by Tamburini et al. (2010). In another study, xylose reductase activity of *P. tannophilus* and *P. stipitis* was found to be induced by using a combination of D-glucose and L-arabinose, yielded about 110 ± 10% of the xylose reductase activity compared with D-xylose (100%). However, very low xylose reductase activity was induced with other carbon sources such as cellobiose, D-mannose, D-glucose, D-galactose, D-fructose, or glycerol (Bicho et al. 1988). In *C. guilliermondii*, D-glucose and glycerol repressed the xylose reductase activity to a greater extent, while with D-fructose, greater repression was seen with xylitol dehydrogenase activity. The increase in relative xylose reductase activity was found when

D-xylose and cellobiose (121 ± 5.7%) were used in combination as compared to D-xylose (100%) alone (Lee et al. 1996).

The maximum activity of xylose reductase (0.582 IU/mg of protein) with *C. guilliermondii* was reported in the medium containing glucose, under the lowest glucose/xylose ratio of 1:25 with sugarcane bagasse hydrolysate after 48 h of fermentation (Silva and Felipe 2006). The volumetric activity of xylose reductase was decreased with *C. guilliermondii* when the initial concentration of xylose was increased from 30 to 70 g/L (Gurpilhares et al. 2009). The enzyme activity of NADPH-dependent xylose reductase from *Hansenula polymorpha* was found to be induced with D-xylose and L-arabinose. Low enzymatic activity was found with other substrates, such as D-glucose and D-mannose, whereas no enzymatic activity was found with D-galactose (Ahmed et al. 2011). A study on effect of different carbon sources on xylose reductase activity from *D. hansenii* has been reported by Biswas et al. (2012). Among all the sugars tested like D-xylose, xylitol, D-arabinose, D-ribose, L-rhamnose, D-galactose, and sucrose, the maximum affinity of xylose reductase enzyme was found with D-xylose (5.3 ± 2 mM). In another study, the xylose reductase activity with *C. tenuis* was induced (951 U/L) when yeast extract (9.2 g/L) was provided in the fermentation media as an organic nitrogen source compared to other organic sources like peptone and urea. However, with inorganic sources of nitrogen, the enzyme activity was found to be higher with NH_4Cl (217 U/L) compared to NH_4NO_3 (2.8 g/L) (Kern et al. 1998). The cells of *P. tannophilus* showed higher xylose reductase activity at 30 °C by utilizing xylose as carbon source (Alexander 1985). It was reported that flavonoids, such as quercetin and rutin, inhibit xylose reductase in non-sigmoidal fashion in *P. stipitis* and the extent of inhibition increased with increasing inhibitor concentration (Webb and Lee 1991). But the maximum activity of XR obtained from *K. marxianus* IMB strain was reported at 45 °C, and xylose reductase activity with NADPH was found to be 322% higher than with NADH (Mueller et al. 2011). Xylose reductase was produced from *C. guilliermondii*, and the maximum enzymatic activity was found at an initial pH of media varying from 4.0 to 6.0 (Sene et al. 2000).

Different microorganisms show different behavior toward xylose metabolism under various aeration conditions due to differences in the nature of initial steps involved during xylose metabolism by different coenzymes. The effect of aeration conditions for xylose metabolism in *C. utilis*, *P. tannophilus*, and *P. stipitis* was reported in a study given by Bruinenberg et al. (1984). It was found that during the metabolism of xylose under anerobic conditions, *C. utilis* did not show the fermentation of xylose, whereas a very little rate of ethanol production as by-product was found with *P. tannophilus*; however, a rapid fermentation of xylose has been reported with *P. stipitis*. Another study with *P. tannophilus* showed the NADPH-dependent xylose reductase activity of 0.27 ± 0.05 IU/mg in both mineral and complex media. In the production media, the introduction of oxygen limitation increased the enzyme activity to 0.39 ± 0.04 IU/mg (Verduyn et al. 1985a). Large aeration conditions inhibit the activity of xylose reductase (both NADH and NADPH dependent) from *P. tannophilus*. Under oxygen-limited conditions, the specific enzyme activity for NADPH- and NADH-linked xylose reductase was found to be 0.222

and 0.169 IU/mg of protein, respectively. However, under anerobic conditions, the specific enzyme activity for NADPH- and NADH-linked xylose reductase was reduced to 0.051 and 0.058 IU/mg of protein, respectively (Cauwenberge et al. 1989). Under oxygen transfer rates (OTR) from 10 to 30 mmol/L/h, the xylose reductase activity was observed with both cofactors, NADH and NADPH in *C. boidinii* NRRL Y-17213 cells. The enzyme activity was found to be higher with NADH compared to NADPH. However, the maximum yield of xylitol (0.48 g/g) was achieved at OTR of 14 mmol/L/h (Vandeska et al. 1995). The effect of dissolved oxygen on enzyme activity of xylose reductase from *C. parapsilosis* was also observed by Kim et al. (1997). In another study, xylitol was produced from detoxified hemicellulose hydrolysate by *P. tannophilus* under microaerophilic conditions. The production of xylitol was found to be about 90% with the optimal pH values of 6.0–7.7 (Converti et al. 1999). *C. maltose* Xu316 showed highest specific xylose reductase activity of 0.590 IU/mg with NADPH under microaerobic conditions in a medium containing xylose as carbon source at 30 °C under stirring conditions of 220 rpm after 72 h of fermentation (Guo et al. 2006). Xylose reductase from *C. guilliermondii* FTI 20037 showed specific xylose reductase activity of 1.45 ± 0.21 IU/mg of protein with the optimized lowest k_{La} value of 12 h^{-1} (Branco et al. 2009). The maximum xylose reductase production (2.5 IU/mg of protein) was obtained with the negative level of dissolved oxygen (20%) and that the enzymatic activity decreased with high values of dissolved oxygen in *C. guilliermondii* (Milessi et al. 2011).

7.5   Production of Recombinant Xylose Reductase

The attempts for producing xylose reductase using recombinant DNA technology has also been made by various workers (Takuma et al. 1991; Dahn et al. 1996; Handumrongkul et al. 1998; Klimacek et al. 2001; Lee et al. 2003; Woodyer et al. 2005; Fernandes et al. 2009; Zhang et al. 2009; Jeon et al. 2012; Hong et al. 2014; Zhang et al. 2015a, b; Kogje and Ghosalkar 2016). The *P. stipitis* XR gene (*xyl1*) was inserted into an autonomous plasmid pXOR to maintain in multicopy. When grown on xylose under aerobic conditions, the strain with pXOR had up to 1.8-fold higher XR activity than the control strain with no plasmid insertion (Dahn et al. 1996). XR gene (*xyl1*) from *C. guilliermondii* was cloned and expressed in *P. pastoris*, and the recombinant strain was able to accumulate 7.8 g/L xylitol under aerobic conditions (Handumrongkul et al. 1998). The amino acids Tyr^{51}, Lys^{55}, and Lys^{80} of xylose reductase from *C. tenuis* were replaced by site-directed mutagenesis. The purified Tyr^{51}CPhe and Lys^{80}CAla mutants showed higher turnover number and catalytic efficiency between 2500- and 5000-fold than wild-type levels, suggesting a catalytic role of both residues (Klimacek et al. 2001). In another study, NADH preferring XR *xyl1* gene was isolated from *C. parapsilosis* and functionally expressed in *C. tropicalis* under the control of alcohol dehydrogenase promoter to enhance gene expression with higher yield and productivity of enzyme (Lee et al. 2003). From the whole genome of *N. crassa*, XR gene was expressed heterogeneously into *E.coli*, and purification product of expressed gene showed high yield (Nyyssola et al. 2005). The catalytic efficiency of the double mutant xylose reductase from thermophilic fungus *Talaromyces emersonii* (k_{cat}/k_m) increased 1.4-fold with NADH and decreased 10.8-fold with NADPH relative to the wild-type enzyme (Fernandes et al. 2009). *xyl1* gene was isolated from *C. tropicalis* SCTCC 300249 strain, and this gene was cloned and functionally expressed in *Escherichia coli* BL21 (DE3) and *S. cerevisiae* W303-1A. The functionally expressed xylose reductase exhibited dual coenzyme specificity, tolerance to wide pH range, and good thermal stability (Zhang et al. 2009). The heterologous expression of NADPH-dependent XR from *A. oryzae* was induced with xylose in the yeast *S. cerevisiae* through plasmid p416TDH3-*xyrA*, and the enzyme activity was found to be 0.498 ± 0.087 IU/mg (Kaneda et al. 2011). To enhance the expression of XR gene, *N. crassa* xylose reductase (NcXR) codons were changed to those preferred in *C. tropicalis*. This codon-optimized NcXR gene (*NXRG*) was placed under control of a constitutive glyceraldehyde-3-phosphate dehydrogenase promoter derived from *C. tropicalis* and integrated into the genome of XDH (*xyl2*)-disrupted *C. tropicalis*. High expression level of *NXRG* was confirmed by determining xylose reductase activity in cells grown on glucose medium. The resulting recombinant strain LNG2 showed high xylose reductase activity whereas parent strain BSXDH-3 showed no enzymatic activity (Jeon et al. 2012). The overexpression of XR gene (*xyl1*) and antisense inhibition of xylulokinase gene (*xyiH*) by genetic engineering showed the increase in xylitol production from 22.8 mM to 24.8 mM with *Trichoderma reesei* (Hong et al. 2014). In another study, NADPH-dependent XR gene from *Rhizopus oryzae* AS 3.819 (*RoXR*) was cloned and expressed in *P. pastoris* GS115. Site-directed mutagenesis at the two coenzyme binding sites Thr^{226} → Glu^{226} and Val^{274} → Asn^{274} was performed, which increased the coenzyme specificity of RoXR^{T226E} and RoXR^{V274N} for NADH by 18.2-fold and 2.4-fold, respectively. These genetic modifications enhanced the preference of enzyme for NADH (Zhang et al. 2015b). The three heterologous and one endogenous xylose reductase were overexpressed in *S. cerevisiae*, and endogenous *GRE3* gene overexpression resulted in the highest xylitol productivity by the recombinant strain (Kogje and Ghosalkar 2016).

7.6 Purification, Characterization, and Immobilization of Xylose Reductase

Xylose reductase was purified from different fungal cultures through multistep purification strategies. NADP-dependent xylose reductase from *P. tannophilus* CBS4044 was resolved into two different enzyme isoforms, and the enzymes A and B both had molecular weight lower than usually observed with xylose reductases of yeasts (Verduyn et al. 1985a). Purification of xylose reductase from *P. tannophilus* was carried by using two affinity chromatography steps, and NADPH- and NADH-linked xylose reductase-specific activity increased approximately 26-fold and 52-fold, respectively. The molecular weight of this multimeric enzyme was found to be 36,500 daltons (Da) using SDS-PAGE (Bolen et al. 1986). Xylose reductase was purified from *P. tannophilus* by Western blot and immunoprecipitation, and two isoforms of the enzyme were found, out of which one isoform requires either NADPH or NADH as cofactor with pI of 5.1; however, the second isoform requires only NADPH as cofactor with pI of 6.4. The molecular weight of both the isoforms was found to be 36,000 Da (Cauwenberge et al. 1989).

Three isoforms (XR1, XR2, XR3) of NADPH-dependent xylose reductases were purified from *C. tropicalis* IFO 0618. They had respective k_m values of 37, 30, and 34 mM for D-xylose and 14, 18, and 9 μM for NADPH, but NADH did not act as a cofactor. Further, gel filtration and cross-linking analysis showed both XR1 and XR2 were dimers with identical monomeric subunits. The pI values of XR1 and XR2 were found to be 4.15 and 4.10, respectively, with molecular weight of 36,497.91 Da and 36,539.68 Da, respectively, as observed from mass spectra analysis (Yokoyama et al. 1995). Xylose reductase from some of the common xylose-metabolizing yeasts such as *C. tropicalis* (Yokoyama et al. 1995), *C. flavus* (Mayr et al. 2003), and *Galdieria sulphuraria* (Gross et al. 1997) has also been reported to be homodimeric. However, xylose reductase from other yeasts such as *P. tannophilus* (Ditzelmuller et al. 1984), *S. cerevisiae* (Kuhn et al. 1995), and *C. tenuis* (Neuhauser et al. 1997) was found to be monomeric. Xylose reductase from fungus *N. crassa* has been purified by using affinity chromatography, HPLC (high performance liquid chromatography), SDS-gradient ultrathin PAGE (sodium dodecyl sulfate polyacrylamide gel electrophoresis), and IEF (isoelectric focusing), and two isoenzymes of xylose reductase were found with molecular mass of 30 KDa and 27 KDa with pI 5.3 and 5.6, respectively. The k_m values against xylose for these two isoforms were 2.3 mM and 1.1 mM (Zhao et al. 1998). Handumrongkul et al. (1998) purified recombinant xylose reductase from *P. pastoris*, and its molecular mass was found to be 36 KDa through SDS-PAGE. Xylose reductase was purified with DEAE Sepharose Fast Flow, Phenyl Sepharose 6 Fast Flow, and Sephadex G-75 columns. Further, ultracentrifugation was carried out by PM 10 (Amicon) with a cutoff value of 10,000 Da. The molar mass of the purified enzyme was detected as 36,000 g/mol and the pI value of 4.5 (Granstrom et al. 2002). The molecular mass of xylose reductase enzyme from *C. mogii* was found to be in the range of 24–36 KDa through SDS- PAGE (Mayerhoff et al. 2004). The xylose reductase gene from *N. crassa* was expressed in *E. coli*, and after transformation, the gene was expressed and purified by size exclusion chromatography and high performance liquid chromatography with a Bio-Sil SEC-250 column. From the retention time, the molecular weight of xylose reductase was found to be 53 KDa, and in the presence of 15% SDS, a single peak was obtained with the molecular weight of 34 KDa (Woodyer et al. 2005). Cortez et al. (2006) has purified xylose reductase by using CTAB (cetyltrimethylammonium bromide)-reverse micelles from *C. guilliermondii* FTI 20037. The $(k_m)_{xylose}$ for the enzyme was about 35% higher than before extraction; however,

$(k_m)_{NADPH}$ was about 30% lower after than before extraction. The v_{max} value against both xylose and NADPH for the enzyme before and after extraction was about 6%. In another study, the molecular mass of XR from *C. tropicalis* was found to be 36.48 KDa, and amino acids, glutamic acid, and aspartic acid were found to be present in high percentage, whereas tyrosine and methionine were present in low values; however, cysteine was not present in the active site of enzyme (Zeid et al. 2008). Xylose reductase was partially purified from *C. guilliermondii* FTI 20037 by agarose column, a membrane reactor, and an Amicon Ultra centrifugal device. The yield of xylose reductase with the three methods was 40%, 7%, and 67%, respectively, with purification factor of 2.84, 3.05, and 13.7, respectively (Tomotani et al. 2009). The xylose reductase activity from *C. tropicalis* was studied with purified, free, and immobilized enzyme, and the enzyme activity was reported to be greatly reduced by different inhibitors like $ZnCl_2$, $FeCl_3$, $CuSO_4$, and dithiothreitol (DTT), and maximum enzymatic inhibition was found with $FeCl_3$ for both for free and immobilized enzyme. Free and immobilized xylose reductase showed higher activity at 45 °C and 50 °C; however, the optimum pH for free and immobilized enzyme was 4.5, but immobilized enzyme showed higher activity with a broader pH range (4.0–6.0). The molecular mass of His tagged xylose reductase fusion protein through SDS- PAGE was observed to be 57 KDa. The k_m values for free and immobilized enzyme were 30.3 mM and 20.1 mM, respectively (Su et al. 2010). Xylose reductase has been purified from *Debaryomyces nepalensis* NCYC 3413, and the molecular weight of enzyme was reported as 74 KDa with monomeric subunits of 36.4 KDa (MALDI-TOF/MS) and pI of 6.0. This enzyme was reported as moderately halotolerant and showed its economical importance for the production of xylitol under conditions where salts are present (Kumar and Gummadi 2011). Xylose reductase from *C. tropicalis* was purified using reverse micelles, and its molecular mass was reported to be 32.42 KDa by SDS-PAGE (Ronzon et al. 2012). The inhibitory effect of tosyl phenylalanyl chloromethyl ketone, phenylmethanesulfonyl fluoride, phenylglyoxal hydrate, and pepstatin A with a concentration ranging from 1 to 10 mM on xylose reductase purified from *Kluyveromyces* sp. IIPE453 has been reported by Dasgupta et al. (2016).

There are scarce reports on immobilization of purified xylose reductase because most of the work on xylitol production was carried out by using whole cells of different microorganisms containing intracellular XR. The immobilization of xylose reductase from *C. tropicalis* on different supports like calcium alginate gel embedding, alginate/chitosan microcapsule, chitosan cross-linked with glutaraldehyde, and polyacrylamide gelatin embedding has been studied by Su et al. (2010). The best results were obtained with chitosan bead, which increased the NADH-dependent xylose reductase activity by 17.15-fold. Xylose reductase immobilized on p-(methyl methacrylate-ethylene dimethyl acrylate-methacrylic acid) (MMA-EDMA-MAA) nanoparticles has shown increase in the stability of enzyme by 18 folds, and the enzyme activity was found to be 5.1 IU/mg (Zhang et al. 2011).

7.7 Patent Status of XR

A number of patents have been granted related with XR production from both native and recombinant fungal strains and expression of XR encoding genes for production of industrial product xylitol and ethanol. Hallborn et al. (1999) patented xylose utilization by new recombinant yeast strain, and the transformed yeast showed enhanced expression of XR gene resulting in higher xylitol production. Aldose reductase inhibitor and pharmaceutically acceptable derivatives from cultures of *A. niger* CFR1046 was patented by Sattur et al. (2003). XR encoding genes from *N. crassa* was heterologously expressed in *E. coli* for enhanced production of xylitol, and XR was shown to have high turnover rate, catalytic efficiency, stability, broad pH profile, and preference of NADPH over NADH (Zhao et al. 2006). Further Zhao et al. (2008) showed the utilization of recombinant *E. coli* for production of xylitol and other alcoholic sugars. The metabolic enhancement of organism was also reported for fermentation of plant biomass into ethanol. Further, the engineering of native strain of XR from *N. crassa* was performed by amino acid mutations, and the mutant XR demonstrated higher preference to xylose over arabinose with enhanced production of xylitol and ethanol (Zhao and Nair 2010). The industrial application of *AKR* from *Z. mobilis* has been patented by Chen and Agrawal (2012).

7.8 Applications of Xylitol

Xylitol has many promising applications in food, pharmaceuticals, and odontological industries. A graphic view of the same is presented in Fig. 7.2.

7.8.1 Food Industry

Xylitol shows many advantages when used as a food ingredient by maintaining the nutritional value of proteins, which is achieved through inhibiting any undesirable changes in the food properties during storage due to its heat stability in Maillard reaction (Parajo et al. 1998; Cunha et al. 2006). Moreover, the sweetening power of xylitol is similar to sucrose (Lourenco et al. 2014). Xylitol also improves the color and taste of food preparations and also limits the tendency of obesity. It extends the shelf life of food by inhibiting the growth of saprophytes such as *Clostridium butyricum* and *Salmonella typhi* (Emodi 1978). This polyol sugar is also used in jams, frozen desserts (Abril et al. 1982), and yoghurts (Hyvonen and Slotte 1983) because it provides texture, color, taste, and long-term storage of these food products. It is also used for thin coating in soft drinks, beverages, chewing gum, and bakery products (Makinen 1992; Lakshmi et al. 2014). Processed cheese is one of the leading

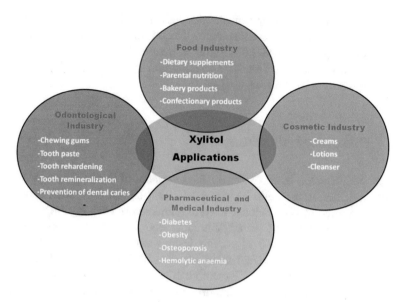

Fig. 7.2 Industrial applications of xylitol

varieties of cheese in the world, and xylitol is used as an ingredient in various processed cheese (Kapoor and Metzger 2008; Kommineni et al. 2012). The prebiotic effect of xylitol has already been approved (Lugani and Sooch 2017).

7.8.2 Pharmaceutical Industry

Xylitol has been shown to provide a wide range of applications in the pharmaceutical industry for the treatment of many diseases like glucose-6-phosphate dehydrogenase deficiency of red blood cells and Meniere's disease. It is also used for the restoration of heart muscle adenine nucleotide levels and preservation of red blood cells (Makinen 2000). The use of xylitol as a protein-sparing and vitamin B-sparing agent was also reported by Makinen (2000). The other major application of xylitol is its use in parental nutrition due to the inability of this sugar to react with amino acids (Parajo et al. 1998). The International Society of Rare Sugars (ISRS) has defined xylitol as a rare sugar with beneficial health properties (Granstrom et al. 2004). There is slow absorption of xylitol in the body with little rise in blood glucose as the metabolism of this polyol is insulin independent; therefore, it acts as a sugar substitute for diabetics (Parajo et al. 1998; Lugani et al. 2017). This property of xylitol makes it useful for postoperative or post-traumatic states when the excessive secretion of stress hormone cortisol, catecholamine, glucagon, growth hormones, etc. causes insulin resistance and reduces the efficient utilization of glucose (Makinen 1976). It is used in combination with mannitol, sorbitol, and citric or adipic acid as a part of coating of pharmaceutical products (Pepper and Olinger 1998). Xylitol also possesses antiketonic (Kinami and Kitagawa 1969) and anti-infection effect (Brown et al. 2004) with skin smoothing properties (Zeid et al. 2008). In a recent study, the protective effect of dietary xylitol on influenza A virus infection is reported by Yin et al. (2014).

7.8.3 Odontological Industry

The human consumption of xylitol for the treatment of dental plaque was firstly reported by Turku sugar studies (Scheinin et al. 1976). From these studies, decayed, missing, and filled (dmf) incidence in teeth with sucrose chewing gum was found to be 2.92 compared to 1.04 with xylitol gum group. The intake of xylitol results in enhanced remineralization of carious lesions and stimulation of the flow of saliva with a decrease in pH (Decker and Loveren 2003; Makinen et al. 2008; Llop et al. 2010; Nayak et al. 2014; Clementine et al. 2016). Xylitol is used in tooth-paste formulations because of its important ability to retain moisture (Chi et al. 2014). The sweetness of xylitol is similar to sucrose, but xylitol does not cause tooth decay because it inhibits the growth of tooth-decaying bacterium *Streptococcus mutans* in the saliva by reducing the level of lactic acid produced by these bacteria (Makinen 1992). Ly et al. (2006) reported that xylitol consumption of less than 5 g per day is more effective for caries prevention than consumption of sorbitol. The consumption of xylitol reduces the mother-to-child transmission of *S. mutans* and thus prevents caries chances in their children (Soderling 2009). The use of xylitol also reduces the amount of supragingival plaque due to its anti-bacterial properties (Twetman 2009). Some studies have also suggested that chewing process enhances the caries inhibitory effect of xylitol chewing gum (Scheie et al. 1998; Machiulskiene et al. 2001; Loveren 2004; Makinen 2009). Pereira et al. (2012) reported the importance of xylitol in reducing the risk of dental caries. The protective effect of xylitol against dental caries has also been reported by Lakshmi et al. (2014).

7.8.4 Other Industries

Xylitol is also used in cosmetic industry as humectants. Therefore, it is used in cleanser, beauty creams, and lotions to maintain moisture. In many industries, xylitol is also used as an intermediate for manufacturing organic chemicals (Velazquez Pereda et al. 2011).

7.9 Conclusion and Future Scope

Xylose reductase is an industrially important enzyme, present in almost all the microorganisms, and is involved in the conversion of xylose to xylitol. Xylitol has many applications in food, pharmaceutical, medical, cosmetic, and odontological sectors. Various reports are available on microbial XR production, purification, and characterization. Some reports on utilization of different agriculture waste as xylose source are available. The production profile of XR is greatly influenced by the availability of oxygen, inducers, and inhibitors. There is an ample scope for utilization of agro-waste as source of xylose for production of xylitol using XR. Most of the previous work on xylitol production has been done with whole microbial cells containing XR. Moreover, very less information is available with the catalytic mechanism, substrate binding sites, and coenzyme specificity of this enzyme. The use of whole cells for xylitol production has many disadvantages like cell recycling problems, clogging of column, product instability, less enzyme to substrate ratio, less purity of product due to interference of other enzymes, etc. Hence, there is an urgent need to make use of XR for xylitol production at industrial scale. There is a need to develop new efficient microbial strains which can utilize agriculture waste and keep the oxygen and energy requirement as low as possible. Many efforts are required to maximize the XR production by using multiple strategies by tailoring the nutritional and process requirements of efficient XR-producing microorganisms. The other constraint in XR system for production of xylitol is the requirement of coenzyme. Hence, the development of a coenzyme regeneration system is recommended for a better economical technology. Several attempts have already been done to improve XR production by recombination DNA technology and site-directed mutagenesis. More research initiatives are required to develop efficient XR-producing strains with new recombinant techniques like microarray and robotic-aided directed evolution. Modern and rapid enzyme purification techniques should also be adapted to improve the purity and recovery of enzyme. Further, initiatives are required to develop immobilization processes using different nanoparticles. Hence, the industrial production of xylitol through XR system requires multidiscipline approaches like microbial technology, molecular and recombination technology, enzyme technology, and proteomics along with medium downstream processes.

Acknowledgments The authors are thankful to the Department of Biotechnology, Punjabi University, Patiala, India, for providing necessary facilities and also thankful to Bhai Kahn Singh Nabha Library of the university for providing access to scientific literature available with them.

References

Abril JR, Stull JW, Taylor RR, Angus RC, Daniel TC (1982) Characteristics of frozen desserts sweetened with xylitol and fructose. Food Sci 47(2):472–475
Agrawal M, Chen RR (2011) Discovery and characterization of a xylose reductase from *Zymomonas mobilis* ZM4. Biotechnol Lett 33(11):2127–2133

Aguiar C, Oetterer M, Menezes TJB (1999) Caracterizacao e aplicacoes do xylitol na industria alimenticia. Boletim SBCTA 33(2):184–193

Ahmed YM, Ibrahim IH, Khan JA, Kumosani TA (2011) Oxidation and reduction of D-xylose by cell-free extract of *Hansenula polymorpha*. Aust J Basic Appl Sci 5(12):95–100

Alexander NJ (1985) Temperature sensitivity of the induction ofxylose reductase in *Pachysolen tannophilus*. Biotechnol Bioeng 27(12):1739–1744

Attfield PV, Bell PJ (2006) Use of population genetics to derive nonrecombinant *Saccharomyces cerevisiae* strains that grow using xylose as a sole carbon source. FEMS Yeast Res 6(6):862–868

Banta S, Boston M, Jarnagin A, Anderson S (2002) Mathematical modeling of in vitro enzymatic production of 2-keto-L-gulonic acid using NAD(H) or NADP(H) as cofactors. Metab Eng 4(4):273–284

Barski OA, Tipparaju SM, Bhatnagar A (2008) The aldo-keto reductase superfamily and its role in drug metabolism and detoxification. Drug Metab Rev 40(4):553–624

Bicho PA, Runnals PL, Cunningham JD, Lee H (1988) Induction of xylose reductase and xylitol dehydrogenase activities in *Pachysolen tannophilus* and *Pichia stipitis* on mixed sugars. Appl Environ Microbiol 54(1):50–54

Billard P, Menart S, Fleer R, Fukuhara MB (1995) Isolation and characterization of the gene encoding xylose reductase from *Kluyveromyces lactis*. Gene 162(1):93–97

Biswas D, Pandya V, Singh AK, Mondal AK, Kumaran S (2012) Co-factor binding confers substrate specificity to xylose reductase from *Debaryomyces hansenii*. PLoS One 7(9):1–11

Bolen PL, Detroy RW (1985) Induction of NADPH-linked D-xylose reductase and NAD-linked xylitol dehydrogenase activities in *Pachysolen tannophilus* by D-xylose, L-arabinose or D-galactose. Biotechnol Bioeng 27(3):302–307

Bolen PL, Bietz JA, Detroy RW (1985) Aldose reductase in the yeast *Pachysolen tannophilus*: purification, characterization and N-terminal sequence. Biotechnol Bioeng Symp 15:129–148

Bolen PL, Roth KA, Freer SN (1986) Affinity purifications of aldose reductase and xylitol dehydrogenase from the xylose fermenting yeast *Pachysolen tannophilus*. Appl Environ Microbiol 52(4):660–664

Boonmee A (2012) Hydrolysis of various Thai agricultural biomasses using the crude enzyme from *Aspergillus aculeatusiizuka* FR60 isolated from soil. Braz J Microbiol 43(2):456–466

Boontham W, Srisuk N, Kokaew K, Treeyoung P, Limtong S, Thamchaipenet A, Yurimoto H (2014) Xylitol production by thermotolerant methylotrophic yeast *Ogataea siamensis* and its xylose reductase gene (*xyl1*) cloning. Chiang Mai J Sci 41(3):491–502

Branco RF, Santos JC, Pessoa A, Silva SS (2009) Profiles of xylose reductase, xylitol dehydrogenase and xylitol production under different oxygen transfer volumetric coefficient values. J Chem Technol Biotechnol 84(3):326–330

Branco RF, Santos JC, Silva SS (2011) A solid and robust model for xylitol enzymatic production optimization. J Bioproces Biotechniq 1(4):1–6

Branden CI (1991) The TIM barrel- the most frequently occurring folding motif in proteins. Curr Opin Struct Biol 1(6):978–983

Brown CL, Graham SM, Cable BB, Ozer EA, Taft PJ, Zabner J (2004) Xylitol enhances bacterial killing in the rabbit maxillary sinus. Laryngoscope 114(11):2021–2024

Bruinenberg PM, But PHM, Dijken JP, Scheffers WA (1984) NADH-linked aldose reductase: the key to anaerobic alcoholic fermentation of xylose by yeasts. Appl Microbiol Biotechnol 19(4):256–260

Cauwenberge JE, Bolen PL, McCracken DA, Bothast RJ (1989) Effect of growth conditions on cofactor linked xylose reductase activity in *Pachysolen tannophilus*. Enzym Microb Technol 11(10):662–667

Chen RR, Agrawal M (2012) Industrial applications of a novel aldo/keto reductase of *Zymomonas mobilis*. US Patent 0,196,342, 2 Aug 2012

Chi DL, Tut OK, Milgrom P (2014) Cluster-randomized xylitol toothpaste trial for early childhood caries prevention. J Dent Child (Chic) 81(1):27–32

Clementine T, Yue CC, Xiaoling W, Marine P, Alex H, Larry M, Daniel W, Laetitia GD (2016) Maltitol and xylitol sweetened chewing gums could modulate salivary parameters involved in dental caries prevention. J Interdiscipl Med Dent Sci 4(2):1–8

Converti A, Perego P, Dominguez JM (1999) Microaerophilic metabolism of *Pachysolen tannophilus* at different pH values. Biotechnol Lett 21(8):719–723

Cortez EV, Pessoa A, Felipe MGA, Roberto IC, Vitolo M (2006) Characterization of xylose reductase extracted by CTAB-reversed micelles from *Candida guilliermondii* homogenate. Braz J Pharm Sci 42(2):251–257

Costanzo L, Penning TM, Christianson DW (2009) Aldo-keto reductases in which the conserved catalytic histidine is substituted. Chem Biol Interact 178(1–3):127–133

Cunha MAA, Converti A, Santos JC, Silva SS (2006) Yeast immobilization in Lentikats: a new strategy for xylitol bioproduction from sugarcane bagasse. World J Microbiol Biotechnol 22(1):65–72

Dahn KM, Davis BP, Pittman PE, Kenealy WR, Jeffries TW (1996) Increased xylose reductase activity in the xylose-fermenting yeast *Pichia stipitis* by overexpression of *xyl1*. Appl Biochem Biotechnol 57-58:267–276

Dasgupta D, Ghosh D, Bandhu S, Agarwal D, Suman SK, Adhikari DK (2016) Purification, characterization and molecular docking study of NADPH dependent xylose reductase from thermotolerant *Kluyveromyces* sp. IIPE453. Process Biochem 51(1):124–133

Decker RT, Loveren C (2003) Sugars and dental caries. Am J Clin Nutr 78(l):881–892

Dijken JP, Scheffers WA (1986) Redox balances in the metabolism of sugars by yeasts. FEMS Microbiol Lett 32(3–4):199–224

Ditzelmuller G, Kubicek CP, Wohrer W, Rohr M (1984) Xylose metabolism in *Pachysolen tannophilus* purification and properties of xylose reductase. Can J Microbiol 30(11):1330–1336

Ellis EM (2002) Microbial aldo-keto reductases. FEMS Microbiol Lett 216(2):123–131

Emodi A (1978) Xylitol: its properties and food applications. Food Technol 32:20–32

Erdei B, Barta Z, Sipos B, Reczey K, Galbe M, Zacchi G (2010) Research ethanol production from mixtures of wheat straw and wheat meal. Biotechnol Biofuels 3(1):1–9

Feldmann SD, Sahm H, Sprenger GA (1992) Pentose metabolism in *Zymomonas mobilis* wild-type and recombinant strains. Appl Microbiol Biotechnol 38:354–361

Fernandes S, Tuohy MG, Murray PG (2009) Xylose reductase from the thermophilic fungus *Talaromyces emersonii*: cloning and heterologous expression of the native gene (*TeXR*) and a double mutant (*TeXRK*[271R + N273D]) with altered coenzyme specificity. J Biosci 34(6):881–890

Granstrom T, Aristidou AA, Leisola M (2002) Metabolic flux analysis of *Candida tropicalis* growing on xylose in an oxygen-limited chemostat. Metab Eng 4(3):248–256

Granstrom TB, Takata G, Tokuda M, Izumori K (2004) A novel and complete strategy for bioproduction of rare sugars. J Biosci Bioeng 97(2):89–94

Granstrom TB, Izumori K, Leisola M (2007) A rare sugar xylitol. Part II: biotechnological production and future applications of xylitol. Appl Microbiol Biotechnol 74(2):273–276

Gross W, Seipold P, Schnarrenberger C (1997) Characterization and purification of an aldose reductase from the acidophilic and thermophilic red alga *Galdieria sulphuraria*. Plant Physiol 114(1):231–236

Guo C, Zhao C, He P, Lu D, Shen A, Jiang N (2006) Screening and characterization of yeasts for xylitol production. J Appl Microbiol 101(5):1096–1104

Gurpilhares DB, Hasmann FA, Pessoa A, Roberto IC (2009) The behavior of key enzymes of xylose metabolism on the xylitol production by *Candida guilliermondii* grown in hemicellulosic hydrolysate. J Ind Microbiol Biotechnol 36(1):87–93

Hagerdal BH, Jeppsson H, Olsson L, Mohagheghi A (1994) An interlaboratory comparison of the performance of ethanol producing microorganisms in a xylose rich acid hydrolysate. Appl Microbiol Biotechnol 41(1):62–72

Hallborn J, Walfridsson M, Airaksinen U, Ojamo H, Hahnhagrbdal B, Penttila M, Kerasnen S (1991) Xylitol production by recombinant *Saccharomyces cerevisiae*. Biotechnology 9(11):1090–1095

Hallborn J, Penttila M, Ojamo H, Walfridsson M, Airaksinen U, Keranen S, Hagerdal BH (1999) Xylose utilization by recombinant yeasts. US Patent 5,866,382, 2 Feb 1999

Handumrongkul C, Ma DP, Silva JL (1998) Cloning and expression of *Candida guilliermondii* xylose reductase gene (*xyl1*) in *Pichia pastoris*. Appl Microbiol Biotechnol 49(4):399–404

Hong Y, Dashtban M, Kepka G, Chen S, Qin W (2014) Overexpression of D-xylose reductase (*xyl1*) gene and antisense inhibition of D-xylulokinase (*xyiH*) gene increase xylitol production in *Trichoderma reesei*. Biomed Res Int 2014:1–8

Horitsu H, Yahashi Y, Takamizawa K, Kawai K, Suzuki T, Watanabe N (1992) Production of xylitol from D-xylose by *Candida tropicalis*: optimization of production rate. Biotechnol Bioeng 40(9):1085–1091

Hyatt MP, Lickteig AJ, Klaassen CD (2013) Tissue distribution, ontogeny and chemical induction of aldo-keto reductases in mice. Drug Metab Dispos 41(8):1480–1487

Hyvonen L, Slotte M (1983) Alternative sweetening of yoghurt. J Food Technol 18(1):97–112

Jeon WY, Yoon BH, Ko BS, Shim WY, Kim JH (2012) Xylitol production is increased by expression of codon-optimized *Neurospora crassa* xylose reductase gene in *Candida tropicalis*. Bioprocess Biosyst Eng 35(1–2):191–198

Jez JM, Penning TM (2001) The aldo-keto reductase (*AKR*) superfamily: an update. Chem Biol Interact 130-132(1–3):499–525

Kaneda J, Sasaki K, Gomi K, Shintani T (2011) Heterologous expression of *Aspergillus oryzae* xylose reductase and xylitol dehydrogenase genes facilitated xylose utilization in the yeast *Saccharomyces cerevisiae*. Biosci Biotechnol Biochem 75(1):168–170

Kapoor R, Metzger LE (2008) Process cheese: scientific and technological aspects: a review. Compr Rev Food Sci Food Saf 7(2):194–214

Karimi K, Kheradmandinia S, Taherzadeh MJ (2006) Conversion of rice straw to sugar by dilute acid hydrolysis. Biomass Bioenergy 30(3):247–253

Kauko K, Makinen KK (2010) Sugar alcohols, caries incidence and remineralization of caries lesions: a literature review. Int J Dent 2010:1–23

Kavanagh KL, Klimacek M, Nidetzky B, Wilson DK (2002) The structure of apo and holo forms of xylose reductase, a dimeric aldo-keto reductase from *Candida tenuis*. Biochem J 41(28):8785–8795

Kavanagh KL, Klimacek M, Nidetzky B, Wilson DK (2003) Structure of xylose reductase bound to NAD⁺ and the basis for single and dual co-substrate specificity in family 2 aldo-keto reductases. Biochem J 373(2):319–326

Kern M, Nidetzky B, Kulbe KD, Haltrich D (1998) Effect of nitrogen sources on the levels of aldose reductase and xylitol dehydrogenase activities in the xylose fermenting yeast *Candida tenuis*. J Ferment Bioeng 85(2):196–202

Khoury GA, Fazelinia H, Chin JW, Pantazes RJ, Cirino PC, Maranas CD (2009) Computational design of *Candida boidinii* xylose reductase for altered cofactor specificity. Protein Sci 18(10):2125–2138

Kim S, Kim J, Oh D (1997) Improvement of xylitol production by controlling oxygen supply in *Candida parapsilosis*. J Ferment Bioeng 83(3):267–270

Kim MD, Jeun YS, Kim SG, Ryu YW, Seo JH (2002) Comparison of xylitol production in recombinant *Saccharomyces cerevisiae* strains harboring *xyl1* gene of *Pichia stipitis* and *GRE3* gene of *S. cerevisiae*. Enzym Microb Technol 31(6):862–866

Kim SR, Park YC, Jin YS, Seo JH (2013) Strain engineering of *Saccharomyces cerevisiae* for enhanced xylose metabolism. Biotechnol Adv 31(6):851–861

Kinami Y, Kitagawa I (1969) Fluctuation of blood sugar, urine sugar and ketone body levels in surgical stress and application of xylitol. Shujutsu 23(11):1487–1491

Klimacek M, Szekely M, GrieMler R, Nidetzky B (2001) Exploring the active site of yeast xylose reductase by site-directed mutagenesis of sequence motifs characteristic of two dehydrogenase/reductase family types. FEBS Lett 500(3):149–152

Kogje A, Ghosalkar A (2016) Xylitol production by *Saccharomyces cerevisiae* overexpressing different xylose reductases using non-detoxified hemicellulosic hydrolysate of corncob. 3 Biotech 6(2):1–10

Kokaew K, Srisuk N, Limtong S, Thamchaipenet A (2009) Cloning and nucleotide sequence analysis of xylose reductase (XR) gene from thermotolerant methylotrophic yeast *Ogataea siamensis* N22. Thai J Genet 2(1):66–71

Kommineni A, Amamcharla J, Metzger LE (2012) Effect of xylitol on the functional properties of low-fat process cheese. J Dairy Sci 95(11):6252–6259

Kuhn A, Zyl C, Tonder AV, Prior BA (1995) Purification and partial characterization of an aldo-keto-reductase from *Saccharomyces cerevisiae*. Appl Environ Microbiol 61(4):1580–1585

Kumar S, Gummadi SN (2011) Purification and biochemical characterization of a moderately halotolerant NADPH dependent xylose reductase from *Debaryomyces nepalensis* NCYC 3413. Bioresour Technol 102(20):9710–9717

Ladisch MR, Lin KW, Voloch M, Tsao GT (1983) Process considerations in enzymatic hydrolysis of biomass. Enzym Microb Technol 5(2):82–102

Lakshmi SV, Yadav HKS, Mahesh KP, Raizaday A, Manne N, Ayaz A, Nagavarma NBV (2014) Medicated chewing gum: an overview. Res Rev J Dent Sci 2(2):50–64

Laskowski RA, MacArthur MW, Moss DS, Thornton JM (1993) Procheck: a program to check the stereochemical quality of protein structures. J Appl Crystallogr 26:283–291

Lee J (1997) Biological conversion of lignocellulosic biomass to ethanol. J Biotechnol 56(1):1–24

Lee H (1998) The structure and function of yeast xylose (aldose) reductases. Yeast 14(11):977–984

Lee H, Sopher CR, Yau KYF (1996) Induction of xylose reductase and xylitol dehydrogenase activities on mixed sugars in *Candida guilliermondii*. J Chem Technol Biotechnol 65(4):375–379

Lee JK, Koo BS, Kim SY (2003) Cloning and characterization of the *xyl1* gene, encoding an NADH-preferring xylose reductase from *Candida parapsilosis* and its functional expression in *Candida tropicalis*. Appl Environ Microbiol 69(10):6179–6188

Llop MR, Jimeno FG, Acien RM, Dalmau LJB (2010) Effects of xylitol chewing gum on salivary flow rate, pH, buffering capacity and presence of *Streptococcus mutans* in saliva. Eur J Paediatr Dent 11(1):9–14

Lourenco MVM, Andreote FD, Vildoso CIA, Basso LC (2014) Biotechnological potential of *Candida* sp. for the bioconversion of D-xylose to xylitol. Afr J Microbiol Res 8(20):2030–2036

Loveren C (2004) Sugar alcohols: what is the evidence for caries-preventive and caries-therapeutic effects? Caries Res 38(3):286–293

Luccio E, Elling RA, Wilson DK (2006) Identification of a novel NADH-specific aldo-keto reductase using sequence and structural homologies. Biochem J 400(1):105–114

Lugani Y, Sooch BS (2017) Xylitol, an emerging prebiotic: a review. Int J Appl Pharm Biol Res 2(2):67–73

Lugani Y, Oberoi S, Sooch BS (2017) Xylitol: a sugar substitute for patients of diabetes mellitus. World J Pharm Pharm Sci 6(4):741–749

Lunzer R, Mamnun Y, Haltrich D, Kulbe KD, Nidetzky B (1998) Structural and functional properties of a yeast xylitol dehydrogenase, a Zn^{2+}-containing metalloenzyme similar to medium-chain sorbitol dehydrogenases. Biochem J 336(1):91–99

Ly KA, Milgrom P, Rothen M (2006) Xylitol, sweeteners and dental caries. Pediatr Dent 28(2):154–163

Machiulskiene V, Nyvad B, Baelum V (2001) Caries preventive effect of sugar-substituted chewing gum. Community Dent Oral Epidemiol 29:278–288

Makinen KK (1976) Possible mechanisms for the cariostatic effect of xylitol. In: Ritzel G, Brubacher G (eds) Monosaccharides and polyalcohols in nutrition, therapy and dietetics. Huber, Bern, pp 368–380

Makinen KK (1992) Dietary prevention of dental caries by xylitol-clinical effectiveness and safety. J Appl Nutr 44:16–28

Makinen KK (2000) The rocky road of xylitol to its clinical application. J Dent Res 79(6):1352–1355

Makinen KK (2009) An end to crossover designs for studies on the effect of sugar substitutes on caries? Caries Res 43(5):331–333

Makinen KK, Alanen P, Isokangas P, Isotupa K, Soderling E, Makinen PL, Wenhui W, Weijian W, Xiaochi C, Yi W, Boxue Z (2008) Thirty-nine-month xylitol chewing-gum programme in initially 8-year-old school children: a feasibility study focusing on mutans *Streptococci* and *Lactobacilli*. Int Dent J 58(1):41–50

Markets and Markets (2016) Industrial enzymes market. http://www.marketsandmarkets.com/PressReleases/industrial-enzymes.asp. Accessed 21 April 2017

Mayerhoff ZDVL, Roberto IC, Franco TT (2004) Purification of xylose reductase from *Candida mogii* in aqueous two-phase systems. Biochem Eng J 18(3):217–223

Mayr P, Bruggler K, Kulbe KD, Nidetzky B (2000) D-xylose metabolism by *Candida intermedia*: isolation and characterization of two forms of aldose reductase with different coenzyme specificities. J Chromatogr B Biomed Sci Appl 737(1–2):195–202

Mayr P, Petschacher B, Nidetzky B (2003) Xylose reductase from the basidiomycete fungus *Cryptococcus flavus*: purification, steady-state kinetic characterization and detailed analysis of the substrate binding pocket using structure-activity relationships. J Biochem 133(4):553–562

Melaja J, Hamalainen L (1977) Process for making xylitol. US Patent 4008285, 15 Feb 1977

Milessi TSS, Chandel AK, Branco RF, Silva SS (2011) Effect of dissolved oxygen and inoculum concentration on xylose reductase production from *Candida guilliermondii* using sugarcane bagasse hemicellulosic hydrolysate. Food Nutr Sci 2(3):235–240

Moyses DN, Reis VC, Almeida JR, Moraes LM, Torres FA (2016) Xylose fermentation by *Saccharomyces cerevisiae*: challenges and prospects. Int J Mol Sci 17(3):1–18

Mueller M, Wilkins MR, Banat IM (2011) Production of xylitol by the thermotolerant *Kluyveromyces marxianus* IMB strains. J Bioprocess Biotechniq 1(2):1–5

Nayak PA, Nayak UA, Khandelwal V (2014) The effect of xylitol on dental caries and oral flora. Clin Cosmet Investig Dent 6:89–94

Neuhauser W, Haltrich D, Kulbe KD, Nidetzky B (1997) NAD(P)H-dependent aldose reductase from the xylose-assimilating yeast *Candida tenuis*. Biochem J 326(3):683–692

Nigam P, Singh D (1995) Processes for fermentative production of xylitol- a sugar substitute. Process Biochem 30(2):117–124

Nyyssola A, Pihlajaniemi A, Palva A, Weymarn N, Leisola M (2005) Production of xylitol from D-xylose by recombinant *Lactococcus lactis*. J Biotechnol 118(1):55–66

Parajo JC, Dominguez H, Dominguez JM (1998) Biotechnological production of xylitol Part 2: operation in culture media made with commercial sugars. Bioresour Technol 65(3):203–212

Pepper T, Olinger PM (1998) Xylitol in sugar-free confections. Food Technol 42:98–106

Pereira AFF, Silva TC, Silva TL, Caldana ML, Baston JRM, Buzalaf MAR (2012) Xylitol concentrations in artificial saliva after application of different xylitol dental varnishes. J Appl Oral Sci 20(2):146–150

Rafiqul ISM, Sakinah AMM (2012) A perspective: bioproduction of xylitol by enzyme technology and future prospects. Int Food Res J 19(2):405–408

Rafiqul SM, Sakinah AM (2015) Biochemical properties of xylose reductase prepared from adapted strain of *Candida tropicalis*. Appl Biochem Biotechnol 175(1):387–399

Rehman A, Gulfraz M, Raja GK, Haq MI, Anwar Z (2015) Comprehensive approach to utilize an agricultural pea peel (*Pisum sativum*) waste as a potential source for bio-ethanol production. Rom Biotechnol Lett 20(3):10422–10430

Ronzon YC, Zaldo MZ, Lozano MLC, Uscanga MGA (2012) Preliminary characterization of xylose reductase partially purified by reversed micelles from *Candida tropicalis* IEC5-ITV, an indigenous xylitol-producing strain. Adv Chem Eng Sci 2(1):9–14

Rosa SMA, Felipe MGA, Silva SS, Vitolo M (1998) Xylose reductase production by *Candida guilliermondii*. Appl Biochem Biotechnol 70(72):127–135

Russo JR (1977) Xylitol: anti-carie sweetener? Food Eng 79:37–40

Saha BC, Bothast RJ (1997) Microbial production of xylitol. In: Saha BC, Woodward J (eds) Fuels and chemicals from biomass. American Chemical Society, Washington, DC, pp 307–319

Saha BC, Bothast RJ (1999) Production of xylitol by *Candida peltata*. J Ind Microbiol Biotechnol 22(6):633–636

Sattur AP, Rao KC, Babu KN, Soundar D, Karanth NG, Tumkur RS (2003) Aldose reductase inhibitor and process for preparation thereof. US Patent 0,134,399, 17 Jul 2003

Scheie AA, Fejerskov O, Danielsen B (1998) The effects of xylitol containing chewing gums on dental plaque and acidogenic potential. J Dent Res 77:1547–1542

Scheinin A, Makinen KK, Ylitalo K (1976) Turku sugar studies. V. Final report on the effect of sucrose, fructose and xylitol diets on the caries incidence in man. Acta Odontol Scand 34(4):179–216

Schneider H (1989) Conversion of pentoses to ethanol by yeasts and fungi. Crit Rev Biotechnol 9(1):1–40

Sene L, Vitolo M, Felipe MGA, Silva SS (2000) Effects of environmental conditions on xylose reductase and xylitol dehydrogenase production by *Candida guilliermondii*. Appl Biochem Biotechnol 84(1):371–380

Sharma A (2014) Production of xylitol by catalytic hydrogenation of xylose. Pharm Innov 2(12):1–6

Silva DDV, Felipe MGA (2006) Effect of glucose:xylose ratio on xylose reductase and xylitol dehydrogenase activities from *Candida guilliermondii* in sugarcane bagasse hydrolysate. J Chem Technol Biotechnol 81(7):1294–1300

Sirisansaneeyakul S, Staniszewski M, Rizzi M (1995) Screening of yeasts for production of xylitol from D-xylose. J Ferment Bioeng 80(6):565–570

Soderling EM (2009) Xylitol, mutans streptococci and dental plaque. Adv Dent Res 21(1):74–78

Su Y, Li W, Zhu W, Yu R, Fei B, Wen T, Cao Y, Qiao D (2010) Characterization of xylose reductase from *Candida tropicalis* immobilized on chitosan bead. Afr J Biotechnol 9(31):4954–4965

Takuma S, Nakashima N, Tantirungkij M, Kinoshita S, Okada H, Seki T, Yoshida T (1991) Isolation of xylose reductase gene of *Pichia stipitis* and its expression in *Saccharomyces cerevisiae*. Appl Biochem Biotechnol 28-29:327–340

Tamburini E, Bianchini E, Bruni A, Forlani G (2010) Cosubstrate effect on xylose reductase and xylitol dehydrogenase activity levels and its consequence on xylitol production by *Candida tropicalis*. Enzym Microb Technol 46(5):352–359

Tomotani EJ, Arruda PVD, Vitolo M, Felipe MGA (2009) Obtaining partial purified xylose reductase from *Candida guilliermondii*. Braz J Microbiol 40(3):631–635

Twetman S (2009) Consistent evidence to support the use of xylitol and sorbitol containing chewing gum to prevent dental caries. Evid Based Dent 10(1):10–11

Uhari M, Tapiainen T, Kontiokari T (2000) Xylitol in preventing acute otitis media. Vaccine 19(1):144–147

Vandeska E, Kuzmanova S, Jeffries TW (1995) Xylitol formation and key enzyme activities in *Candida boidinii* under different oxygen transfer rates. J Ferment Bioeng 80(5):513–516

Velazquez Pereda MDC, Polezel MA, Dieamant GC, Cecilia Nogueira C, Mussi L, Rossan MR, Carlos Correia RD, Camilo NS (2011) Xylitol esters and ethers applied as alternative emulsifier, solvents, co-emulsions and preservative systems for pharmaceutical and cosmetic products. US Patent 0,251,415, 13 Oct 2011

Verduyn C, Jzn JF, Dijken JPV, Scheffers WA (1985a) Multiple forms of xylose reductase in *Pachysolen tannophilus* CBS4044. FEMS Microbiol Lett 30(3):313–317

Verduyn C, Kleef RV, Frank J, Schreuder H, Dijken JPV, Scheffers WA (1985b) Properties of the NAD(P)H-dependent xylose reductase from the xylose fermenting yeast *Pichia stipitis*. Biochem J 226(3):669–677

Vogl M, Kratzer R, Nidetzky B, Brecker L (2011) *Candida tenuis* xylose reductase catalysed reduction of acetophenones: the effect of ring-substituents on catalytic efficiency. Org Biomol Chem 9(16):5863–5870

Vongsuvanlert V, Tani Y (1988) Purification and characterization of xylose isomerase of a methanol yeast, *Candida boidinii*, which is involved in sorbitol production from glucose. Agric Biol Chem 52(7):1817–1824

Webb SR, Lee H (1991) Inhibition of xylose reductase from the yeast *Pichia stipitis*. Appl Biochem Biotechnol 30:325–337

Woodyer R, Simurdiak M, Donk WA, Zhao HM (2005) Heterologous expression, purification, and characterization of a highly active xylose reductase from *Neurospora crassa*. Appl Environ Microbiol 71(3):1642–1647

Yablochkova EN, Bolotnikova OI, Mikhailova NP, Nemova NN, Ginak AI (2003) Specific features of fermentation of D-xylose and D-glucose by xylose-assimilating yeasts. Appl Biochem Biotechnol 39(3):265–269

Ye Q, Hyndman D, Green NC, Li L, Jia Z, Flynn TG (2001) The crystal structure of an aldehyde reductase Y50F mutant NADP complex and its implications for substrate binding. Chem Biol Interact 130-132(1–3):651–658

Yin SY, Kim HJ, Kim HJ (2014) Protective effect of dietary xylitol on influenza: a virus infection. PLoS One 9(1):1–7

Yokoyama S, Suzuki T, Kawai K, Horitsu H, Takamizawa K (1995) Purification, characterization and structure analysis of NADPH-dependent D-xylose reductases from *Candida tropicalis*. J Ferment Bioeng 79(3):217–223

Yoshitake J, Ohiwa H, Shimamura M, Imai T (1971) Production of polyalcohol by a *Corynebacterium* sp. Part I Production of pentitol from aldopentose. Agric Biol Chem 35(6):905–911

Yoshitake J, Ishizaki H, Shimamura M, Imai T (1973) Xylitol production by an *Enterobacter* species. Agric Biol Chem 37(10):2261–2266

Zeid AAA, Fouly MZ, Zawahry YA, Mongy TM, Aziz ABA (2008) Bioconversion of rice straw xylose to xylitol by a local strain of *Candida tropicalis*. J Appl Sci Res 4(8):975–986

Zhang F, Qiao D, Xu H, Lio C, Li S, Cao Y (2009) Cloning, expression and characterization of xylose reductase with higher activity from *Candida tropicalis*. J Microbiol 47(3):351–357

Zhang Y, Gao F, Zhang SP, Su ZG, Ma GH, Wang P (2011) Simultaneous production of 1,3-dihydroxyacetone and xylitol from glycerol and xylose using a nanoparticle supported multi-enzyme system with in situ cofactor regeneration. Bioresour Technol 102(2):1837–1843

Zhang B, Zhang J, Wang D, Gao X, Sun L, Hong J (2015a) Data for rapid ethanol production at elevated temperatures by engineered thermotolerant *Kluyveromyces marxianus* via the NADP(H)-preferring xylose reductase-xylitol dehydrogenase pathway. Data Brief 5:179–186

Zhang M, Jiang ST, Zheng Z, Li XJ, Luo SZ, Wu XF (2015b) Cloning, expression, and characterization of a novel xylose reductase from *Rhizopus oryzae*. J Basic Microbiol 55(7):907–921

Zhao H, Nair NU (2010) Xylose reductase mutants and uses thereof. US Patent 0,291,645, 18 Nov 2010

Zhao X, Gao P, Wang Z (1998) The production and properties of a new xylose reductase from fungus. Appl Biochem Biotechnol 70–72(1):405–414

Zhao H, Woodyer R, Simurdiak M, Donk WA (2006) Highly active xylose reductase from *Neurospora crass*. US Patent 0,035,353, 16 Feb 2006

Zhao H, Woodyer R, Simurdiak M, Donk WA (2008) Highly active xylose reductase from *Neurospora crass*. US Patent 7,381,553, 3 Jun 2008

Zheng Y, Yu X, Li T, Xiong X, Chen S (2014) Induction of D-xylose uptake and expression of NAD(P)H-linked xylose reductase and NADP -linked xylitol dehydrogenase in the oleaginous microalga Chlorella sorokiniana. Biotechnol Biofuels 7(1):1–8

Chapter 8
Single Cell Oils (SCOs) of Oleaginous Filamentous Fungi as a Renewable Feedstock: A Biodiesel Biorefinery Approach

Mahesh Khot, Gouri Katre, Smita Zinjarde, and Ameeta RaviKumar

Abstract Single cell oils (SCOs) from oleaginous fungi are fast occupying centre stage as biodiesel feedstock and offer many advantages over plant- and algal-based oils. The biorefinery concept involves the production of SCOs along with other coproducts by these fungi when grown on waste agro-residue biomass. Filamentous fungi, in general, are able to effectively utilize this waste biomass as they have the capacity to produce lignocellulosic enzymes, namely, cellulase, xylanase, etc. The utilization of these wastes as growth substrates would not only solve the problem of waste disposal but also help in reducing the production cost of biodiesel. This chapter deals with production of SCOs from various filamentous fungi as feedstock for biodiesel when grown on lignocellulosic wastes. Two important parameters to be considered for biodiesel production are feedstock selection and fuel properties of biodiesel which are strain and growth substrate specific. Approaches to improve the process efficiency like optimization of fermentation conditions, one-step transesterification and metabolic engineering as well as the physico-chemical properties of biodiesel are also discussed.

8.1 Introduction

Extensive utilization of coal and petroleum has forced reserves of these non-renewable fuels to near depletion resulting in these fuel sources becoming increasingly limited and expensive to acquire. Therefore, there exists an acute need for alternative fuels along with sustainable methodologies for their production. These methodologies should be cost-effective and environment-friendly and yet produce high yields of quality biodiesel. Furthermore co-synthesis of other valuable

M. Khot · G. Katre · S. Zinjarde
Institute of Bioinformatics and Biotechnology, Pune, India

A. RaviKumar (✉)
Institute of Bioinformatics and Biotechnology, Pune, India

Department of Biotechnology, Savitribai Phule Pune University, Pune, India
e-mail: ameeta@unipune.ac.in

© Springer International Publishing AG, part of Springer Nature 2018
S. Kumar et al. (eds.), *Fungal Biorefineries*, Fungal Biology,
https://doi.org/10.1007/978-3-319-90379-8_8

145

products would be desirable as it would lead to reduction in production costs. For this, in recent times a biorefinery approach has been suggested. Biorefinery refers to a facility that can generate multiple products, such as biofuels, chemicals and other valuable products from waste biomass. The United States is leading with 213 biorefinery facilities directed at producing a range of coproducts with ethanol. In Europe, the Netherlands supports 5 commercial biorefineries and 12 demo and pilot facilities (REN21 2016). The development and scale-up of biorefineries have gained importance with growing efforts in emerging economies such as India. For example, Godavari Biorefineries (India) succeeded in securing more than USD 14 million towards ethanol production while also adding specialty chemical production capacity.

The UN-backed Intergovernmental Panel on Climate Change (IPCC) under the United Nations Framework Convention on Climate Change (UNFCCC) in its report (http://ipcc-wg2.gov/AR5/) has also recommended the complete phase out of fossil fuels used in power generation by the year 2100. As a part of contribution to climate mitigation efforts, it has urged all nations to avoid emissions of greenhouse gases (GHGs) and increase the share of clean/renewable energy from existing 30–80% by the year 2050. This includes biofuels as an important option for transition towards a low-carbon economy. In 2016, renewable energy contributed 4% of global fuel to the transport sector. For road transport (light- and heavy-duty road vehicles), liquid and gaseous biofuels are used. Advances in new markets and applications such as aviation biofuels have been observed in recent years. A study (2011) by the UN's Food and Agriculture Organization (UNFAO) showed that biofuel production can boost food and energy security and reduce poverty by increasing job opportunities in developing countries like India.

In this chapter, we discuss the strategies to produce biodiesel and chemicals of biotechnological interest by fermentation of waste agro-residues using filamentous fungi. Further, the yields and physico-chemical properties of the biodiesel produced are also considered.

8.2   Biodiesel

World liquid biofuel production falls into two major groups—ethanol and biodiesel [includes fatty acid methyl esters (FAMEs) and hydrotreated vegetable oil (HVO)]. Biodiesel is an environmentally friendly alternative fuel to petro-diesel, consisting of mono-alkyl esters of long-chain fatty acids (mostly FAMEs). It is an alternative fuel that reduces net greenhouse effects and is in use in many countries. The combustion characteristics of biodiesel are similar to petro-diesel, and its use for diesel engines actually predates the use of petroleum. When Sir Rudolf Diesel presented the first compression ignition engine at the world exhibition in Paris in 1898, peanut oil fuel was used as fuel. The use of vegetable oils had been common until 1920, when diesel engines were modified for use of petroleum. Today biodiesel can be used without any major modifications for the diesel engines in vehicles, and it is

Fig. 8.1 Advantages of biodiesel over petro-diesel

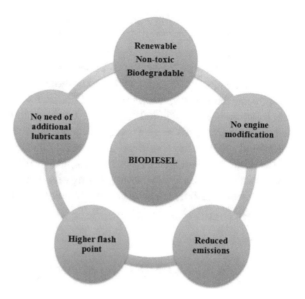

compatible with the current fuel infrastructure. Biodiesel is currently being produced on commercial scale from edible vegetable oils such as rapeseed, soybean, sunflower, palm, etc. and can be used as transportation fuel with multiple benefits over petro-diesel (Fig. 8.1).

Global biofuel production increased to 135 billion litres in 2016 with biodiesel accounting for ~23% of the total at 30.8 billion litres (REN21 2017) with the United States, Brazil, Germany and Argentina leading in biodiesel production worldwide. In India, 148 million litres of biodiesel fuel were produced in 2016 from used cooking oil, animal fats and tallow (GAIN report 2017). Currently, in India five to six large plants (10,000–250,000 metric tons per year) are present with 28% capacity utilization to produce 130–140 million litres of biodiesel from a variety of feedstocks such as inedible vegetable oils, edible oil waste and animal fats. The biodiesel consumers include brick kilns, cellular communication towers, progressive farmers, state transport corporations, automobiles and transport companies, institutions and privately owned retail outlets that run diesel generators as source of power (GAIN Report 2016).

8.3 Feedstock for Biodiesel

Biodiesel is mainly obtained by the transesterification of fat and vegetable oils in the presence of a catalyst by a primary alcohol (usually methanol) leading to a fatty acid methyl ester (FAME), which is used as a biofuel. Vegetable oils produced from edible plants such as rapeseed, palm, soybean, sunflower and other oil (oleaginous) crops are the most common feedstock used in biodiesel production in different parts of the world. However, the use of edible oils for biodiesel has received criticism due to its low sustainability, conflict with food for the use of arable land along with high water and fertilizer requirements. The feedstock/raw material price is the governing factor for the economic viability of biodiesel market as revealed from several studies and accounts for 70–95% of the total biodiesel production cost (Dorado et al. 2006; Krawczyk 1996; Zhang et al. 2003). Hence, nonedible cost-effective feedstocks are now being studied to produce biodiesel in a sustainable and economical manner. Recent trends in feedstocks for biodiesel production have been reviewed by Pinzi et al. (2013). Most of these are at research stage and have not been commercialized but are projected to enter the market in the coming years. These include nonedible plant oils, waste cooking/frying oils (yellow grease), animal fats (tallow, lard and chicken fat), soapstocks, insect oils and microbial lipids also known as single cell oils (SCOs).

8.4 Single Cell Oils (SCOs) of Fungi

Microorganisms which can accumulate >20% (w/w) or more of their biomass as lipid (single cell oil, SCO) are known from the early 1940s and have been termed "oleaginous". The oleaginicity is limited to a relatively small number of microorganisms and is distributed among specific genera and species of microalgae, yeasts, filamentous fungi as well as bacteria (Ratledge 1989).

Triacylglycerols (TAGs) are the major component of the SCOs followed by free fatty acids, monoacylglycerols, diacylglycerols and steryl esters, sterols and polar lipids (glyco-, sphingo-, phospholipids). Lipid accumulation is not observed in oleaginous microbes under balanced nutrient conditions. When one or more of the nutrients, such as N, P, K, Mg, S or Fe, required for cell proliferation are limiting, the available carbon is channelized into lipid synthesis. Most of the times, N (nitrogen) is the limiting factor for lipid accumulation. Under these conditions, the existing cells continue to assimilate the carbon available to them which gets converted into lipid by specific lipid biosynthetic pathway(s). The prospects of microbial oils (SCOs) from oleaginous microalgae, fungi and bacteria for biodiesel production have been discussed by Li et al. (2008). A number of advantages of using oleaginous microbes for the production of biodiesel as compared to their plant counterparts were enlisted, namely, short life cycle, less labour and lower land resources required, easy scale-up and independent of venue, season or climate.

Although algae have unique advantages of that being able to use sunlight and carbon dioxide for lipid synthesis, large-scale cultivation is limited by problems related to photobioreactor designing and high-density cultivation. Subramaniam et al. (2010) have also cited suitability of SCOs as alternative sources for biodiesel production over conventional sources such as oilseeds and animal fats. These include no conflict/competition with food, potential for large-scale production and net energy gain.

The metabolism and engineering of lipids in order to attain higher productivity have been studied in heterotrophic microbes (Kosa and Ragauskas 2011). For this, process parameters using oleaginous heterotrophic organisms need to be developed. Optimization of the processes using cheap, renewable carbon sources such as agro-industrial residues leading to high SCO yields needs to be explored. Further for industrial level scale-up, reproducible, sustainable and high-quality production requires standardization. Biodiesel composition can be optimized to contain methyl esters of palmitic acid, oleic acid and decanoic acid by controlling the chain length and unsaturation degree. This can be achieved through environmental factors such as temperature, substrate and control of key enzymes such as fatty acid synthase and desaturases via genetic modification. The optimization of process parameters and growth conditions with respect to C/N ratio, nitrogen source and oxygen demand with novel process configurations such as solid-state/semi-solid-state fermentation (SSF) has been shown to enhance SCO accumulation (Kosa and Ragauskas 2011).

Fungi represent the second largest and diverse group among living organisms with the species number estimated to be 1.5 million on earth (Hawksworth 1991). Fungi are chemoorganotrophs and display simple nutritional requirements. Oleaginous fungi are limited to the individual strains and may not be present in the whole species or even genus. SCOs of oleaginous filamentous fungi and some yeast strains have long been known as a source of polyunsaturated fatty acids (PUFAs) of food and medical interest (Certik and Shimizu 1999; Ratledge and Wynn 2002). PUFA-rich lipids are highly demanded in the food industry for use as food additives to fortify and/or amend fatty acid composition of specific foods (e.g. infant foods). The first commercially viable SCO process was the production of an oil rich in γ-linolenic acid (GLA) using oleaginous mould *Mucor circinelloides* (Cohen and Ratledge 2005). Solid- or semi-solid-state fermentation with fungi-producing PUFAs rich in specific fatty acids can be augmented to improve agricultural products (e.g. cereals) and by-products (e.g. orange peels, apple or pear pomace) for their direct use as food and/or feed supplements. Bellou et al. (2016) have recently reviewed the sources of PUFAs including fungi, the biochemical pathways of PUFA synthesis and the current biotechnological research concerning the production of polyunsaturated SCOs for use as food additives. Industrial production of PUFA-rich SCOs from filamentous fungi has been launched by various industries enabling commercially available food supplements, e.g. GLA from *Mucor circinelloides* (J and E. Sturge) and arachidonic acid from *Mortierella alpina* (CABIO).

The potential of oleaginous fungi for biodiesel production has been less explored and is being studied only recently. In these fungi, the major lipid component has been found to be triacylglycerols (TAGs) which account for 85–90% of the total accumulated lipid. Among different growth conditions, temperature and carbon source are known to affect the TAG content. In fact oleaginous fungi possess a number of advantages for SCO production over their plant and algal counterparts as illustrated in Fig. 8.2.

Fig. 8.2 Advantages of fungal single cell oils (SCOs) over plant and algal sources

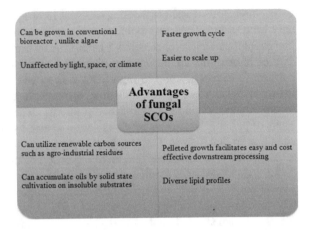

Can be grown in conventional bioreactor , unlike algae

Faster growth cycle

Unaffected by light, space, or climate

Easier to scale up

Advantages of fungal SCOs

Can utilize renewable carbon sources such as agro-industrial residues

Pelleted growth facilitates easy and cost effective downstream processing

Can accumulate oils by solid state cultivation on insoluble substrates

Diverse lipid profiles

8.5 Biochemistry of Lipid Accumulation in Oleaginous Fungi

Most of the information on the biochemistry responsible for lipid accumulation in oleaginous fungi has been obtained from the non-oleaginous yeast *Saccharomyces cerevisiae* (Kohlwein 2010) and *Yarrowia lipolytica*, which represent a model for bio-oil production (Beopoulos et al. 2009). The oleaginous microorganisms follow either the "ex novo" or "de novo" lipid accumulation pathways when provided with hydrophobic and hydrophilic substrates, respectively (Papanikolaou and Aggelis 2011a). Principal differences occur between the two pathways, and "de novo" process commences with depletion in the medium of an essential nutrient (usually nitrogen) and occurs in the presence of hydrophilic substrates, while lipid accumulation from hydrophobic substrates via the "ex novo" pathway is a growth-associated process and requires no nitrogen-limiting conditions (Aggelis et al. 1995; Papanikolaou et al. 2001, 2002).

A wide variety of hydrophilic substrates have been investigated for "de novo" SCO production from oleaginous fungi. For example, sugar-based substrates including both simple (glucose, fructose, xylose, lactose, sucrose) and complex sugars (starch, pectin), whey, glucose-enriched wastes, molasses, ethanol, glycerol, low-molecular-weight organic acids such as citric acid and acetic acid and other low-cost renewable substrates including starch hydrolysate and tomato waste hydrolysate have been investigated (Huang et al. 2013). Similarly a variety of hydrophobic substrates produced during industry processes have been used for SCO production as substrate and co-substrate (along with a hydrophilic substrate). Examples of fatty material used as substrates for lipid accumulation are vegetable oils, fatty alkyl esters (methyl, ethyl, butyl or vinyl esters of fatty acids), soapstocks, pure free fatty acids, glycerol, waste cooking oil, fish oils and industrial fats composed of free fatty acids of animal or vegetable origin.

The "de novo" biochemical pathway of lipid biosynthesis does not differ in oleaginous and non-oleaginous fungi. A single- or multiple-nutrient limitation (such as nitrogen or phosphorus) and presence of an excess carbon source in the medium

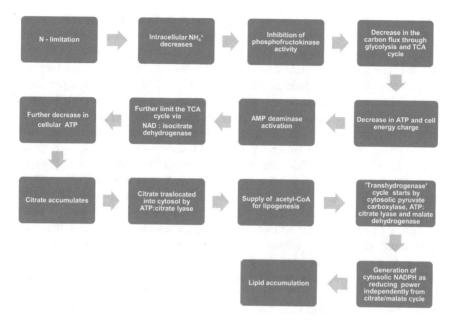

Fig. 8.3 Biochemical events involved in de novo lipogenesis of oleaginous fungi consolidated bioprocessing (CBP) microorganism capable of directly growing on cellulosic substrates and producing lipids. The presence of two secretory endoglucanases and one β-D-glucosidase, along with a set of accessory cell wall-degrading enzymes, was demonstrated in the fungus through genomic, proteomic and biochemical analyses (Wei et al. 2013)

were found to induce lipid accumulation in fungi (Papanikolaou et al. 2004b). The most efficient condition for inducing lipogenesis has been found to be nitrogen limitation. The high lipid-accumulating ability of oleaginous fungi depends mostly on the regulation the biosynthetic pathway and the supply of the precursors (i.e. acetyl-CoA, malonyl-CoA and glycerol-3-phosphate) and the cofactor NADPH. The presence of exogenous aliphatic chains from hydrophobic substrates may inhibit fatty acid synthetase and ATP-citrate lyase, and therefore when fatty material is used as sole carbon and energy source, de novo lipid biosynthesis does not occur from the acetyl-CoA. In some oleaginous organisms such as *Pichia methanolica*, de novo biosynthesis could potentially occur from glucose or glycerol, in spite of the presence of significant exogenous quantities of fatty materials (Aoki et al. 2002). Both oleaginous yeasts and filamentous fungi show the presence of all key enzymes responsible for higher levels of storage lipid. However, in the course of few studies on lipid-accumulating moulds, it has become apparent that the regulation of lipogenesis may differ between filamentous fungi and yeasts.

Wynn et al. (2001) studied the biochemical events associated with the onset of lipid accumulation in two oleaginous moulds under growth conditions of nitrogen limitation. The results of this study on *Mucor circinelloides* and *Mortierella alpina* have put forward a revised and more concerted mechanism for the initiation of storage lipid accumulation for filamentous fungi (Fig. 8.3). In fact the biochemical mechanism found to differ in many fundamental ways from those described in oleaginous yeasts. For example, the source of reducing power for fatty acid biosynthesis, i.e. NADPH, is believed to be the malic enzyme in *Mucor circinelloides* and *Mortierella alpina* and is therefore considered as the rate-limiting step for fatty acid biosynthesis (Wynn and Ratledge 1997; Wynn et al. 1999). However, according to Ratledge (2014), though malic enzyme plays a role in majority of species, the provision of NADPH for lipid biosynthesis in oleaginous microorganisms is still not completely understood. It has been suggested that the possible route for generation of NADPH is the pentose phosphate pathway and cytosolic isocitrate dehydrogenase reaction. Chen et al. (2015) used the oleaginous fungus *Mortierella alpina* as a model system and demonstrated that a significant role is played by the pentose phosphate pathway during fungal lipogenesis. The increase in levels of NADPH produced by the pathway, especially glucose-6-phosphate dehydrogenase, is one of the critical determinants for efficient fatty acid synthesis in oleaginous microbes.

Multiple malic enzyme isoforms have been identified in filamentous fungi with up to 2, 5 and 11 separate genes being involved in *Mortierella alpina*, *Mucor circinelloides* and *Pythium splendens*, respectively (Song et al. 2001; Zhang et al. 2007). Rodríguez-Frómeta et al. (2013) indicated leucine metabolism as one of the pathways involved in the generation of the acetyl-CoA required for fatty acid biosynthesis. In this study, the genetically modified strain of *M. circinelloides* overexpressing gene coding for cytosolic malic enzyme isoforms III/IV showed similar lipid content to that of the control strain despite higher enzyme activity. On the other hand, a 2.5-fold increase in lipid accumulation was observed by the prototrophic strains in comparison with leucine auxotrophic strains. The results suggested that leucine metabolic pathway is necessary for lipid accumulation as the endogenously produced leucine is degraded for generation of acetyl-CoA for incorporation into fatty acids. Similarly in *Mortierella alpina*, it has been suggested that while malic enzyme may be relevant in fatty acid synthesis, it is not the sole rate-limiting enzyme, and leucine metabolism may have a critical role (Hao et al. 2014). It thus suggests that lipid biosynthesis in oleaginous fungi could be influenced by multiple steps.

Tang et al. (2015) compared the biochemical analysis of lipid accumulation between low and high lipid-producing strains of *Mucor circinelloides*. Substantially higher activities of glucose-6-phosphate dehydrogenase and 6-phosphogluconate dehydrogenase along with ATP-citrate lyase and fatty acid synthase were observed in high lipid-producing strain. These activities may provide more NADPH and cooperatively regulate the fatty acid biosynthesis. On the other hand, lower activities of NAD^+- and $NADP^+$-dependent isocitrate dehydrogenase were observed in high lipid-producing strain suggesting inhibition of TCA cycle which in turn may lead ATP-citrate lyase to produce more acetyl-CoA.

Besides the biochemistry responsible for oleaginicity, cellulolytic capacity of *Mucor circinelloides* was also elucidated through biochemical analyses to evaluate its potential as a.

The "ex novo" lipid accumulation process involves bio-modification of fats and oils, and thus it is possible to "tailor-make" lipids of high-added value via an improvement and upgrade of the fatty materials utilized as substrates (Papanikolaou and Aggelis 2011a, b). Thus, the biotechnological valorization of low-cost fatty materials as substrates for SCO production is an eco-friendly strategy and holds great potential for the production of cocoa butter substitutes and other tailor-made high-added-value lipids (Papanikolaou and Aggelis, 2010).

The free fatty acids transported inside the cell or produced from lipase-induced intracellular triacylglycerol hydrolysis are the initial substrates for SCO production. The incorporated fatty acids may get dissimilated for growth purposes. Alternatively intracellular bio-transformations may take place resulting in the synthesis of new fatty acid profiles which did not exist previously in the original substrate (Papanikolaou and Aggelis 2011a). The dissimilated free fatty acids undergo degradation to generate smaller chain acyl-CoAs and acetyl-CoA via β-oxidation catalysed by various acyl-CoA oxidases (Aox) (Fickers et al. 2005; Beopoulos et al. 2009). These biochemical events supply the energy for cell growth and maintenance and also for the formation of intermediary metabolites.

The enzymatic and molecular aspects of degradation of hydrophobic substrates have been studied in detail for the strains of the yeast *Y. lipolytica* (Papanikolaou and Aggelis 2011a). An extracellular lipase—Lip2p—encoded by the LIP2 gene is secreted by this yeast when cultivated on triacylglycerol-like substrates (Pignede et al. 2000). This enzyme is a 301-amino-acid glycoprotein belonging to the TAG hydrolase family (EC 3.1.1.3). The cell eventually takes up the released free fatty acids, produced after lipase-catalysed hydrolysis of the triacylglycerols, and the biochemical process is a multistep reaction requiring different enzymatic activities of five acyl-CoA oxidase isozymes (Aox1p through Aox5p), encoded by the POX1 through POX5 genes.

8.6 Growth, SCO Production and Fatty Acid Profile of Oleaginous Fungi

Among different renewable sources and waste substrates available for microbial oil production, lignocellulosic biomass holds promise as it constitutes the most abundant and locally available inexpensive source of carbohydrates in the form of waste agro-industrial residues. As there are still technical challenges in production of biodiesel using lignocellulose, the process has been restricted to the laboratory stage. The studies reported till date involve two key steps to process lignocellulosic substrate for growth medium preparation: (a) thermochemical pretreatment of substrate resulting in hydrolysates and (b) hydrolysis mediated by enzyme cocktails. The examples of lignocellulosic substrates used as hydrolysates include wheat straw, pitch pine, corn cob, rice hulls, sugarcane bagasse, rice straw, tomato waste, corn stalk and tree (*Populus euramevicana*) leaves (Yousuf 2012). These studies involved submerged cultivation of oleaginous yeasts and fungi in media based on detoxified hydrolysates generated after different pretreatment methods.

The oleaginous fungus, *Trichoderma reesei*, when grown on 80 g L^{-1} glucose showed the presence of total saturated fatty acids (49.7%), total monounsaturated fatty acids (36.6%) and total polyunsaturated fatty acids (9.8%) (Bharathiraja et al. 2017). *Mortierella elongata* strain PFY could directly use the cellulose in rice straw as substrate. The yield of total lipid in dried material was found to be 7.07% (g/gds or gramme per gramme dry substrate) on the seventh day. The fatty acid profile revealed the presence of C16:0, C16:1, C17:1, C18:1 and C18:2 fatty acids to be 39.93%, 0.61%, 2.31%, 35.87% and 17.58%, respectively (Yao et al. 2012). *Thamnidium elegans* strain CF-1465 was cultivated as a shake flask culture using low-cost sugars from a sugar refinery plant (glucose, fructose and sucrose) as carbon sources under high C/N ratio (Papanikolaouet al.2010). Under increasing initial sugar concentrations, lipid production and content increased with maximum SCO concentration of >9 g L^{-1} and 70% (w/w) lipid content. The produced SCO contained oleic acid as the major fatty acid followed by palmitic and linoleic acids and thus found to be suitable for biodiesel production.

Cunninghamella echinulata on tomato waste hydrolysate (TWH) yielded 7.8 g L^{-1} of reserve lipid of which 44% was consumed to yield 3.2 g L^{-1} of lipid-free biomass. TAG proportion decreased from 26.2 to 6.9% indicating that it was consumed preferentially, while γ-linolenic acid increased from 9.2 to 15.3%, and palmitic acid (C16:0) decreased from 22.7% to 15.6%. Membrane polar lipids were produced during lipid turnover phase (Fakas et al. 2007). When the fungus was grown on TWH supplemented with glucose containing no assimilable organic nitrogen, the lipid produced was in low quantities and contained mainly neutral lipids, which were used up during lipid turnover phase. The lipid yield coefficient was 0.48 per gramme biomass wherein the lipid yield was 8.7 g L^{-1} and the γ-linolenic acid yield was more than 1 gL^{-1} (Fakas et al. 2008).

The biomass, total lipid, substrate consumption and fatty acid profile of two oleaginous *Mucorales*, *Mortierella isabellina* ATHUM 2935 and *Cunninghamella echinulata* ATHUM 441, were studied using xylose, raw glycerol and glucose under nitrogen limitation. Glucose-containing media favoured the production of 27gL^{-1} biomass and 44.6% lipid content for *M. isabellina* and 15gL^{-1} biomass and 46% lipid content for *C. echinulata*. In xylose-containing medium, *M. isabellina* and *C. echinulata* yielded lipid contents of 65.5% and 57.7%, respectively, with *C. echinulata* producing 6.7 g L^{-1} lipid and 1119 mg L^{-1} of γ-linolenic acid (Fakas et al. 2009a, b).

Meeuwse et al. (2013) have recently demonstrated the suitability of solid-state fermentation (SSF) for production of SCO as biodiesel feedstock in terms of cultivation costs and energy input both of which are lower as compared to submerged fermentation processes. Submerged and SSF processes were evaluated for biodiesel production from an oleaginous yeast and/or filamentous fungus grown on sugar beet pulp as substrate for lipid yield and energy use. Kinematic models were used to predict lipid yields, and energy use was estimated in different steps such as cultivation, substrate pretreatment and downstream processing. Asadi et al. (2015) have also highlighted the potential benefits of SSF for production of SCO by fungi belonging to the genus *Mortierella*. The filamentous fungi are the most preferred microorganisms for SSF because their hyphal growth pattern assists them in colonization and penetration of the solid substrate resulting in better utilization of nutrients. They are also well adapted to low water activity present in SSF systems which facilitates the use of bound water of substrate allowing their growth in the absence of free water (Cheirsilp and Kitcha 2015). Reports are available on SCO production from filamentous fungi employed in SSF of both pretreated and untreated substrates such as pear pomace, wheat straw, wheat bran, rice straw, palm by-products (palm empty fruit bunches, palm pressed fibre, palm kernel cake), soybean hulls, sweet sorghum and SPORL (sulphite pretreatment to remove recalcitrance of lignocellulose) and pretreated Douglas fir (Dey et al. 2011; Economou et al. 2010; Kitcha and Cheirsilp 2014; Lin et al. 2010; Peng and Chen 2008; Zhang and Hu 2012; Harde et al. 2016). Few of these studies have attempted the direct conversion of lignocellulosic biomass into SCO. For example, endophytic fungi with cellulase activity have been isolated from oleaginous plants by Peng and Chen (2007). The process yielded 19–42 mg SCO/g of the substrate, and yield was further improved up to 74 mg/g dry substrate composed of steam exploded wheat straw mixed with wheat bran (Peng and Chen 2008).

In a study by Economou et al. (2010), semi-solid-state fermentation process was developed with 92% water content for lab-scale production of SCO from *Mortierella isabellina* with yield of 11 g/100 g dry weight of sweet sorghum stems used as substrate. The fatty acids comprised mostly oleic (49–55%) and palmitic (24–35%), making it suitable feedstock for biodiesel. The direct conversion of wheat and rice straws into fungal lipids by cellulolytic fungi has also been attempted (Dey et al. 2011; Lin et al. 2010). Soybean hulls without any pretreatment have also been used for SSF by an oleaginous fungus *Mortierella isabellina* to produce lipid up to 47.9 mg/g of substrate (Zhang and Hu 2012). With the improved yield and productivity, the fungal SCOs produced by SSF hold the potential to be promising candidates for source of biodiesel in the near future (Meeuwse et al. 2013). Table 8.1 describes the biomass, lipid production, yield coefficient per gramme biomass and fatty acid profile of different fungi grown on various substrates.

A lipid yield of 36.6 mg/gds (gramme dry substrate) was obtained from the solid-state fermentation of a cellulosic waste, wheat straw and bran mixture, by the cellulolytic fungus *Aspergillus oryzae* A-4. On the sixth day of fermentation, a cellulase activity of 1.82 FPU/gds resulted in 25.25% utilization of holocellulose in the substrate. The study indicated that cellulose-degrading fungi could be potential feedstocks for biodiesel production (Lin et al. 2010). When *M. isabellina* ATHUM 2935 was grown on high glucose (100 g L^{-1}), nitrogen-limited medium, a biomass of up to 35.9 g L^{-1}, was produced along with a lipid yield of 18.1 g L^{-1}. The lipid content was 50% (w/w) of the dry biomass. The fatty acid profile indicated that it produced 51% oleic acid and 17% linoleic acid (Papanikolaou et al. 2004a). *M. isabellina*, when grown on acid- and alkali-pretreated corn stover hydrolysate, yielded a biomass of 16.8 g L^{-1} and a lipid yield of 5.1 g L^{-1} with a lipid content of 30% (w/w). The process was scaled up to 7.5 L in a bioreactor using the combined hydrolysate with glucose (28.6 g L^{-1}), xylose (16.1 g L^{-1}) and acetate (3.4 g L^{-1}). 18.7 g L^{-1} of

Table 8.1 Biomass, lipid production, yield coefficient and fatty acid profile of different fungi grown on various substrates

Organism/substrate used	Time (h)	X (g/L)	L (g/L)	Y$_{LX}$ (g/g)	C16:0	C16:1	C18:0	C18:1	C18:2	C18:3	C20:0	Others	Reference
Mortierella isabellina ATHUM 2935 Glucose—45 g L^{-1}	145	14.9	8.2	0.55	22	3	1.5	51	17	4.4	–	–	Papanikolaou et al. (2004a)
Cunninghamella echinulata ATHUM 4411 glucose—40 g L^{-1}	310–400	9.4	4.4	0.47	15.9	–	5	50.5	13.1	11	–	–	Papanikolaou et al. (2004b)
M. isabellina ATHUM 2935 Glucose—40 g L^{-1}	95–140	13.5	7.6	0.56	20	3	1.5	52	16.5	3.5	–	–	
M. isabellina ATHUM 2935													Papanikolaou et al. (2007)
Starch—30 g L^{-1}	170	10.4	3.7	0.36	26.5	–	2.1	56.5	10.5	4	–	–	
Pectin—30 g L^{-1}	180	8.4	2	0.24	22.8	–	6.8	42.5	14.5	6.1	–	–	
Lactose—30 g L^{-1}	190	9.5	3.5	0.37	23	–	1.9	56.5	11	4.1	–	–	
C. echinulata ATHUM 4411													
Starch—30 g L^{-1}	456	13.5	3.8	0.28	18	–	4.1	47.7	13.5	14.2	–	–	
Pectin—30 g L^{-1}	180	4.1	0.45	0.1	24.1	–	1.5	32	12.9	16	–	–	
Mucor circinelloides-glucose—10 g L^{-1}	72	3.73	0.71	0.19	20	2.3	2	37	14.3	18.5	–	–	Vicente et al. (2009)
Aspergillus niger LFMB 1	140	4.4	1.4	0.32	13.9	–	6.1	29.9	45.1	–	–	–	André et al. (2010)
A. niger NRRL 364	165	6.4	2.5	0.39	19.1	–	8.9	31.5	38.1	–	–	–	
Geotrichum robustum strain G9	192	40.25	28.63	0.71	23.61	–	26.47	37.36	4.46	–	–	–	Cao et al. (2010)
M. isabellina ATHUM 2935 160 g L^{-1} glucose													Chatzifragkou et al. (2010)

														Reference
Commercial glucose—60 g L⁻¹	237	13.2	9.9	0.74	28.1	3.3	3.1	53.3	7.2	3.5	–	–	–	Lin et al. (2010)
Commercial fructose—60 g L⁻¹	405	12.1	7.4	0.61	28.5	3.8	2.2	51.3	9.6	2.5	–	–	–	
Molasses—60 g L⁻¹	150	9.5	3.1	0.54	29.1	2.1	3.8	51.2	10.1	2.1	–	–	–	
Sweet sorghum extract		100 g DW	100 g DW											
A. oryzae A-4, wheat straw	144	305	36.6	0.12	32.95	–	9.96	22.64	27.74	–	–	–	–	Lin et al. (2010)
To wheat bran 1:9		Mg/gds	Mg/gds											
M. isabellina ATHUM 2935	195	32	8.1	0.25	24.5	2.5	4.2	49.1	14.3	3.8	–	–	–	Vamvakaki et al. (2010)
Colletotrichum sp. glucose—100 g L⁻¹	240	11.4	5.3	0.49	5.06	0.26	3.28	58.14	23.21	2.16	0.17	7.72		Dey et al. (2011)
Alternaria sp. glucose—100 g L⁻¹	240	12.1	7.1	0.58	6.12	0.56	5.15	60.15	23.51	3.6	0.27	0.64		
M. isabellina ATHUM 2935	20	4.31	2.2	0.51	–	–	–	–	–	–	–	–		Economou et al. (2011)
Reducing sugars—13.03 g L⁻¹ Sorghum extract														
Epiococcum purpurascens AUMC 5615 commercial molasses 40 g L⁻¹	336	15.63	12.5	0.8	45.35	–	12.24	–	–	–	–	38.39		Koutb and Mohamed (2011)
Aspergillus niger NRRL 363	122	14.4	7.1	0.49	8.9	–	3	78.8	6.6	–	0.7	–		Papanikolaou et al. (2011c)
A. niger NRRL 364	95	8.4	3.5	0.42	12.1	–	3.1	78.7	5.6	–	0.5	–		
A. niger LFMB 1	73	9.1	5.1	0.56	11.2	–	3.4	75.3	7.4	–	2.7	–		
Aspergillus sp. ATHUM 3482	191	12	7.7	0.64	10.4	–	19	62.4	6	–	2.2	–		

(continued)

Table 8.1 (continued)

Organism/substrate used	Time (h)	X (g/L)	L (g/L)	Y_LX (g/g)	FA profile (%) C16:0	C16:1	C18:0	C18:1	C18:2	C18:3	C20:0	Others	Reference
A. niger LFMB 2	96	10.6	3.9	0.37	17.4	–	6.8	71.3	1	–	3.5.	–	
Penicillium expansum NRRL 973	94	15.4	9.7	0.31	7.3	–	1	76.3	15.4	–	–	–	
Waste cooking oil 15 g L^{-1}													Venkata Subhash and
Aspergillus sp. corncob waste liquor	48	2	0.44	0.22	–	–	–	–	–	–	–	–	Venkata Mohan (2011)
Aspergillus terreus IBB M1													Khot et al. (2012)
Glucose—30 g L^{-1}	72	5.52	2.95	0.54	20.1	0.4	23.6	30.1	22.3	0.4	0.8	1.8	
M. isabellina ATCC 24613													Ruan et al. (2012)
Glucose—82.5 g L^{-1}	408	22.3	10	0.45	20.6	2.51	2.53	59.5	12.9	0.36	–	1.65	
Xylose—39.1 g L^{-1}	408	11.4	5	0.43	24.6	2.85	3.59	50.9	13	3.41	–	1.65	
M. isabellina M2	144	17.52	8.84	0.5	16.59	2.35	4.05	39.57	28.85	3.85	0.34	2.48	Xing et al. (2012)
Corn fibre hydrolysate													
M. isabellina IFO7884	168	–	47.9 Mg/gds	–	24.3	3.3	10.3	22.2	20.3	1.1	3.2	15.3	Zhang and Hu (2012)
Soybean hull													
M. isabellina glucose—30 g L^{-1}	–	7.3	4.88	0.67	22	–	11	50	18	3	–	–	Zheng et al. (2012)
Xylose—30 g L^{-1}	–	5	2.52	0.5	22	–	1	52	14	3	–	8	
A. oryzae DSM 1862	120	8.75	3.5	0.4	11.6	15.6	19.3	30.3	6.5	5.5	2	6.3	Muniraj et al. (2013)
Potato-processing wastewater													
M. isabellina ATCC 42613	93	16.8	5.1	0.3	–	–	–	–	–	–	–	–	Ruan et al. (2014)
Corn Stover hydrolysate													
A. awamori MTCC 11639	72	–	–	0.31	–	–	–	–	–	–	–	–	Venkata subhash and Venkata Mohan (2014)
Glucose—40 g L^{-1}													

M. isabellina IFO7884 Corn Stover hydrolysate	144	0.84	0.48	0.57	16	2	7	43	15	11.5	2	3.5	Zhang and Hu (2014)
Aspergillus tubingensis TSIP9 Palm empty fruit bunches (EFB) and palm kernel (PK)	120	–		79.9 (mg/gds)	–	–	–	–	–	–	–	–	Cheirsilp and Kitcha (2015)
Mortierella isabellina NRRL 1757 SPORL-pretreated Douglas fir	168	–	0.21 (g/gs)	–	–	–	–	–	–	–	–	–	Harde et al. (2016)
Mucor fragilis AFT7-4 Glucose—86 g L^{-1}	120	27.5	12.86	0.46	19.41	2.28	6.69	66.77	–	–	0.89	3.03	Huang et al. (2016)
Cunninghamella echinulata LFMB 5	262	2.9	1.61	0.55	21.7	–	6.0	42.5	20.8	7.5	–	–	Papanikolaou et al. (2010)
Mortierella isabellina ATHUM 2935	222	8.1	5.4	0.66	22.0	–	3.9	51.2	18.8	4.0	–	–	
Mortierella ramanniana MUCL 9235	142	7.6	3.30	0.43	21.4	–	5.4	48.9	17.8	5.4	–	–	
Mucor sp. Glycerol—30 g L^{-1}	310	6.8	2.93	0.43	25.0	–	4.0	26.4	27.0	10.0	–	–	
Penicillium brevicompactum NRC 829 Sunflower oil cake—8 g L^{-1}	144	13.91	8.01	0.57	0.78	15.72	–	–	61.83	7.97	0.23	13.47	Ali et al. (2017)

Not mentioned, X biomass, L lipid yield, Y_{LX} lipid yield coefficient per gramme biomass, DW dry weight, gds gramme dry substrate, SPORL sulphite pretreatment to remove recalcitrance of lignocellulose, gs gramme sugar biomass was generated along with 6.9 g L^{-1} total lipid. The lipid content was 36.8% (w/w), indicating that the lipid production could be scaled up effectively from shake flask to bioreactor (Ruan et al. 2014).

The growth of microbes and the SCO produced have also been investigated on hydrophobic substrates as the kind or type of fatty acids produced is known to vary with the raw substrate used for fermentation. The fatty acid profile of *Mucor fragilis* AFT7–4 when grown on sunflower oil cake revealed the presence of prominently C16:0 (19.41), C16:1 (2.28), C18:0(6.69) and C18:1(66.77), respectively (Huang et al. 2016). Aggelis and Sourdis (1997) proposed a numerical model for the prediction of lipid accumulation-degradation by oleaginous yeasts and moulds growing on vegetable oils. *Yarrowia lipolytica*, the model system of lipid accumulation, was shown to accumulate some cocoa butter-like lipids (Papanikolaou et al. 2003; Papanikolaou and Aggelis 2010) from mixtures of saturated free fatty acids (stearin, tallow and its industrial derivative), technical glycerol and glucose. CBE is mainly composed of three different kinds of triacylglycerols (TAGs): POP (C16:0–C18:–C16:0), POS (C16:0–C18:1–C18:0) and SOS (C18:0–C18:1–C18:0). The storage lipids of yeasts, mainly TAGs, also contain relative high level of C16 and C18 fatty acids and might be used as CBE (Wei et al. 2017).

Thus, predetermined lipids with composition similar to cocoa butter could be synthesized using low-cost substrates. The levels of CBE depended on the intracellular level of stearic acid when *Y. lipolytica* was grown in the presence of stearin along with oleic acid as co-substrate. An important factor to be noted for industrial production of CBE is the low oxygen saturation required for CBE by the yeast. *Trichosporon oleaginosus* DSM11815 produced 27.8% potential POP and POS at levels of 378 mg TAGs/g dry cell weight with,3% as SOS on nitrogen-limited glucose media (Wei et al. 2017). Large-scale (pilot) production of a cocoa butter equivalent (CBE) fat was also carried out using wild-type strain of yeast, *Candida curvata* (now *Cryptococcus curvatus*). The process was scaled to 250 m^3 to produce palm oil equivalent using lactose as feedstock originating from the cheese creamery processes in New Zealand (Davies 1988).

In another study, two strains of *Aspergillus niger*, LFMB 1 and NRRL 364, were used for the production of lipid (up to 3.5 g L^{-1}) and oxalic acid (up to 21.5 g L^{-1}) in nitrogen-limited medium with 60 g L^{-1} biodiesel-derived glycerol. Two strains of *Lentinula edodes*, AMRL 119 and AMRL 121, were also grown on carbon-limited medium (crude glycerol at 20 g L^{-1}) to produce 5.2 g L^{-1} biomass, which contained 10% lipids (w/w). The fatty acid profile revealed the predominance of linoleic acid (70.8–75.4%, w/w). Thus, the biodiesel-derived glycerol was valorised to yield value-added products in a biorefinery approach (André et al. 2010). Chatzifragkou et al. (2011) evaluated various yeast and zygomycetous fungal strains to assimilate biodiesel-derived waste glycerol and convert it into value-added metabolic products. The fungi tested were able to accumulate higher levels of lipids (18.1–42.6% w/w of dry biomass) containing the medically important GLA in varying amount. *Thamnidium elegans* was further cultivated on increased initial glycerol concentrations and produced 11.6 gL^{-1} of SCO, with 71.1% (w/w) lipid content, while the maximum concentration of GLA was 371 mg L^{-1}. The phospholipid fraction of *T. elegans*-derived SCO was found to be rich in PUFA. Zygomycetous fungal strains, namely, *Cunninghamella echinulata* ATHUM 4411, *Mortierella manniana* MUCL 9235, *Mucor* sp. LGAM 36, *Thamnidium elegans* CCF-1465 and *Zygorhynchus moelleri* MUCL 143, were cultivated on pure glycerol (Bellou et al. 2012). The changes of the major lipid fractions implicated in the PUFA biosynthesis, as well as their fatty acid compositional shifts, were investigated. Neutral lipid was the major fraction in all the growth phases of fungi tested than phospho- and glycolipids. The linoleic and γ-linolenic acid concentration showed increment with time in all lipid fractions of *C. echinulata*. However, during growth of *M. ramanniana*, PUFA concentration gradually decreased in all lipid fractions, while n *C. echinulata*, PUFA biosynthesis was observed even after growth ceased. In *M. ramanniana*, PUFA concentrations decreased even in low temperature conditions indicating association of PUFA biosynthesis with primary metabolism. In another study on different yeasts and fungi with glucose and industrial (biodiesel-derived) glycerol by Papanikolaou et al. (2017), *Mortierella isabellina*, *M. ramanniana* and *Mucor* sp. accumulated high lipid quantities than yeasts. *M. ramanniana* MUCL 9235 preferred glycerol over glucose for lipid accumulation yielding SCO titre of 4 g L^{-1} and yield of 0.16–0.22gg^{-1} glycerol consumed. *M. isabellina* ATHUM 2935 displayed highly efficient SCO production on both glucose and glycerol, with yield of 0.22–0.25 g g^{-1} of substrate consumed and GLA at titre of 350 mg L^{-1}.

Waste cooking oil (WCO) is another important hydrophobic substrate that can be used for SCO production. The biomass, lipid yield and yield coefficient of lipid/biomass of the oleaginous yeast *Y. lipolytica* NCIM 3589 were found to be 10.1gL^{-1}, 4.3 gL^{-1} and 0.43 gg^{-1}, respectively, when grown on 100 gL^{-1} WCO (Katre et al. 2012). In another study, Ayadi et al. (2016) have reported that when *Candida viswanathii* Y-E4 was grown on 200 gL^{-1} WCO, the yield coefficient of lipid/biomass was 0.45 gg^{-1}.

8.7 Statistical Optimization Methods for SCO Production

The traditional method of optimizing one variable at a time (OVAT) is time-consuming and laborious. The technique suffers from an additional disadvantage that it does show the interaction between two or more variables. Statistical optimization strategies like Plackett-Burman design (PBD), Taguchi orthogonal array (Taguchi OA) and response surface methodology (RSM) help to not only increase the product yield, but also RSM helps in studying the interaction between two or more variables.

Plackett-Burman is a type of fractional factorial design which involves the selection of the significant variables that affect the process. Two levels, one high (+1) and another low (−1), are allowed for each variable. The Taguchi orthogonal array is another factorial design that involves studying the effect of large number of variables using a smaller set of experiments. Response surface methodology (RSM) is the most used technique for optimization. It involves the study of the interaction factors and usually generates a second-order quadratic equation by which the process can be understood. A three-dimensional model is generated which shows the interaction between variables. This technique has been widely used in biotechnology to optimize media components and cultivation conditions (Wang and Lu 2005). A Box-Behnken design (BBD) was used to increase the lipid yield to 12.86 g L^{-1} in *Mucor fragilis* AFT7–4 grown on glucose under the optimum conditions (glucose 86 g/L, ammonium sulphate 2.5 g/L, inoculum (v/v) 11% and fermentation time 5 days). This represented a 33.63% improvement in the lipid yield (Huang et al. 2016). Optimization studies employing CCD resulted in maximum lipid/dry biomass of 57.6% when the fungus was grown on sunflower oil cake, NaNO$_3$ and KCl at final concentrations of 8 gL^{-1}, 0.75 gL^{-1} and 0.25 gL^{-1}, respectively (Ali et al. 2017).

Aspergillus awamori MTCC 11639 was investigated for various parameters like glucose, pH, incubation temperature, nitrogen, phosphorous, proteins and sodium chloride concentration. A Taguchi OA was used to optimize the lipid production of the fungus. Glucose, pH and incubation temperature were found to be most significant factors affecting lipid yield. Validation experiments were performed, and an increased lipid productivity of 31% accompanied by 90% of substrate degradation was noted (Venkata Subhash and Venkata Mohan 2014).

M. isabellina IFO 7884 was investigated for various factors like moisture content, inoculum size, fungal spore age and nutrient supplements when grown on soybean hull. The final yield of total lipids was 47.9 mg/g; soybean hull was obtained when the moisture content, spore inoculum size and age were 75%, 10^4 and 7 days per g soybean hull. The total of C16 and C18 fatty acids was around 80.4% of the total fatty acids (Zhang and Hu 2012).

Aspergillus tubingensis strain TSIP9 was grown on lignocellulosic palm by-products, palm kernel cake (PK) and alkaline-pretreated empty fruit branches (EFB); it yielded cellulolytic enzymes. A three-factor CCD (central composite design) was used for the solid-state fermentation (SSF) of alkali-pretreated EFB to which PK was added as the alternate nitrogen source. The variables chosen were ratio of PK (20–80%), moisture content (40–80%) and the inoculum size (10^6–10^8 spores/gds). The experiments resulted in enhancement of the production of lipid and coproducts, cellulose and xylanase to 88.5 ± 4.9 mg/gds, 26.1 ± 0.1 U/gds and 59.3 ± 0.3 U/gds, respectively (Kitcha and Cheirsilp 2014). Optimizing the nutrient requirements to increase the lipid productivity and the lipid content would help to reduce the cost of the feedstock required for biodiesel production (Chang et al. 2015).

8.8   Biodiesel Production Technologies from Lipid/Oils

Biodiesel can be produced from lipids or oils by a number of techniques and methodologies such as dilution (direct mixing), microemulsion, pyrolysis and transesterification. The use of vegetable oils directly in diesel engine is not recommended due to their inherent high viscosity, low volatility, free fatty acid content, gum formation due to oxidation and polymerization during storage and combustion. They are unsuitable for diesel engine causing carbon deposits and gelling of engine oil over long-term use.

Transesterification is still the preferred and widely used technique at industrial scale for biodiesel production being the most promising solution to the high viscosity problem with simplicity, low cost and high conversion efficiency (Lin et al. 2011). Vegetable oils or animal fats are chemically triacylglycerols (TAGs) which during transesterification react with alcohol in the presence of a catalyst. It takes place via multiple consecutive reversible reactions in which diglycerides are produced from triglycerides and then diglycerides are converted to monoglycerides followed by their conversion to glycerol. In each step a fatty ester is liberated, and thus three fatty acid alkyl ester molecules are produced from one molecule of triglyceride. Glycerol is a by-product with a commercial value. Methanol is the preferred alcohol for transesterification because of its relatively low cost. Other alcohols studied include ethanol, isopropanol, butanol, tert-butanol, branched alcohols and octanol.

Catalytic transesterification methods involve acids, alkali or enzymes as catalysts. The SCO of *Mucor circinelloides* was transesterified with ethanol using an immobilized lipase from *Candida antarctica* (Novozym® 435), and it was found that 93% of the lipids were transesterified to fatty acid ethyl esters (FAEEs) (Carvalho et al. 2015). Non-catalysed transesterification processes have been developed as well, viz. the BIOX (cosolvent) process and the supercritical alcohol (methanol) process (Demirbas 2009; Leung et al. 2010). Aransiola et al. (2014) have recently reviewed technologies that have been used for biodiesel production till date, and an attempt has been made to compare commercial suitability of these methods on the basis of available feedstocks and associated challenges.

In another approach instead of the conventional three-step transesterification, a direct one-step process of in situ transesterification had been employed for FAME production. It was seen that under optimum conditions (8 h, 65 °C), dry biomass of *Mucor circinelloides* yielded 18.9% FAME by acid-catalysed one-step transesterification (Vicente et al. 2010). The biomass of *Aspergillus candidus* grown on whey was directly converted to FAME (fatty acid methyl ester) to yield 320 mg of FAME (Kakkad et al. 2015a). The biomass of *Y. lipolytica* NCIM 3589 grown on 100 gL^{-1} WCO was optimized using a PBD and a one variable design (OVD) for the in situ transesterification, and a FAME yield of 0.88 g was obtained (Katre et al. 2018). The conventional three-step procedure was thus avoided which would help to save time and energy.

Varying mathematical models have been developed to study fungal growth and lipid production. A detailed mathematical model that could predict biomass, growth and lipid accumulation and sugar and nitrogen consumption was developed for *Mortierella isabellina* which was grown on sweet sorghum. This model could be used as a tool to design bioreactor for the production of biofuel (Economou et al. 2011).. In chemostat cultures of *Umbelopsis isabellina* grown on mineral medium containing glucose and NH_4^+, model parameters were obtained which showed that dilution rate (D) influenced lipid production. The maximum specific lipid production rate was found to be independent of the actual specific growth rate. In solid-state fermentation, when the supply rate of substrate monomers which are released from the solid substrate is low, the lipid yield is low because the condition is somewhat similar to a chemostat with single-carbon limitation or dual (carbon and nitrogen) limitation (Meeuwse et al. 2011). Similarly, when a model was developed to predict growth, lipid accumulation and substrate consumption in different oleaginous yeasts, it was found that the first priority was given to maintenance, second priority to growth and third priority to lipid production. The maximum specific lipid production rate was independent of the growth rate (Meeuwse and Tramper 2011).

For submerged fermentation, a model was developed that described growth, lipid production and subsequent lipid turnover in three phases. The first is an exponential phase, wherein both the carbon and nitrogen sources are available, the second is carbohydrate and lipid production phase wherein the nitrogen source is depleted, and the third is lipid turnover phase wherein the carbon source is depleted. Batch and chemostat cultures were compared for *Umbelopsis isabellina*, and it was found that the initial specific lipid production rate for batch cultures was almost four times higher than for chemostat cultures which decreased exponentially with time. In the case of chemostat cultures, the specific lipid production rate was independent of residence time indicating that different mechanisms were active for batch and chemostat cultures (Meeuwse et al. 2012a).

A model describing the kinetics and diffusion of glucose, alanine and oxygen containing κ-carrageenan gel was validated empirically. The model described the different phases of culture very well: exponential growth, linear growth due to oxygen limitation, accumulation of lipids and carbohydrates after depletion of nitrogen lipid turnover after carbon depletion (Meeuwse et al. 2012b). The model indicated that oxygen limitation is very important in solid-state cultures using monomers explaining the difference in production rate with submerged cultures.

Studies on the relationship between lipid accumulation and morphology development in *Mortierella isabellina* revealed that the free dispersed mycelia form was desirable to obtain maximum lipid productivity. Different concentrations of magnesium silicate microparticles were added to develop different morphologies of the fungus, viz. pellets of different sizes, free dispersed mycelia and broken hyphal fragments. The study provided insight into molecular mechanisms of biosynthesis of lipid and its relation with morphological development in addition to design and optimization of bioprocess to enhance lipid-based biofuels (Gao et al. 2014).

Mutagenesis and genetic engineering techniques have often been used to obtain mutants exhibiting increased in production of SCOs. A filamentous fungus, *Ashbya gossypii*, widely used for industrial production of riboflavin has recently been engineered to overproduce lipids suitable for biodiesel production. The fatty acyl-CoA pool was altered and the Δ9 desaturase gene manipulated to yield fourfold increase in lipid production (Lozano-Martínez et al. 2017). The engineered strain could utilize and convert molasses to lipids. The same fungus was also engineered to utilize xylose, a common pentose present in lignocellulosic biomass and the overexpression of an endogenous xylose utilization pathway resulted in increased conversion rates of xylose to xylitol (up to 97% in 48 h). Using a heterologous phosphoketolase pathway, metabolic flux channelling resulted in increased lipid content (54%) in the xylose-utilizing strain. The lipid content was further enhanced to 69% by blocking the β-oxidation pathway. A novel biocatalyst was thus generated with promising properties in consolidated bioprocessing (Díaz-Fernández et al. 2017). A molecular breeding approach was employed in *Mortierella alpina* to achieve overproduction of linoleic and oleic acids. The linoleic acid production rate as to total fatty acid increased five times when the *Δ12 desaturase* (DS) gene from *Coprinopsis cinerea* was expressed in the 6DSΔ activity-defective mutant. The oleic acid accumulation rate to total fatty acids reached 68% by suppressing the endogenous *Δ6I* gene using RNAi in the 12DSΔ activity-defective mutant (Sakamoto et al. 2017).

Random mutagenesis as a means of strain improvement can also be used to increase the SCO production. Two stable mutants of *Y. lipolytica* NCIM 3589 grown on WCO were generated by treatment with a chemical mutagen, *N*-methyl-*N'*-nitro-*N*-nitrosoguanidine (MNNG), as well as an additional treatment with cerulenin, a fatty acid synthase inhibitor. The two mutants, YlC7 and YlE1, were capable of accumulating 60 and 67% lipid (Katre et al. 2017). When compared to the wild-type strain, the increase in lipid contents of the mutants was 1.33- and 1.49-fold for YlC7 and YlE1, respectively. Similar results have been noted for mutant of *Colletotrichum* sp. DM06, wherein a ~1.7-fold increase in lipid content was observed after its genetic transformation with the CtDGAT2b gene from *Candida tropicalis* SY005. In another study, after physical mutagenesis of *Rhodosporidium toruloides* np11 using atmospheric and room temperature plasma method and chemical mutagenesis with MNNG, the mutant strain *R. toruloides* XR-2, 450 accumulated 41% lipids as compared to 23% of the wild type (Zhang et al. 2016).

It has been found that free fatty acids (FFAs) were released form cells of the filamentous fungus *Aspergillus oryzae* when grown on non-ionic surfactants such as Triton X-100. When a *faaA* disruptant of this fungus was grown in the presence of 1% Triton X-100, 50% of the FFAs that were synthesized de novo were released (Tamano et al. 2017). This helped to develop a cost-effective technique to release FFAs accumulated in the cytosol without the need for energy-intensive cell disruption which would be useful in metabolic engineering studies in which microbes are exploited to produce FFAs.

Another significant development in the field is the development of newer and efficient methods for estimating the efficiency of lipid extraction. An FTIR spectroscopy method for estimating lipid extraction has been established which can be used to identify chitin, glucuronans, polyphosphates and other compounds that may affect the lipid extraction process (Forfang et al. 2017). A major disadvantage in using oleaginous fungi and development of efficient biosystems is lack of high-throughput method to grow and screen strains capable of increased lipid accumulation. In view of this, a study conducted recently has shown the potential for using a Duetzmicrotiter plate system for growth of oleaginous fungi and a rapid, high-throughput FTIR spectroscopy for estimation of lipid (Kosa et al. 2017).

Large-scale production of oleaginous fungi for biodiesel has been previously evaluated for its feasibility by a techno-economic study using *Mucor plumbeus*

(Ahmad et al. 2016). Oil palm empty fruit bunch (EFB), a readily available ligno-cellulosic substrate, was pretreated using dilute acid treatment to give the liquid hydrolysate which was detoxified to give empty fruit bunch liquid hydrolysate (EFBLH). The solid residue was hydrolysed enzymatically to give empty fruit bunch enzymatic hydrolysate (EFBEH). The fungus was grown on both EFBLH and EFBEH. It was found that a total of 93.6 kg of microbial oil was produced from 1 tonne dry EFB. These results were comparable with the estimated amount of microbial oil produced per tonne of wheat straw (103 kg) by *Mortierella isabellina* in an earlier study (Zheng et al. 2012). The outcomes of these studies prove the potential of integrating the microbial oil production with the existing processes that generate these lignocellulosic wastes and the advantages of using these renewable wastes as growth substrates.

8.9 Biodiesel Fuel Standards, Fuel Properties and Fungal Lipids as Feedstock

In 1991, rapeseed oil methyl esters were the first commercially to be used as bio-diesel fuel in Austria. A biodiesel standard was therefore set to define and approve the standards for rapeseed oil methyl esters as a diesel fuel. In 1997, a German standard, DIN 51606, was formalized. Eventually, the successful introduction and commercialization of biodiesel around the world resulted in the development of biodiesel standards in many countries. An American Society for Testing and Materials (ASTM) standard, D6751, was set in the United States for biodiesel in the form of the lower alkyl esters of fatty acids and published in 2002. This was soon followed by a new Europe-wide biodiesel standard, DIN EN14214, in 2003. ASTM D6751 and EN14214 have served in the development of other standards around the world. Various specifications that a biodiesel fuel must meet are contained in these biodiesel standards. Biodiesel is most commonly used as a blend with petroleum diesel. Pure biodiesel is designated as B100, and for its blends, a number following the "B" indicates the percentage of biodiesel in a gallon of fuel; the remainder of the gallon can be No. 1 or No. 2 diesel, kerosene, jet A, JP8, heating oil or any other distillate fuel (NREL 2009). The biodiesel properties included as specifications in different biodiesel standards comprise of a range of physico-chemical properties and quality parameters, viz. cetane number (CN), density, kinematic viscosity (KV), flash point, acid number, iodine value (IV), ester content, copper corrosion, water and sediment, distillation range, sulphur and phosphorus content, total sulphur and oxidative stability. Ramos et al. (2009) have shown that the biodiesel quality depends upon the fatty acid composition of the oil feedstock. For any oleaginous fungus to be considered as a suitable feedstock for biodiesel, besides the total lipid content (>20%), the type of fatty acids (long-chain saturated and/or monounsatu-rated fatty acids) also constitutes an important criterion. Lipid content and fatty acid composition of SCOs varies in response to environmental factors such as type of

Table 8.2 Select fuel properties of biodiesel (B100) listed as specifications in two major fuel standards

Fuel property	Units	Limits in fuel standards	
		ASTM D6751-12[a]	DIN EN14214:2012[b]
Density (15 °C)	Gcm^{-3}	NS	0.86–0.9
Kinematic viscosity (40 °C)	Mm^{2}s^{-1}	1.9–6.0	3.5–5.0
Cetane number	N/A	47 min	51 min
Flash point	°C	93 min	101 min
Acid number	N/A	0.5 max	0.5 max
Distillation temperature (T90)	°C	360 max	NS
Iodine value	N/A	NS	120 max
Cu strip corrosion	N/A	No. 3 max	Class 1 max
Acid number	Mg KOH/g	0.50 max	0.50 max
Water and sediment/water	Vol. %/mg/ kg	0.050 max	500 max
Concentration of linolenic acid (C18:3)	Wt. %	NS	12 max
FAME having ≥4 double bonds	Wt. %	NS	1 max
Visual test	N/A	Clear and free	NS

N/A not applicable, *NS* not specified
[a]http://www.afdc.energy.gov/fuels/biodiesel_specifications.html
[b]http://www.dieselnet.com/tech/fuel_biodiesel_std.php#spec

carbon source, pH and temperature and is species- and strain-specific (Subramaniam et al. 2010; Venkata Subhash and Venkata Mohan 2011). In *Mortierella* species, *M. alpina* was able to produce 40% (w/w) SCO, while *M. isabellina* ATHUM 2935 gave 50.4% (w/w) oil (Wynn et al. 2001; Papanikolaou et al. 2004a) when grown on glucose. The lipid production (w/w wet biomass) in n-hexadecane- and glucose-grown cells were 4.43 and 0.62%, respectively, by *Aspergillus terreus* MTCC 6324 (Kumar et al. 2010).

Since the accumulation of lipids by oleaginous fungi varies, not all oleaginous fungal cells can be used as a feedstock for biodiesel production. Therefore, careful selection of the oleaginous strains of the fungal species and characterization of lipid composition needs to be performed to ascertain their suitability for biodiesel production. Since biodiesel fuel properties such as ignition quality, cold flow, oxidative stability and iodine value are determined by the fatty acid profiles of the oil feedstock, it is important to evaluate the fatty acid compositions of fungal SCOs in terms of carbon chain length and unsaturation degree. As FAME composition determines the quality of biodiesel, a higher content of saturated fatty esters imparts desirable fuel properties to biodiesel, viz. high CN and better oxidative stability. For example, higher CN helps ensure good cold-start properties and reduces smoke formation (NREL 2009). However, saturated FAMEs display higher melting points than unsaturated ones, and hence higher amounts of saturated fatty esters may result in biodiesel with poor cold flow properties (Knothe 2005). A recent study suggests that while unsaturated (Table 8.2) fatty acid esters are required for better cold flow properties, the levels of saturated FAMEs should be below 26% in biodiesel for flow properties at low temperature (Moser and Vaughn 2012). On the other hand, the percentage of unsaturated fatty acid methyl esters affects oxidative stability and quality of biodiesel during extended storage (Knothe 2005). A high total PUFA methyl ester content also results in increased viscosity and low CN, which is undesirable for biodiesel. As a selection criterion for biodiesel production, the range for the iodine value (54–120) has been fixed as to simultaneously eliminate FAMEs with poor oxidative stability and poor cold flow properties (Moser and Vaughn 2012). Biodiesel with high monounsaturated fatty acid content has better characteristics with respect to ignition quality, nitrogen oxide emissions, fuel stability and flow properties, and hence methyl esters of palmitoleic (C16:1) and oleic (C18:1) acids are warranted as they are liquid at room temperature. In fact, monounsaturated fatty acid methyl esters are considered to be desirable for CN and IV, without any adverse effect on biodiesel cold properties. Thus, an ideal biodiesel is made mainly of monounsaturated with balanced levels of saturated and polyunsaturated methyl esters (Ramos et al. 2009).

Direct measurement of fuel properties of biodiesel is quite complex with high cost and high error in reproducibility and requires a considerable amount of fuel sample (Tong et al. 2011). Specific linear mixing rules can be used to predict values of some biodiesel properties from individual properties of their component FAMEs, while empirical equations can be used to predict few other physico-chemical fuel properties (Demirbas 1998; Azam et al. 2005). The following general expression is used to estimate/predict the fuel properties of fungal SCOs.

$$f_b = \sum_{i=1}^{n} z_i \cdot f_i$$

where f is a function representing a physical property (the subscripts b and i refer to the biodiesel and the pure ith FAME, respectively) and z_i is the mass or mole fraction of the ith FAME. The function f_b must be replaced by the variables specifying cetane number, natural logarithm of kinematic viscosity, density and higher heating value of biodiesel, while the function f_i must be interchanged by the respective variables in order to specify the properties of the individual ith FAME (Ramírez-Verduzco et al. 2012). Predicting density, KV and CN from values of individual neat esters has been accomplished (Ramos et al. 2009; Tong et al. 2011; Pratas et al. 2011; Knothe and Steidley 2011) and shown in Table 8.3.

Ramírez-Verduzco et al. (2012) provided four new empirical correlations to estimate the CN, KV, density and HHV of individual neat FAMEs from two structural features of molecules, namely, molecular weight and degree of unsaturation (Table 8.4).

Where:

c_i is the concentration, and ρ_i is the density of component i.
ν_{mix} is kinematic viscosity of the sample of biodiesel, A_c is the relative amount of the fatty acid methyl ester, and ν_c is the kinematic viscosity of the individual FAME.

Table 8.3 Mixing rules and empirical correlations used in estimation of biodiesel fuel properties for fungal SCOs

Fuel property/quality parameter of biodiesel	Mixing rule/empirical formula
Density	$\rho = \sum c_i \rho_i$ Pratas et al. (2011)
Kinematic viscosity (KV)	$\nu_{mix} = \sum (A_c \nu_c)$ Knothe and Steidley (2011)
Cetane number (CN)	$CN = 1.068 \sum (CN_i W_i) - 6.747$ Tong et al. (2011)
	$CN = \sum X_{ME}(wt\%)$. CN_{ME} Ramos et al. (2009)
Saponification number (SN)	$SN = (560 \times A_i/MW_i)$ Azam et al. (2005)
Iodine value (IV)	$IV = (254 \times D \times A_i/MW_i)$ Azam et al. (2005)
Higher heating value (HHV)	$HHV = 49.43 - [0.041(SN) + 0.015(IV)]$ Demirbaş (1998)

Table 8.4 Empirical correlations for prediction of physical fuel properties of FAMEs (Ramírez-Verduzco et al. 2012)

Property	Empirical correlation N = number of double bonds in a given FAME; M_i = molecular weight of the ith FAME
Density	$\rho_i = 0.8463 + \dfrac{4.9}{M_i} + 0.0118.N$
KV	$\ln(\eta_i) = -12.503 + 2.496. \ln(M_i) - 0.718. N$
CN	$\phi_i = -7.8 + 0.302. M_i - 20. N$
HHV	$\delta_i = 46.19 - \dfrac{1794}{M_i} - 0.21.N$

CN is the cetane number of the biodiesel, CN_i is the cetane number of the individual neat methyl ester, and W_i is the weight fraction of the fatty acid methyl ester.

CN is the cetane number of the biodiesel, X_{ME} is the weight percent of the individual methyl ester, and CN_{ME} is the cetane number of the individual neat methyl ester.

A_i is the percentage, D is the number of double bonds, and MW_i is the molecular mass of each fatty acid methyl ester.

Where:

ρ_i is the density of the ith FAME at 20 °C in g/cm³.
η_i is the KV at 40 °C of the ith FAME in mm²/s.
ϕ_i is the CN of the ith FAME.
δ_i is the HHV of the ith FAME in MJ/kg.
M_i is the molecular weight of the ith FAME.

8.9.1   Cetane Number (CN)

CN serves as an important ignition quality indicator for diesel fuel equivalent to octane number of petrol. Higher octane number of a fuel compound is associated with lower CN and vice versa. Upon injection of fuel into the engine, the combustion chamber experiences an ignition delay time (the time between fuel injection into engine cylinder and its ignition) which is measured as CN. The higher the CN, the shorter the ignition delay time and vice versa (Knothe 2005, 2008; Moser 2009). Higher CN of diesel results in better ignition quality leading to easier starting and quieter operation.

The CN is included in the biodiesel standards of different countries and organizations with a prescribed minimum value of 47 (ASTM D6751) or 51 (EN14214) for B100. The term lipid combustion quality number has been suggested for biodiesel components (FAMEs) because of their high CNs which exceed the cetane scale (Knothe 2005). A standard scale has been established using hydrocarbons and accepted worldwide for assigning CNs to diesel fuel. Hexadecane ($C_{16}H_{34}$; trivial name cetane) is the reference compound with highest CN of 100 on the cetane scale, and 2,2,4,4,6,8,8-heptamethylnonane (HMN, also $C_{16}H_{34}$) is the low-quality standard with an assigned CN of 15. This scale also explains the suitability of fatty acid alkyl esters as diesel fuel because of the presence of long hydrocarbon chain in both of these esters and straight-chain alkanes. However, as compounds have been identified with CN >100 or CN < 15, arbitrary nature of the cetane scale has to be noted (Moser 2009). The CN of fatty acid alkyl esters is dependent on both saturation and chain length of these compounds and is known to increase with increasing saturation and increasing chain length. Palm oil ethyl esters were first evaluated for CN tests and displayed a high CN, while FAMEs such as methyl palmitate and stearate have cetane numbers >80. Methyl oleate displays the CN in the range of 55–58, while methyl linoleate and linolenate exhibit CNs around 40 and 25, respectively (Knothe 2008). Biodiesel derived from oils of soy, sunflower and rapeseed possesses CNs closer to 47. CN of 60 or higher is shown by biodiesel derived from highly saturated feedstocks (animal fat, waste frying oil).

In order to determine the potential of fungal SCOs as biodiesel feedstock, CN values have been calculated using prediction model equations or empirical formula either based on fatty acid composition or other fuel properties. FAMEs derived from lipid containing biomass of filamentous fungus *Mortierella isabellina* were estimated using known empirical formulas (Liu and Zhao 2007;Azam et al. 2005). The calculated CN of 56.4 is in accordance with known biodiesel specifications of ASTM D6751 (USA), EN14 214 (Europe) and DIN V51 606 (Germany). When SCOs of oleaginous fungi isolated from the mangrove wetlands of the Indian west coast were analysed for their potential as biodiesel feedstock, the calculated CNs were found to be between 56 and 61 and thus in the acceptable range of international biodiesel standard norms suggesting their possible suitability as a fuel. In fact, transesterified oils of these mangrove fungi displayed higher levels of methyl esters of long-chain saturated fatty acids, namely, stearic acid (C18:0) and palmitic acid (C16:0) (Khot et al. 2012). In other studies, oil storing mangrove fungal strain *Aspergillus terreus* IBB M1 produced SCO with calculated CN of 52 and 50.4, respectively, when grown on copra cake and sugarcane bagasse residue left over after bagasse pretreatment (Kamat et al. 2013; Khot et al. 2015). The SCOs of *A. terreus* IBB M1 displayed CN in good agreement with international biodiesel fuel specifications as it was rich in saturated fatty acids (60%) comprised of lauric (C12:0), myristic (C14:0), palmitic (C16:0) and stearic acids (C18:0) (Khot et al. 2015), while for sugarcane bagasse residue, 46.7% saturated fat was present in SCO comprised of heneicosanoic (C21:0), behenic (C22:0), undecanoic (C11:0), tridecanoic (C13:0), palmitic (C16:0), heptadecanoic (C17:0) and stearic (18:0) acids (Kamat et al. 2013).

Zheng et al. (2012) estimated CNs of lipids produced by six different filamentous fungi, namely, *Aspergillus terreus, Cunninghamella elegans, Mortierella isabellina, Mortierella vinacea, Rhizopus oryzae* and *Thermomyces lanuginosus*, grown on wheat straw hydrolysate. The calculated values vary from 54.8 to 61.6 and all being above 51 and thus matched the criteria of EN14214 biodiesel specifications for CN. The biomass of *Aspergillus candidus* IBB G4 was grown on whey to produce FAMEs with predicted CN of 54 and thus satisfies the criteria of both ASTM D6751 and EN14214 (Kakkad et al. 2015a). The CN was predicted to be 53.11 for *Trichoderma reesei* and satisfied both ASTM and EN specifications (Bharathiraja et al. 2017).

8.9.2 Viscosity

The engine fuel viscosity is known to play a critical role in the fuel spray, mixture formation and combustion process. The fuel atomization is affected by its viscosity upon fuel injection into the combustion chamber and in turn determines the formation of engine deposits. A much higher diesel viscosity is not desirable due to a number of reasons. It increases the tendency of the fuel to form engine deposits. The high viscosity results in insufficient fuel atomization and early fuel injection due to high line pressure. Besides, increase in fuel viscosity leads to increase in the mean diameter of fuel droplets from the injector and their penetration (Canakci and Sanli 2008). These operational problems have been found to be associated with the use of highly viscous vegetable oils containing triglycerides, and therefore alkyl esters were produced to form biodiesel by transesterification of vegetable oils (Knothe 2009).

Biodiesel viscosity is about an order of magnitude lower than that of the oil feedstock causing better fuel atomization in the engine combustion chamber. Biodiesel production can be monitored by means of the difference in viscosity of the oil feedstock and their alkyl ester (Knothe 2005). As the chain length (the number of $-CH_2$ moieties in the fatty ester chain) increases, viscosity increases. However, viscosity is observed to decrease with an increasing unsaturation with double bonds in *cis*-configuration causing a significant reduction of viscosity. However, fatty esters with trans-double bonds and their saturated counterparts exhibit almost similar viscosity values.

The kinematic viscosity values for a variety of short-, medium- and long-chain fatty esters with varying saturation levels can be found. It is important to note that temperature is known to affect viscosity to a greater extent in such a way that it decreases exponentially with an increasing temperature. Hence, lower viscosity facilitates the handling of biodiesel fuel at lower temperatures. The reason behind this is the flow properties which are influenced in turn by viscosity making it a critical parameter to consider when using biodiesel, at low temperatures (Knothe2008). Biodiesel standards contain specifications for kinematic viscosity (1.9–6.0 mm^2/s in ASTM D6751, 3.5–5.0 mm^2/s in EN14214, 5.0 mm^2/s ISO3104).

SCOs of oleaginous fungi isolated from the mangrove wetlands of the Indian west coast were analysed for potential as biodiesel feedstock, and their calculated KVs were found to be in the acceptable range of international biodiesel standard norms suggesting their possible suitability as a fuel (Khot et al. 2012). *Aspergillus terreus* IBB M1, an oleaginous mangrove fungal strain, produced SCO with predicted KV of 3.32 and 4.85 mm^2/s, respectively, when grown on copra cake and sugarcane bagasse residue left over after bagasse pretreatment (Kamat et al. 2013; Khot et al. 2015). Zheng et al. (2012) calculated the KV values for the lipids obtained from six moulds, viz. *Aspergillus terreus*, *Cunninghamella elegans*, *Mortierella isabellina*, *Mortierella vinacea*, *Rhizopus oryzae* and *Thermomyces lanuginosus*. The estimated KV varied from 4.2 to 4.5 mm^2/s and thus satisfied the EN14214 criterion. An oleaginous mangrove fungal strain *Aspergillus candidus* produced FAMEs with calculated KV of 4.27 mm^2/s in case of whey as substrate (Kakkad et al. 2015a) while it ranged from 4.12 to 4.83 mm^2/s when grown on glucose and solid agro-industrial residues (Kakkad et al. 2015b). For the oleaginous fungi *Aspergillus oryzae* and *Mucor plumbeus*, the kinematic viscosities were 4.42 and 4.27 mm^2/s, respectively, when grown on EFBLH and 4.50 mm^2/s and 4.51 mm^2/s, respectively, when grown on EFBEH (Ahmad et al. 2016).

8.9.3 Density

Density is another important physical fuel property of biodiesel which depends on the fatty acid profile of alkyl esters present which in turn is affected by the raw materials used in fuel production. The importance of the density lies in the fact that it affects the amount of mass injected at the injection system because a precise fuel quantity is essential for proper combustion. Density data play an important role in a variety of unit operations involved in biodiesel production such as designing of reactors, distillation units, storage tanks and piping. Besides, the total energy consumption can be estimated with fuel density data because the injection system of diesel engine injects a specific fuel volume into the cylinder determined by the Electronic Control Unit (ECU) depending on the driving conditions. The ECU of a vehicle in general is standardized for a constant, fixed density value. When biodiesel is used as an alternative fuel in the diesel engine (as B100 or blended), mismatching of the engine parameters may take place due to the density differences leading to an increase in the emissions compared to those of the optimized conditions. Therefore, diesel and biodiesel norms (EN-590 and EN-14214, respectively) have put limits for fuel density with EN-3675 (hydrometer method) or EN-12158 (oscillating U-tube method) as the standard test methods for the density measurement at a temperature of 15 °C (Pratas et al. 2011, 2010). In addition, density helps in determination of a range of other important fuel properties, and thus correct estimation/measurement of this property is relevant. For example, the CN and viscosity of biodiesel are highly dependent on its density (the higher the density, the lower the viscosity and the CN) (Lapuerta et al. 2010).

Density values have been determined both experimentally and using prediction models for the fungal SCOs in order to evaluate their potential as biodiesel feedstock. Density of transesterified lipids from isolated filamentous fungus *Aspergillus* sp. was found to vary from 0.80 to 0.83 g/cm^3 depending on growth substrates and transesterification methods (Venkata Subhash and Venkata Mohan 2011). In a study on tropical mangrove oleaginous fungi, experimentally determined and calculated density values of transesterified oils of three isolates vary from 0.860 to 0.875 g/cm^3 and thus were within the range of EN14214 specification (Kamat et al. 2013; Khot et al. 2012, 2015). *Aspergillus awamori* MTCC 11639 produced lipids with good biodiesel properties and possess density of 0.8241 g/cm^3 after transesterification which could be attributed to the higher number of saturated fatty acids (Venkata Subhash and Venkata Mohan 2014). The suitability of lipids obtained from six moulds, viz. *Aspergillus terreus*, *Cunninghamella elegans*, *Mortierella isabellina*, *Mortierella vinacea*, *Rhizopus oryzae* and *Thermomyces lanuginosus*, was studied by calculating their density values (Zheng et al. 2012). The results demonstrated that the fungal lipids matched the EN14214 specification criterion of density very well being in the range of 0.86 to 0.90 g/cm^3. Similar results were obtained for the FAMEs produced from mangrove oleaginous fungal isolate, *Aspergillus candidus* IBB G4 (Kakkad et al. 2015a, b).

8.9.4 Higher Heating Value (HHV)

Higher heating value (HHV) is also known as the gross calorific value or gross heat of combustion and represents the amount of heat released during the combustion of 1 gramme of fuel to produce CO_2 and H_2O at its initial temperature. The energy content of the fuel is defined with the help of its calorific value. HHV of biodiesel is in the range of 39.57–41.33 MJ/kg and thus around 12% lesser compared to that of petro-diesel (46 MJ/kg). Calorific value of FAMEs and TAG is approximately 1300 kg cal/mol and 3500 kg cal/mol for fatty acids and their esters. This physical fuel property of biodiesel increases with chain length and ratio of carbon and hydrogen to oxygen and nitrogen, while unsaturation has a negative effect on heating value. It helps in determining fuel consumption in such a way that the lower the gross calorific value, the higher the fuel consumption. Thus, the lowest heating values have been observed for the short-chain saturated fatty esters with higher fuel consumption (Knothe 2005, 2008; Ramírez-Verduzco et al. 2012).

HHV was theoretically calculated for SCOs of two endophytic fungal cultures, namely, *Colletotrichum* sp. (isolate DM06) and *Alternaria* sp. (isolate DM09), taking the following five fatty acids into consideration, namely, C16:0, C18:0, C18:1, C18:2 and C18:3. The predicted HHV of the lipids of these two fungal endophytes was compared with rapeseed, jatropha, sunflower, palm and soybean oils and found to be close to those of above-mentioned plant oils which are most commonly used lipid raw materials for biodiesel production (Gopinath et al. 2009; Dey et al. 2011). The HHVs were determined experimentally and also calculated empirically for the SCOs of five lipid-accumulating filamentous fungal strains isolated from mangrove wetlands of the Indian west coast, and the values of ~40 MJ/kg found were similar to those of vegetable oils used for biodiesel production (Khot et al. 2012). The calculated HHVs were 39.7–39.8 MJ/kg for the lipids produced from *Aspergillus terreus*, *Cunninghamella elegans*, *Mortierella isabellina*, *Mortierella vinacea*, *Rhizopus oryzae* and *Thermomyces lanuginosus* (Zheng et al. 2012). Kakkad et al. (2015a, b) have estimated HHVs to be around 40 MJ/kg for the FAMEs produced from *Aspergillus candidus* IBB G4 biomass grown on different substrates. The HHVs for *Aspergillus oryzae* and *Mucor plumbeus* were 40.11 MJ/kg and 39.10 MJ/kg, respectively, when grown on EFBLH and 39.79 MJ/kg and 39.84 MJ/kg, respectively, when grown on EFBEH (Ahmad et al. 2016).

8.9.5 Iodine Value (IV)

IV is a chemical fuel property of biodiesel and acts as a crude measure of the degree of unsaturation of the fuel sample. It is used in connection with oxidative stability of biodiesel fuel with an implication that fuel with low IV will possess high oxidative stability than that with a high IV. Therefore, European biodiesel standard EN14214 has included IV in their specifications. Due to dependence of IV on molecular weight, FAMEs have higher IV than their ethyl ester counterparts (Azam et al. 2005; Knothe 2009).

In a study on biodiesel production from oleaginous fungus *M. circinelloides*, the calculated IV for the saponifiable lipids and the free fatty acids extracted from fungus was 107.6 mg I_2/g, being far below the specified limit of 120 mg I_2/g in the European Union Standards (Vicente et al. 2009). In comparison to the vegetable oils, the IV was very similar to the one obtained in biodiesel from rapeseed oil (107.7 mg of I_2/g), which is the preferred raw material for biodiesel production in Europe (Vicente et al. 2010). In a study by Dey et al. (2011), the calculated IVs (g I_2/100 g of lipid) were found to be 95.43 for *Colletotrichum* sp. and 100.28 for *Alternaria* sp., and these values are well below the European biodiesel standards, i.e. 120–130. Khot et al. (2012) reported the experimentally determined and the predicted IVs of the transesterified oils from five oleaginous mangrove fungal isolates to be below the EN14214 specification (120 max) suggesting good oxidative stability. Calculated IVs of the lipids of six filamentous fungi (*Aspergillus terreus, Cunninghamella elegans, Mortierella isabellina, Mortierella vinacea, Rhizopus oryzae, Thermomyces lanuginosus*) were ranging from 68.3 to 97.1 and matched the EN14214 criterion (Zheng et al. 2012). IV of fungal lipids from isolated filamentous fungus *Aspergillus* sp. was found to vary from 12 to 16 depending upon the carbon source and transesterification method employed. The low IVs obtained in any case indicate the less risk of fuel polymerization, and low values were seen for lipids from *Aspergillus awamori* MTCC 11639 (Venkata Subhash and Venkata Mohan 2011, 2014). SCOs derived from mangrove fungal isolate *Aspergillus terreus* IBB M1 were calculated to possess IV of 98.3 with sugarcane bagasse residue as substrate and 32 (experimental) when culture was grown on copra cake. The results are in accordance with EN14214 specifications for IV (Kamat et al. 2013; Khot et al. 2015). In another study, the iodine value was predicted to be 64.34 for *Trichoderma reesei* (Bharathiraja et al. 2017).

8.10 The Biorefinery Concept

The biorefinery concept that involves conversion of the plant biomass to fuels such as biodiesel as one of its products, along with other coproducts (chemicals and metabolites), would be an economically viable option. The onus is on "zero waste" generation during the processing of biomass as the waste generated after the product obtained from one stream would serve as the raw material for the generation of the second product (FitzPatrick et al. 2010). Thus, sustainable harvesting and the waste biomass utilization has no negative impact on the environment and is carbon neutral (Liu et al. 2012). The aim is to reduce industrial waste disposal while using the waste to produce economically viable products. About 60–75% of the total production cost is contributed by the carbon source (Subramaniam et al. 2010). Oleaginous fungi are heterotrophic, and the cost of the carbon source can be cut down by using a zero or negative waste, a waste that could be a threat to the environment if not disposed suitably. The use of the recalcitrant lignocellulosic wastes and other wastes like industrial and municipal waste waters is fast emerging as a viable option.

Aspergillus tubingensis TSIP9 grown on palm empty fruit bunches (EFB) and palm kernel (PK) yielded 86.6 mg/gds (mg per gramme dry substrate) along with the production of 22.1 U/gds of cellulase and 78.9 U/gds of xylanase (Cheirsilp and Kitcha 2015). *Aspergillus* sp. ATHUM 3482 accumulated lipids up to 64% (w/w) after 191 h when grown on 15 g L^{-1} waste olive oil (Table 8.1). This strain was also capable of synthesizing and accumulating oxalic acid (5 g L^{-1}). In the same study, it was found that another strain *A. niger* NRRL 364 produced 645 U ml^{-1} lipase. This study demonstrated the biovalourization potential of these fungal species, which could convert the hydrophobic waste to value-added products like lipase and oxalic acid (Papanikolaou et al. 2011c). Kamat et al. (2013) have used sugarcane bagasse to co-produce biodiesel and high value products like xylitol and cellulose.

A mangrove yeast, *Williopsis saturnus*, yielded 0.51 g xylitol/g xylose consumed in 72 h. The sugarcane bagasse residue obtained after the hydrolysis of sugarcane bagasse was used as a growth substrate for another mangrove fungus, *Aspergillus terreus* IBB M1, for the production of xylanase (12.74 U/ml) and cell biomass (9.8 g/L). The fungal biomass was used as feedstock for SCO production and yielded 0.19 g/g and was transesterified to biodiesel. This novel, green biorefining approach allowed for an efficient and sustainable use of biomass and reduced the competition for different uses of a single source of biomass. *Aspergillus awamori* MTCC 11639 was grown on lignocellulosic wastewaters like corncob waste liquor (CWL), paper mill effluent (PME) and a cellulosic substrate-deoiled algae extract (DAE) to obtain microbial lipid (Venkata Subhash and Venkata Mohan 2015). The removal of colour was efficient (78%) along with reduction in the chemical oxygen demand (COD). Extracellular laccase activity was detected which validated the colour removal and degradation of lignin. The approach was a low-cost alternative for the production of fungal lipid production using lignocellulosic waste as substrate.

Crude glycerol, a by-product of the transesterification reaction, is produced in large quantities and is fast becoming a waste stream. Hence, to reduce waste disposal of glycerol by the biodiesel industry, several alternatives aimed at the use of crude glycerol to produce fuels and chemicals by microbial fermentation have been evaluated and reviewed (Almeida et al. 2012). Different strategies have been employed to produce biofuels and chemicals, namely, ethanol, n-butanol, 1,3-propanediol, 2,3-butanediol, organic acids and polyols, by microbial fermentation of glycerol.

8.11 Future Prospects

Increasing petroleum cost, demand for fuels and chemicals worldwide, awareness of environmental issues and pressure to reduce the emission of pollutants have all led to the development of biomass conversion processes for production of biofuels. Oleaginous fungi are an alternative partial solution to the ever-increasing demand for fossil fuels. Their ability to produce hydrolytic enzymes helps them to grow on recalcitrant wastes which can be further used to produce single cell oils along with other coproducts like organic acids in a biorefinery. Other technologies to improve the overall process yield like pretreatment of substrate, enhanced recovery of lipid and mathematical modelling to study the techno-economics would pave the way for biorefineries which would be more feasible economically. In addition, recent approaches like next-generation sequencing (NGS) which reveals the entire genome of these fungi would be immensely useful to facilitate newer technologies like metabolic engineering, and directed evolution would help to increase the productivity of SCOs.

References

Aggelis G, Sourdis J (1997) Prediction of lipid accumulation-degradation in oleaginous microorganisms growing on vegetable oils. Antonie Van Leeuwenhoek 72:159–165

Aggelis G, Komaitis M, Papanikolaou S, Papadopoulos G (1995) A mathematical model for the study of lipid accumulation in oleaginous microorganisms. I. Lipid accumulation during growth of *Mucor circinelloides* CBS 172-27 on a vegetable oil. Gracas y Aceites 46:169–173

Ahmad FB, Zhang Z, Doherty WOS, Hara IMO, Crops T (2016) Evaluation of oil production from oil palm empty fruit bunch by oleaginous micro-organisms. Biofuels Bioprod Biorefin 10:378–392

Ali TH, El-Gamal MS, El-Ghonemy DH, Awad GE, Tantawy AE (2017) Improvement of lipid production from an oil-producing filamentous fungus, *Penicillium brevicompactum* NRC 829, through central composite statistical design. Ann Microbiol 67:601–613

Almeida JRM, Fávaro LC, Quirino BF (2012) Biodiesel biorefinery: opportunities and challenges for microbial production of fuels and chemicals from glycerol waste. Biotechnol Biofuels 5:48–64

André A, Diamantopoulou P, Philippoussis A, Sarris D, Komaitis M, Papanikolaou S (2010) Biotechnological conversions of bio-diesel derived waste glycerol into added-value compounds by higher fungi: production of biomass, single cell oil and oxalic acid. Ind Crop Prod 31:407–416

Aoki H, Miyamoto N, Furuya Y, Mankura M, Endo Y, Fujimoto K (2002) Incorporation and accumulation of docosahexaenoic acid from the medium by *Pichia methanolica* HA-32. Biosci Biotechnol Biochem 66:2632–2638

Aransiola EF, Ojumu TV, Oyekola OO, Madzimbamuto TF, Ikhu-Omoregbe DIO (2014) A review of current technology for biodiesel production: state of the art. Biomass Bioenergy 61:276–297

Asadi SZ, Khosravi-Darani K, Nikoopour H, Bakhoda H (2015) Evaluation of the effect of process variables on the fatty acid profile of single cell oil produced by *Mortierella* using solid-state fermentation. Crit Rev Biotechnol 35:94–102

Ayadi I, Kamoun O, Trigui-Lahiani H, Hdiji A, Gargouri A, Belghith H, Guerfali M (2016) Single cell oil production from a newly isolated *Candida viswanathii* Y-E4 and agro-industrial by-products valorization. J Ind Microbiol Biotechnol 43:901–914. https://doi.org/10.1007/s10295-016-1772-4

Azam MM, Waris A, Nahar NM (2005) Prospects and potential of fatty acid methyl esters of some non-traditional seed oils for use as biodiesel in India. Biomass Bioenergy 29:293–302

Bellou S, Moustogianni A, Makri A, Aggelis G (2012) Lipids containing polyunsaturated fatty acids synthesized by Zygomycetes grown on glycerol. Appl Biochem Biotechnol 166:146–158

Bellou S, Triantaphyllidou I, Aggeli D, Elazzazy A, Baeshen M, Aggelis G (2016) Microbial oils as food additives: recent approaches for improving microbial oil production and its polyunsaturated fatty acid content. Curr Opin Biotechnol 37:24–35

Beopoulos A, Cescut J, Haddouche R, Uribelarrea J (2009) Progress in lipid research *Yarrowia lipolytica* as a model for bio-oil production. Prog Lipid Res 48:375–387

Bharathiraja B, Sowmya V, Sridharan S, Yuvaraj D, Jayamuthunagai J, Praveenkumar R (2017) Biodiesel production from microbial oil derived from wood isolate *Trichoderma reesei*. Bioresour Technol 239:538–541

Canakci M, Sanli H (2008) Biodiesel production from various feedstocks and their effects on the fuel properties. J Ind Microbiol Biotechnol 35:431–441

Cao Y, Yao J, Chen X, Wu J (2010) Breeding of high lipid producing strain of *Geotrichum robustum* by ion beam implantation. Electron J Biotechnol. https://doi.org/10.2225/vol13-issue6-fulltext-4

Carvalho AKF, Rivaldi JD, Barbosa JC, De Castro HF (2015) Biosynthesis, characterization and enzymatic transesterification of single cell oil of *Mucor circinelloides* – a sustainable pathway for biofuel production. Bioresour Technol 181:47–53

Certik M, Shimizu S (1999) Biosynthesis and regulation of microbial polyunsaturated fatty acid production. J Biosci Bioeng 87:1–14

Chang Y-H, Chang K-S, Lee C-F, Hsu C-L, Huang C-W, Jang H-D (2015) Microbial lipid production by oleaginous yeast *Cryptococcus* sp. in the batch cultures using corncob hydrolysate as carbon source. Biomass Bioenergy 72:95–103

Chatzifragkou A, Fakas S, Galiotou-Panayotou M, Komaitis M, Aggelis G, Papanikolaou S (2010) Commercial sugars as substrates for lipid accumulation in *Cunninghamella echinulata* and *Mortierella isabellina* fungi. Eur J Lipid Sci Technol 112:1048–1057

Chatzifragkou A, Makri A, Belka A, Bellou S, Mavrou M, Mastoridou M, Mystrioti P, Onjaro G, Aggelis G, Papanikolaou S (2011) Biotechnological conversions of biodiesel derived waste glycerol by yeast and fungal species. Energy 36:1097–1108

Cheirsilp B, Kitcha S (2015) Solid state fermentation by cellulolytic oleaginous fungi for direct conversion of lignocellulosic biomass into lipids: fed-batch and repeated-batch fermentations. Ind Crop Prod 66:73–80

Chen H, Hao G, Wang L, Wang H, Gu Z, Liu L (2015) Identification of a critical determinant that enables efficient fatty acid synthesis in oleaginous fungi. Sci Rep 5:11247

Cohen Z, Ratledge C (eds) (2005) Single cell oils. AOCS Press, Champaign

Davies R (1988) Yeast oil from cheese whey; process development. In: Moreton R (ed) Single cell oil. Longman, London, pp 99–145

Demirbaş A (1998) Fuel properties and calculation of higher heating values of vegetable oils. Fuel 77:1117–1120

Demirbas A (2009) Progress and recent trends in biodiesel fuels. Energy Convers Manag 50:14–34

Dey P, Banerjee J, Maiti MK (2011) Comparative lipid profiling of two endophytic fungal isolates – *Colletotrichum* sp. and *Alternaria* sp. having potential utilities as biodiesel feedstock. Bioresour Technol 102:5815–5823

Díaz-Fernández D, Martínez PL, Buey RM, Revuelta JL, Jiménez A (2017) Utilization of xylose by engineered strains of *Ashbya gossypii* for the production of microbial oils. Biotechnol Biofuels 10:3

Dorado MP, Cruz F, Palomar JM, Lopez FJ (2006) An approach to the economics of two vegetable oil-based biofuels in Spain. Renew Energy 31:1231–1237

Economou C, Makri A, Aggelis G, Pavlou S, Vayenas DV (2010) Semi-solid state fermentation of sweet sorghum for the biotechnological production of single cell oil. Bioresour Technol 101:1385–1388

Economou C, Aggelis G, Pavlou S, Vayenas DV (2011) Modeling of single-cell oil production under nitrogen-limited and substrate inhibition conditions. Biotechnol Bioeng 108:1049–1055

Fakas S, Galiotou-panayotou M, Papanikolaou S, Komaitis M, Aggelis G (2007) Compositional shifts in lipid fractions during lipid turnover in *Cunninghamella echinulata*. Enzym Microb Technol 40:1321–1327

Fakas S, Papanikolaou S, Galiotou-Panayotou M, Komaitis M, Aggelis G (2008) Organic nitrogen of tomato waste hydrolysate enhances glucose uptake and lipid accumulation in *Cunninghamella echinulata*. J Appl Microbiol 105:1062–1070

Fakas S, Makri A, Mavromati M, Tselepi M, Aggelis G (2009a) Fatty acid composition in lipid fractions lengthwise the mycelium of *Mortierella isabellina* and lipid production by solid state fermentation. Bioresour Technol 100:6118–6120

Fakas S, Papanikolaou S, Batsos A, Galiotou-panayotou M, Mallouchos A, Aggelis G (2009b) Evaluating renewable carbon sources as substrates for single cell oil production by *Cunninghamella echinulata* and *Mortierella isabellina*. Biomass Bioenergy 33:573–580

Fickers P, Benetti P, Wache Y, Marty A, Mauersberger S, Smit M, Nicaud J (2005) Hydrophobic substrate utilisation by the yeast *Yarrowia lipolytica*, and its potential applications. FEMS Yeast Res 5:527–543

FitzPatrick M, Champagne P, Cunningham MF, Whitney RA (2010) A biorefinery processing perspective: treatment of lignocellulosic materials for the production of value-added products. Bioresour Technol 101:8915–8922

Forfang K, Zimmermann B, Kosa G, Kohler A, Shapaval V (2017) FTIR spectroscopy for evaluation and monitoring of lipid extraction efficiency for oleaginous fungi. PLoS One 12(1):e0170611

GAIN Report (2016) Number IN6088, India biofuels annual, 2016, USDA Foreign Agricultural Service

GAIN Report (2017) Number IN7075, India biofuels annual, 2017, USDA Foreign Agricultural Service

Gao D, Zeng J, Yu X, Dong T, Chen S (2014) Improved lipid accumulation by morphology engineering of oleaginous fungus *Mortierella Isabellina*. Biotechnol Bioeng 111:1758–1766

Gopinath A, Puhan S, Nagarajan G (2009) Theoretical modeling of iodine value and saponification value of biodiesel fuels from their fatty acid composition. Renew Energy 34:1806–1811

Hao G, Chen H, Wang L, Gu Z, Song Y, Zhang H, Chen W, Chen Q (2014) Role of malic enzyme during fatty acid synthesis in the oleaginous fungus *Mortierella alpina*. Appl Environ Microbiol 80:2672–2678. https://doi.org/10.1128/AEM.00140-14

Harde SM, Wang Z, Horne M, Zhu JY, Pan X (2016) Microbial lipid production from SPORL-pretreated Douglas fir by *Mortierella isabellina*. Fuel 175:64–74

Hawksworth DL (1991) The fungal dimension of biodiversity: magnitude, significance, and conservation. Mycol Res 95:641–655

Huang C, Chen X, Xiong L, Ma L, Chen Y (2013) Single cell oil production from low-cost substrates: the possibility and potential of its industrialization. Biotechnol Adv 31:129–139

Huang G, Zhou H, Tang Z, Liu H, Cao Y, Qiao D, Cao Y (2016) Novel fungal lipids for the production of biodiesel resources by *Mucor fragilis* AFT7-4. Environ Prog Sustain Energy 35:1784–1792

Kakkad H, Khot M, Zinjarde S, Ravikumar A (2015a) Biodiesel production by direct in situ transesterification of an oleaginous tropical mangrove fungus grown on untreated agro-residues and evaluation of its fuel properties. Bioenergy Res 8:1788–1799

Kakkad H, Khot M, Zinjarde S, Ravikumar A, Ravi Kumar V, Kulkarni BD (2015b) Conversion of dried *Aspergillus candidus* mycelia grown on waste whey to biodiesel by in situ acid transesterification. Bioresour Technol 197:502–507

Kamat S, Khot M, Zinjarde S, RaviKumar A, Gade WN (2013) Coupled production of single cell oil as biodiesel feedstock, xylitol and xylanase from sugarcane bagasse in a biorefinery concept using fungi from the tropical mangrove wetlands. Bioresour Technol 135:246–253

Katre G, Joshi C, Khot M, Zinjarde S, RaviKumar A (2012) Evaluation of single cell oil (SCO) from a tropical marine yeast *Yarrowia lipolytica* NCIM 3589 as a potential feedstock for biodiesel. AMB Express 2:36. https://doi.org/10.1186/2191-0855-2-36

Katre G, Ajmera N, Zinjarde S, RaviKumar A (2017) Mutants of *Yarrowia lipolytica* NCIM 3589 grown on waste cooking oil as biofactories for biodiesel production. Microb Cell Factories 16:176

Katre G, Raskar S, Ravi Kumar V, Kulkarni B, Zinjarde S, RaviKumar A (2018) Optimization of the *in situ* transesterification step for biodiesel production using biomass of *Yarrowia lipolytica* NCIM 3589 grown on waste cooking oil. Energy 142 944e952

Khot M, Kamat S, Zinjarde S, Pant A, Chopade B, Ravikumar A (2012) Single cell oil of oleaginous fungi from the tropical mangrove wetlands as a potential feedstock for biodiesel. Microb Cell Factories 11:71

Khot M, Gupta R, Barve K, Zinjarde S, Govindwar S, RaviKumar A (2015) Fungal production of single cell oil using untreated copra cake and evaluation of its fuel properties for biodiesel. J Microbiol Biotechnol 25:459–463

Kitcha S, Cheirsilp B (2014) Bioconversion of lignocellulosic palm byproducts into enzymes and lipid by newly isolated oleaginous fungi. Biochem Eng J 88:95–100

Knothe G (2005) Dependence of biodiesel fuel properties on the structure of fatty acid alkyl esters. Fuel Process Technol 86:1059–1070

Knothe G (2008) "Designer" biodiesel: optimizing fatty Ester composition to improve fuel properties. Energy Fuels 22:1358–1364

Knothe G (2009) Improving biodiesel fuel properties by modifying fatty ester composition. Energy Environ Sci 2:759–766

Knothe G, Steidley KR (2011) Kinematic viscosity of fatty acid methyl esters: prediction, calculated viscosity contribution of esters with unavailable data, and carbon – oxygen equivalents. Fuel 90:3217–3224

Kohlwein SD (2010) Triacylglycerol homeostasis: insights from yeast. J Biol Chem 285:15663–15667. https://doi.org/10.1074/jbc.R110.118356

Kosa M, Ragauskas AJ (2011) Lipids from heterotrophic microbes: advances in metabolism research FA. Trends Biotechnol 29:53–61

Kosa G, Kohler A, Tafintseva V, Zimmermann B, Forfang K, Afseth NK, Tzimorotas D, Vuoristo KS, Horn SJ, Mounier J, Shapaval V (2017) Microtiter plate cultivation of oleaginous fungi and monitoring of lipogenesis by high – throughput FTIR spectroscopy. Microb Cell Factories 16:101

Koutb M, Mohamed F (2011) A potent lipid producing isolate of *Epicoccum purpurascens* AUMC5615 and its promising use for biodiesel production. Biomass Bioenergy 35:3182–3187

Krawczyk T (1996) Biodiesel. In: International news on fats, oils and related materials. AOCS Press, Champaign, p 801

Kumar AK, Vatsyayan P, Goswami P (2010) Production of lipid and fatty acids during growth of *Aspergillus terreus* on hydrocarbon substrates, Appl. Biochem.Biotechnol !60: 1293–1300

Lapuerta M, Rodríguez-fernández J, Armas O (2010) Correlation for the estimation of the density of fatty acid esters fuels and its implications. A proposed Biodiesel Cetane Index. Chem Phys Lipids 163:720–727

Leung DYC, Wu X, Leung MKH (2010) A review on biodiesel production using catalyzed transesterification. Appl Energy 87:1083–1095

Li Q, Du W, Liu D (2008) Perspectives of microbial oils for biodiesel production. Appl Microbiol Biotechnol 80:749–756

Lin H, Cheng W, Ding HT, Chen XJ, Zhou QF, Zhao YH (2010) Direct microbial conversion of wheat straw into lipid by a cellulolytic fungus of *Aspergillus oryzae* A-4 in solid-state fermentation. Bioresour Technol 101:7556–7562

Lin L, Cunshan Z, Vittayapadung S, Xiangqian S, Mingdong D (2011) Opportunities and challenges for biodiesel fuel. Appl Energy 88:1020–1031

Liu B, Zhao Z (2007) Biodiesel production by direct methanolysis of oleaginous microbial biomass. J Chem Technol Biotechnol 82:775–780

Liu S, Abrahamson LP, Scott GM (2012) Biorefinery: ensuring biomass as a sustainable renewable source of chemicals, materials, and energy. Biomass Bioenergy 39:1–4

Lozano-Martínez P, Buey RM, Ledesma-Amaro R, Jiménez A, Revuelta JL (2017) Engineering *Ashbya gossypii* strains for de novo lipid production using industrial by-products. Microb Biotechnol 10:425–433

Meeuwse P, Tramper J (2011) Modeling lipid accumulation in oleaginous fungi in chemostat cultures. II: validation of the chemostat model using yeast culture data from literature. Bioprocess Biosyst Eng 34:951–961

Meeuwse P, Tramper J, Rinzema A (2011) Modeling lipid accumulation in oleaginous fungi in chemostat cultures: I. Development and validation of a chemostat model for *Umbelopsis isabellina*. Bioprocess Biosyst Eng 34:939–949

Meeuwse P, Akbari P, Tramper J, Rinzema A (2012a) Modeling growth, lipid accumulation and lipid turnover in submerged batch cultures of *Umbelopsis isabellina*. Bioprocess Biosyst Eng 35:591–603

Meeuwse P, Klok AJ, Haemers S, Tramper J, Rinzema A (2012b) Growth and lipid production of *Umbelopsis isabellina* on a solid substrate — mechanistic modeling and validation. Process Biochem 47:1228–1242

Meeuwse P, Sanders JPM, Tramper J, Rinzema A (2013) Lipids from yeasts and fungi: tomorrow's source of biodiesel? Biofuels Bioprod Biorefin 7:521–524

Moser BR (2009) Biodiesel production, properties, and feedstocks. In vitro cellular and developmental biology. Plants 45:229–266

Moser BR, Vaughn SF (2012) Efficacy of fatty acid profile as a tool for screening feedstocks for biodiesel production. Biomass Bioenergy 37:31–41

Muniraj IK, Xiao L, Hu Z, Zhan X, Shi J (2013) Microbial lipid production from potato processing wastewater using oleaginous filamentous fungi *Aspergillus oryzae*. Water Res 47:3477–3483

NREL (2009) Biodiesel Handling and Use Guide 4th edition National Renewable EnergyLaboratory.http://biodiesel.org/docs/using-hotline/nrel-handling-and-use.pdf

Papanikolaou S, Aggelis G (2010) *Yarrowia lipolytica*: a model microorganism used for the production of tailor-made lipids. Eur J Lipid Sci Technol 112:639–654

Papanikolaou S, Aggelis G (2011a) Lipids of oleaginous yeasts. Part I: biochemistry of single cell oil production. Eur J Lipid Sci Technol 113:1031–1051

Papanikolaou S, Aggelis G (2011b) Lipids of oleaginous yeasts. Part II: Technology and potential applications. Eur J Lipid Sci Technol 113:1052–1073

Papanikolaou S, Chevalot I, Komaitis M, Aggelis G, Marc I (2001) Kinetic profile of the cellular lipid composition in an oleaginous *Yarrowia lipolytica* capable of producing a cocoa-butter substitute from industrial fats. Antonie Van Leeuwenhoek 80:215–224

Papanikolaou S, Chevalot I, Komaitis M, Marc I, Aggelis G (2002) Single cell oil production by *Yarrowia lipolytica* growing on an industrial derivative of animal fat in batch cultures. Appl Microbiol Biotechnol 58:308–312

Papanikolaou S, Muniglia L, Chevalot I, Aggelis G, Marc I (2003) Accumulation of a cocoa-butter-like lipid by *Yarrowia lipolytica* cultivated on agro-industrial residues. Curr Microbiol 46:0124–0130

Papanikolaou S, Komaitis M, Aggelis G (2004a) Single cell oil (SCO) production by *Mortierella isabellina* grown on high-sugar content media. Bioresour Technol 95:287–291

Papanikolaou S, Sarantou S, Komaitis M, Aggelis G (2004b) Repression of reserve lipid turnover in *Cunninghamella echinulata* and *Mortierella isabellina* cultivated in multiple-limited media. J Appl Microbiol 97:867–875

Papanikolaou S, Galiotou-Panayotou M, Fakas S, Komaitis M, Aggelis G (2007) Lipid production by oleaginous Mucorales cultivated on renewable carbon sources. Eur J Lipid Sci Technol 109:1060–1070

Papanikolaou S, Diamantopoulou P, Chatzifragkou A, Philippoussis A, Aggelis G (2010) Suitability of low-cost sugars as substrates for lipid production by the fungus *Thamnidium elegans*. Energy Fuel 24:4078–4086

Papanikolaou S, Dimou A, Fakas S, Diamantopoulou P, Philippoussis A (2011) Biotechnological conversion of waste cooking olive oil into lipid-rich biomass using *Aspergillus* and *Penicillium* strains. J Appl Microbiol 110:1138–1150

Papanikolaou S, Rontou M, Belka A, Athenaki M, Gardeli C, Mallouchos A, Kalantzi O, Koutinas A, Kookos I, Zeng A, Aggelis G (2017) Conversion of biodiesel-derived glycerol into biotechnological products of industrial significance by yeast and fungal strains. Eng Life Sci 17:262–281

Peng X, Chen H (2007) Microbial oil accumulation and cellulase secretion of the endophytic fungi from oleaginous plants. Ann Microbiol 57:239–242

Peng X, Chen H (2008) Single cell oil production in solid-state fermentation by *Microsphaeropsis* sp. from steam-exploded wheat straw mixed with wheat bran. Bioresour Technol 99:3885–3889

Pignède G, Wang H, Fudalej F, Seman M, Gaillardin C, Nicaud J (2000) Autocloning and amplification of LIP2 in *Yarrowia lipolytica*. Appl Environ Microbiol 66:3283–3289

Pinzi S, Leiva-Candia D, López-García I, Redel-Macías MD, Dorado MP (2013) Latest trends in feedstocks for biodiesel production. Biofuels Bioprod Biorefin 8:126–143

Pratas MJ, Freitas S, Oliveira MB, Monteiro SC, Lima AS, Coutinho JAP (2010) Densities and viscosities of fatty acid methyl and ethyl esters. J Chem Eng Data 55:3983–3990

Pratas MJ, Freitas SVD, Oliveira MB, Monteiro SC, Lima S, Coutinho JAP (2011) Biodiesel density: experimental measurements and prediction models. Energy Fuel 25:2333–2340

Ramírez-Verduzco LF, Rodríguez-Rodríguez JE, Jaramillo-Jacob ADR (2012) Predicting cetane number, kinematic viscosity, density and higher heating value of biodiesel from its fatty acid methyl ester composition. Fuel 91:102–111. https://doi.org/10.1016/j.fuel.2011.06.070

Ramos MJ, Fernández CM, Casas A, Rodríguez L, Pérez Á (2009) Influence of fatty acid composition of raw materials on biodiesel properties. Bioresour Technol 100:261–268

Ratledge C (1989) Biotechnology of oils and fats. In: Ratledge C, Wilkinson S (eds) Microbial Lipids. Academic, London, pp 567–668

Ratledge C (2014) The role of malic enzyme as the provider of NADPH in oleaginous microorganisms: a reappraisal and unsolved problems. Biotechnol Lett 36:1557–1568

Ratledge C, Wynn J (2002) The biochemistry and molecular biology of lipid accumulation in oleaginous microorganisms. Adv Appl Microbiol 51:1–52

REN21 – Renewable Energy Policy Network for the 21st century (2016) Renewables 2016 Global Status Report. http://www.ren21.net/status-of-renewables/global-status-report/

REN21 – Renewable Energy Policy Network for the 21st century (2017) Renewables 2017 Global Status Report. http://www.ren21.net/status-of-renewables/global-status-report/

Rodríguez-Frómeta RA, Gutiérrez A, Torres-Martínez S, Garre V (2013) Malic enzyme activity is not the only bottleneck for lipid accumulation in the oleaginous fungus Mucor circinelloides. Appl Microbiol Biotechnol 97:3063–3072

Ruan Z, Zanotti M, Wang X, Ducey C, Liu Y (2012) Evaluation of lipid accumulation from lignocellulosic sugars by Mortierella isabellina for biodiesel production. Bioresour Technol 110:198–205

Ruan Z, Zanotti M, Archer S, Liao W, Liu Y (2014) Oleaginous fungal lipid fermentation on combined acid- and alkali-pretreated corn Stover hydrolysate for advanced biofuel production. Bioresour Technol 163:12–17

Sakamoto T, Sakuradani E, Okuda T, Kikukawa H, Ando A (2017) Metabolic engineering of oleaginous fungus Mortierella alpina for high production of oleic and linoleic acids. Bioresour Technol (in press) https://doi.org/10.1016/j.biortech.2017.06.089

Song Y, Wynn JP, Li Y, Grantham D, Ratledge C (2001) A pre-genetic study of the isoforms of malic enzyme associated with lipid accumulation in Mucor circinelloides. Microbiology 147:1507–1515

Subramaniam R, Dufreche S, Zappi M, Bajpai R (2010) Microbial lipids from renewable resources: production and characterization. J Ind Microbiol Biotechnol 37:1271–1287

Tamano K, Miura A, Koike H, Kamisaka Y, Umemura M (2017) High-efficiency extracellular release of free fatty acids from Aspergillus oryzae using non-ionic surfactants. J Biotechnol 248:9–14

Tang X, Chen H, Chen YQ, Chen W, Garre V, Song Y, Ratledge C (2015) Comparison of biochemical activities between high and low lipid-producing strains of Mucor circinelloides: an explanation for the high oleaginicity of strain WJ11. PLoS One 10(6):e0128396

Tong D, Hu C, Jiang K (2011) Cetane number prediction of biodiesel from the composition of the fatty acid methyl esters. J Am Oil Chem Soc 88:415–423

Vamvakaki AN, Kandarakis I, Kaminarides S, Komaitis M, Papanikolaou S (2010) Cheese whey as a renewable substrate for microbial lipid and biomass production by Zygomycetes. Eng Life Sci 10:348–360

Venkata Subhash G, Venkata Mohan S (2011) Biodiesel production from isolated oleaginous fungi Aspergillus sp. using corncob waste liquor as a substrate. Bioresour Technol 102:9286–9290

Venkata Subhash G, Venkata Mohan S (2014) Lipid accumulation for biodiesel production by oleaginous fungus Aspergillus awamori: influence of critical factors. Fuel 116:509–515

Venkata Subhash G, Venkata Mohan S (2015) Sustainable biodiesel production through bioconversion of lignocellulosic wastewater by oleaginous fungi. Biomass Convers Biorefinery 5:215–226

Vicente G, Bautista LF, Rodriguez R, Gutiérrez FJ, Sadaba I, Ruiz-Vazquez RM, Torres-Martinez S, Garre V (2009) Biodiesel production from biomass of an oleaginous fungus. Biochem Eng J 48:22–27

Vicente G, Bautista LF, Gutierrez FJ, Rodríguez R, Martínez V, Rodríguez-Frometa RA, Ruiz-Vazquez RM, Torres-Martínez S, Garre V (2010) Direct transformation of fungal biomass from submerged cultures into biodiesel. Energy Fuels 24:3173–3178

Wang Y, Lu Z (2005) Optimization of processing parameters for the mycelial growth and extracellular polysaccharide production by *Boletus* spp. ACCC 50328. Process Biochem 40:1043–1051

Wei H, Wang W, Yarbrough JM, Baker JO, Laurens L, Van WS, Chen X, Ii LET, Xu Q, Himmel ME, Zhang M (2013) Genomic, proteomic and biochemical analyses of oleaginous mucor circinelloides: evaluating its capability in utilizing cellulolytic substrates for lipid production. PLoS One 8:e71068

Wei Y, Siewers V, Nielsen J (2017) Cocoa butter-like lipid production ability of non-oleaginous and oleaginous yeasts under nitrogen-limited culture conditions. Appl Microbiol Biotechnol (2017) 101:3577–3585

Wynn JP, Ratledge C (1997) Malic enzyme is a major source of NADPH for lipid accumulation by *Aspergillus nidulans*. Microbiology 143:253–257

Wynn JP, Hamidt A, Ratledge C (1999) The role of malic enzyme in the regulation of lipid accumulation in filamentous fungi. Microbiology 145:1911–1917

Wynn JP, Hamid AA, Li Y, Ratledge C (2001) Biochemical events leading to the diversion of carbon into storage lipids in the oleaginous fungi *Mucor circinelloides* and *Mortierella alpina*. Microbiology 147:2857–2864

Xing D, Wang H, Pan A, Wang J, Xue D (2012) Assimilation of corn fiber hydrolysates and lipid accumulation by *Mortierella isabellina*. Biomass Bioenergy 39:494–501

Yao R, Zhang P, Wang H, Deng S, Zhu H (2012) One-step fermentation of pretreated rice straw producing microbial oil by a novel strain of *Mortierella elongata* PFY. Bioresour Technol 124:512–515

Yousuf A (2012) Biodiesel from lignocellulosic biomass - prospects and challenges. Waste Manag 32:2061–2067

Zhang J, Hu B (2012) Solid-state fermentation of *Mortierella isabellina* for lipid production from soybean hull. Appl Biochem Biotechnol 166:1034–1046

Zhang J, Hu B (2014) Microbial lipid production from corn Stover via *Mortierella isabellina*. Appl Biochem Biotechnol 174:574–586

Zhang Y, Dubé MA, McLean DD, Kates M (2003) Biodiesel production from waste cooking oil: 2. Economic assessment and sensitivity analysis. Bioresour Technol 90:229–240

Zhang Y, Adams IP, Ratledge C (2007) Malic enzyme: the controlling activity for lipid production? Overexpression of malic enzyme in *Mucor circinelloides* leads to a 2.5-fold increase in lipid accumulation. Microbiology 153:2013–2025

Zhang C, Shen H, Zhang X, Yu X, Wang H, Xiao S et al (2016) Combined mutagenesis of *Rhodosporidium toruloides* for improved production of carotenoids and lipids. Biotechnol Lett 38:1733–1738

Zheng Y, Yu X, Zeng J, Chen S (2012) Feasibility of filamentous fungi for biofuel production using hydrolysate from dilute sulfuric acid pretreatment of wheat straw. Biotechnol Biofuels 5:50

Chapter 9
Fungal Biorefinery for the Production of Single Cell Oils as Advanced Biofuels

Abu Yousuf, Baranitharan Ethiraj, Maksudur Rahman Khan, and Domenico Pirozzi

Abstract Biofuel production from edible substrate raises a social conflict because it hikes the food price, occupies the arable land and competes with the food industry. Therefore, microbial oils or single cell oils (SCOs) have attracted the attention as suitable alternative of triglycerides. A variety of microbes have been studied such as microalgae, bacteria, yeast and fungi. A comprehensive study for the functionality of fungi in the purpose of biofuel synthesis is not addressed yet. This chapter describes the biorefinery concept of fungi to biofuels, which includes their culture techniques, culture medium, growth, SCO extraction and transesterification of SCO. It also compares the first-generation to third-generation biofuels.

9.1 Introduction

During the nineteenth century, Henry Ford and Rudolf Diesel demonstrated the operation of internal-combustion engines on plant-based feedstocks. As days passed, the attention towards the oil production from plant-based feedstock was decreased due to the availability of petroleum and petrol-based products. But, in the past few decades, there had been a growing interest towards biofuels due to the

A. Yousuf (✉)
Department of Chemical Engineering & Polymer Science, Shahjalal University of Science and Technology, Sylhet, Bangladesh
e-mail: ayousuf-cep@sust.edu

B. Ethiraj
Department of Biotechnology, Bannari Amman Institute of Technology, Sathyamangalam, Erode District, India

M. R. Khan
Faculty of Chemical and Natural Resources Engineering, University Malaysia Pahang, Gambang, Malaysia

D. Pirozzi
Department of Chemical Engineering, Materials and Industrial Production, University Naples Federico II, Naples, Italy

© Springer International Publishing AG, part of Springer Nature 2018
S. Kumar et al. (eds.), *Fungal Biorefineries*, Fungal Biology,
https://doi.org/10.1007/978-3-319-90379-8_9

185

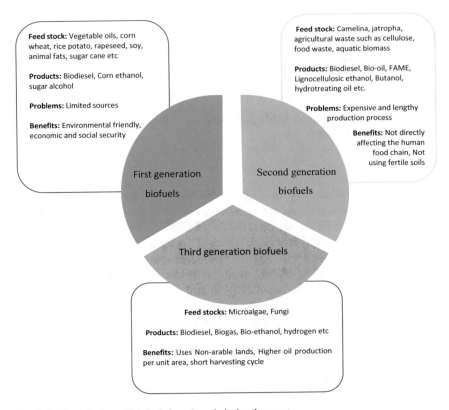

Feed stock: Vegetable oils, corn wheat, rice potato, rapeseed, soy, animal fats, sugar cane etc

Products: Biodiesel, Corn ethanol, sugar alcohol

Problems: Limited sources

Benefits: Environmental friendly, economic and social security

Feed stock: Camelina, jatropha, agricultural waste such as cellulose, food waste, aquatic biomass

Products: Biodiesel, Bio-oil, FAME, Lignocellulosic ethanol, Butanol, hydrotreating oil etc.

Problems: Expensive and lengthy production process

Benefits: Not directly affecting the human food chain, Not using fertile soils

First generation biofuels

Second generation biofuels

Third generation biofuels

Feed stocks: Microalgae, Fungi

Products: Biodiesel, Biogas, Bio-ethanol, hydrogen etc

Benefits: Uses Non-arable lands, Higher oil production per unit area, short harvesting cycle

Fig. 9.1 Classification of biofuels based on their development

strategic and environmental challenges associated with the oil economy (Cherubini 2010; Naik et al. 2010). This is because these biofuels have considerable environmental benefits since they are found to emit less greenhouse gases than conventional fuels (De Gorter and Just 2010). Several research studies have reported that the greenhouse gas balance leans on the way feedstocks are produced, biofuel processing and its distribution (Cherubini et al. 2009; Gnansounou et al. 2009). The first-generation biofuels are obtained from raw materials such as ethanol made from corn or sugarcane, but these feedstocks are basically in competition with food and feed companies. The development of second-generation biofuels (Fig. 9.1) attracted worldwide interest in the last three decades since they are a likely alternative to conventional fuels and also they overcome the limitations of first-generation biofuels. Unfortunately, the technology to make these newer fuels is still in development. Other important factors to be considered are emission of greenhouse gases, energy balances, the impact of biomass production on the environment and the potential rivalry with food production. The common sources of second-generation biofuels are camelina and jatropha, cellulose, etc. The third-generation biofuels are obtained from oleaginous material derived from microorganisms (microalgae, yeasts, bacteria) capable of growing (photo-)heterotrophically on organic waste or phototrophically on inorganic carbon (Muller et al. 2014). The research and development of third-generation biofuels mainly focused on algae and yeasts.

The main advantages offered by third-generation biofuels are as follows:

- Biomass can be cultivated on nonarable land.
- Oil produced 15–300 times more than per unit of area than typical terrestrial crops.
- Short harvesting cycle (approx 1–10 days).
- Cultivation of biomass does not depend on climate and season.

9.1.1 Microbial Oil

Since algae are cultivated in water, these species overcome one of the real issues confronted by different biofuel feedstocks – rivalry for area with terrestrial crop yields. Algae can accumulate up to 50% lipids into their cell, which can later be extracted as oil for biodiesel production. The fast growth rate of algae helps to meet high energy demand in the future. However, excessive production of algae causes massive die-offs, as crowded algae cells may not receive enough light to perform photosynthesis. Currently, there are only few business makers of algal oil since none of the present technology has the capacity to control and maintain the algae growth. On the other hand, from 1980, yeasts and fungi are considered as oleaginous microorganisms (Ratledge 1993). Among the yeasts, few strains like *Lipomyces* sp. and *Rhodosporidium* sp. are able to accumulate lipids up to 70% of their biomass dry weight, while fungus like *Mortierella alliacea* is able to yield around 46 g/L dry cell weight, 7 g/L arachidonic acid and 19.5 g/L total fatty acid (Takeno et al. 2005). The oleaginous yeast could be considered as one of the potential alternative oil resources to meet the growing energy demands in the future. However, not all of them are suitable for oil production.

9.2 Biorefineries

9.2.1 Biorefinery Concept

The biorefinery is a facility providing complete downstream processing solutions to produce biomass-based biofuels, using a variety of different technologies. Consequently, the concept of biorefinery, now widely accepted, is analogous to that of petroleum refinery (Fig. 9.2a), except that it utilizes renewable feedstock (derived from plants), whereas an oil refinery is produced from non-renewable feedstocks.

The technologies based upon the concept of biorefinery can offer many bioproducts (Fig. 9.2b) that include biofuels (bioethanol, biodiesel, biogas and biomethane), biomaterials (fibres, pulp) and a range of bioactive compounds through downstream fermentation and refining processes, thus increasing the value derived from the biomass feedstock.

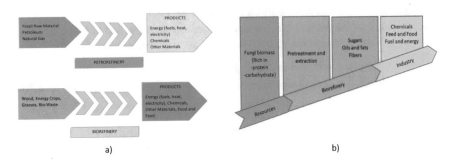

Fig. 9.2 (**a**) Comparison of the basic principles of the petroleum refinery and the biorefinery. (**b**) Biorefinery of fungal biomass

9.2.2 Development of Biorefineries as an Alternative to Petroleum Refineries

Energy security, climate change and rural development are the least three distinct advantages of a biorefinery using biofuel feedstocks compared to petrochemical feedstocks (Cherubini 2010). Plant biomass is a renewable energy source, whereas fossil fuels have a limited quantity of supply, which may deplete during this century based upon its utilization as an energy resource. Plant-based feedstock will therefore increase our energy security and reduce the dependency on crude oil (non-renewable).

Biorefineries give solution to reduce waste streams and lower the pollution, assisting against climate change by reducing the final products of fossil fuel combustion released to the atmosphere (Cherubini 2010). In addition, a biorefinery-based economy will promote the rural development, creating new business ventures and employments for agricultural-based countries.

On comparison with petrol-based biofuels, biomass generally has more oxygen, less carbon and negligible amount of hydrogen. The compositional properties of biofuel feedstock have an advantage together with a disadvantage. An advantage is that biorefineries can generate different bioproducts that petroleum refineries can depend on broad range of raw materials. A disadvantage is that, with the need of relatively advanced processing technologies, most of these technologies are still at a pre-commercial stage, though they are being rapidly developed.

9.2.3 Non-food Agriculture

The recent evolution of the concept of biorefinery has also been affected by the concept of 'non-food agriculture'. With the increasing demand of food, it is a questionable practice to replace the edible feedstocks with non-edible feedstock. Instead, a non-food material such as cellulose offers an alternative feedstock resulting in the production of different products such as bio-based chemicals and biofuels. The commercialization of raw materials from agriculture and forestry resources for the biofuel industry is still in the early stages. Approximately 6 billion tons of biomass generated annually are utilized, from which about 3.0–3.5% is utilized in the non-food area, for example, chemistry (Zoebelin 1996).

9.2.3.1 Biorefining as a New Science

Biorefining is still unexplored to a great extent and requires further research studies and therefore has opportunities to open doors for the commercialization of bio-based products from arable areas. As of now, petroleum chemistry depends mostly to produce simple hydrocarbons and stable chemical compounds in biorefineries. But a complete biorefinery is focused on the production of basic chemicals, inter-mediates and final complex products.

This rule of petroleum refineries must be reassigned to biorefineries. Biofuel feedstocks are able to synthesis products naturally and contain another C: H: O: N proportion than petroleum. The process of biomass origin should be modified to adapt the produced biomass for the use of successive downstream processing and desirable target products already have been formed.

Plant biomass contains the following major components: carbohydrates, lignin, proteins and fats, apart from minor components such as vitamins, dyes, flavours and aromatic compounds. Biorefineries unite the vital technologies between the raw materials, intermediates and final products.

It is noted that there are few technically feasible separation techniques, which would permit the extraction of basic bioactive compounds and subsequent down-stream processing of these basic compounds, existing until now merely in the form of initial studies. Assuming 170 billion tons of biomass produced by biosynthesis annually, this biomass contains 75% carbohydrates (existing mainly in the form of starch, cellulose and saccharose), 20% lignin and 5% remaining compounds such as fats (oils), proteins and few other substances. However, the main focus is shifted to carbohydrates and their subproducts. Glucose, extracted via chemical or microbial method, from starch, sugar or cellulose, is fated for a key position as a fundamental compound, on the grounds that an expansive palette of chemical or biotechnological products can easily be obtained from glucose.

On account of starch, the benefit of enzymatic hydrolysis than conventional chemical hydrolysis is today effectively figured out. But, in the cellulose case, this is not yet figured out. Cellulose hydrolysing proteins can alone act better after pretreatment, to separate the exceptionally stable lignin/cellulose/hemicellulose composites. These pretreatments are still temperature related, and so they require a large quantity of energy. The treatments for microbial change of substances out of glucose are expansive; the final products are vigorously productive. It is important to consolidate the degradation processes by means of glucose to mass chemicals with the building procedures to their ensuing items and materials (Kamm et al. 2006a).

Bio-based items are set up for an economic use by a sensible combination of various techniques and procedures (physical, chemical, thermal and biological). Therefore, biorefining is an interdisciplinary science including the connection between biology (microbiology), chemistry (natural chemistry), engineering (bio/chemical and procedure) and mathematics (material science). It is subsequently much like biotechnology and in this admiration applies the standards of the fundamental sciences and is intensely dependent upon enzyme and fermentation technology. Therefore, it gives off an impression of being sensible to explain the term 'biorefinery design' which implies uniting experimental and technologic basics with initial advancements, products and product offerings inside the biorefineries.

More attention must be given to the blend of chemical and biotechnological substance changeover and the required vitality contribution for the transformation. The fundamental changes of each biorefinery can be outlined as follows:

As of now, four complex biorefinery frameworks are investigated for research and development:

1. The 'green biorefineries' utilizing 'nature-wet' biomasses, for example, green grass, horse feed, clover or juvenile grain (Kromus et al. 2006).
2. The 'lignocellulosic feedstock biorefinery' utilizing 'nature-dry' biomasses, for example, cellulose-containing biomass and wastes (Koutinas et al. 2006).
3. The 'entire crop biorefinery' utilizes crude material, for example, oats or maize (Kamm et al. 2006b).
4. The 'biorefinery two platforms concept' which comprises two platforms: sugar platform and syngas platform (Werpy and Petersen 2004).

9.3 Single Cell Oils (SCOs)

In recent years, single cell oils (SCOs) or microbial oils have been significantly considered among the researchers due to its specific characteristics and functions. SCOs are the triglycerides produced by oleaginous microbes such as bacteria, microalgae, moulds and yeasts (Fig. 9.3). Typically microorganisms that are capable to accumulate more than 20% of their dry weight are considered as oleaginous microorganisms (Ratledge 1991). Until 1980, the biochemistry and lipid metabolism by oleaginous microorganisms were the main focus of many researchers.

During last two decades, biochemical process involved in SCO accumulation was given more attention as the SCOs play an important function on human health through replacing valuable materials, e.g. cocoa butter (Beopoulos et al. 2009). The biochemical mechanisms involved in the accumulation of lipids by microbes were well studied in those years (Evans and Ratledge 1983; Holdsworth and Ratledge 1991; Alvarez and Steinbüchel 2002; Ratledge 2004). Moreover, the key enzymes and the expression pathways for lipid accumulation were also involved in those

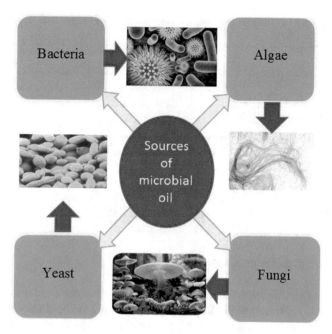

Fig. 9.3   Sources of microbial oils or SCOs

studies (Wynn et al. 1997; Polakowski et al. 1999; Zhang et al. 2007). The SCO attention further increased on medical benefits of polyunsaturated fatty acids (PUFA) such as docosahexaenoic acid (DHA), eicosapentaenoic acid (EPA), arachidonic acid (ARA) and γ-linolenic acid (GLA) (Ward and Singh 2005). It can be used as an alternative lipid substrate for biodiesel production as it is cheap in comparison to traditional vegetable oils (Li et al. 2008). Thus, it clearly shows that, in the current world, the SCO plays a pivotal role in providing solution to depleting fossil fuels.

9.3.1   SCO from Fungi

9.3.1.1   Microbial Types

The application of lignocellulosic hydrolysates by yeast (*Pichia stipitis* and *Saccharomyces cerevisiae)* and fungi *(Trichoderma reesei* and *Aspergillus niger*) as raw material for different industrial fermentations was assessed by Rumbold et al. (2009). Additionally, accumulation of lipids by oleaginous yeast and fungi on distinct substrates such as whey, glycerol, sewage waste and molasses was reassessed (Subramaniam et al. 2010). Recently, *Acinetobacter baylyi* ADP1, a metabolic engineered strain for the enhanced triacylglycerol production, were also studied (Santala et al. 2011). Table 9.1 summarizes some well-known strains, which have been studied for the production of SCO.

Table 9.1 Oil content of microorganisms on a dry weight basis (Meng et al. 2009; Zheng et al. 2012; Yousuf et al. 2014)

Strain	Substrate	Lipid content (%)	Lipid concentration (g L^{-1})	Lipid yield (mg g^{-1})
Aspergillus .niger	Glucose	9.6	0.55	21
	Xylose	8.0	0.37	15
Mortierella isabellina	Glucose	67.0	4.88	195
	Xylose	50.9	2.52	121
Mucor circinelloides	Glucose	23.8	0.95	57
	Xylose	17.3	0.63	36
Cunninghamella elegans	NDLH	17.0	0.80	38
	DLH	23.1	0.88	50
Rhizopus oryzae	NDLH	16.1	0.84	38
	DLH	19.7	0.77	45

9.3.1.2 Performance to Accumulate Lipids

The important specification for an oleaginous microorganism in order to consider as an appropriate substrate for biodiesel production is that the lipid content present in the substrate must be greater than 20% dry weight. Besides that, the fatty acid type must be either long-chain saturated or monounsaturated. The source of carbon type, pH, temperature and the strain type (species specific) have an influence over the fatty acid composition and lipid content (Subramaniam et al. 2010). This was evidently proved in the studies that dealt with effect of glucose and temperature on the composition and degree of unsaturation of fatty acids using psychrophilic oleaginous yeast *Rhodotorula glacialis* (Amaretti et al. 2010). Growth of *Mortierella* species such as *M. alpine and M. isabellina* ATHUM 2935 on glucose substrate is capable to produce 40% (w/w) and 50.4% (w/w) oil, respectively (Wynn et al. 2001; Papanikolaou et al. 2004). The n-hexadecane and lipid production were also found to be 4.43 and 0.62%, respectively, with the similar feedstock (Kumar et al. 2010).

9.3.2 Substrates for Single Cell Oil Production

9.3.2.1 Single Cell Oil Production from Hydrophilic Materials

Molasses

Industrial sugarcane molasses or beet molasses from sugar processing plants generally contain monosaccharides and disaccharides such as glucose, fructose and sucrose. These sugars were found to be suitable raw material for cost-effective media formulation and were extensively used in ethanol fermentation and hydrogen and lactic acid production (Doelle and Doelle 1990; Tanisho and Ishiwata 1995; Wee et al. 2004). Moreover, Bednarski et al. (1986) proved that these substrates can

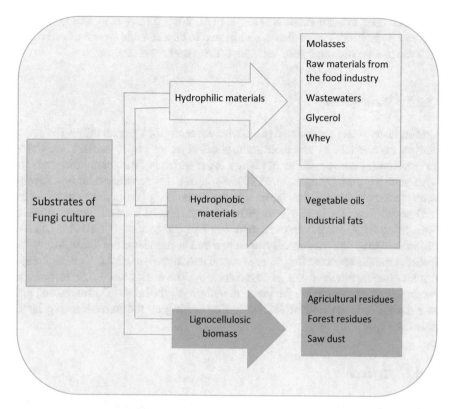

Fig. 9.4 Promising substrates for single cell oil production

also be used for SCO production. Chatzifragkou et al. (2010) used *Cunninghamella echinulata* to demonstrate decolourization and efficient utilization of waste molasses for SCO production. But, high levels of nitrogen in the molasses inhibit its growth and accumulation of lipids in oleaginous microorganisms (Zhu et al. 2008). Potential nutrient sources are summarized in Fig. 9.4.

9.3.3 Raw Materials from the Food Industry

The starch hydrolysates from food industry can be used as sole carbon source for oleaginous microorganisms because they contain distinct fermentable sugars. For instance, Zhu et al. (2003) used *Mortierella alpine strain and* produced 1. 47 g/L of arachidonic acid using starch hydrolysate, and Vega et al. (1988) used the banana juice as a raw material for SCO production. Furthermore, Fakas et al. (2007) examined and proved the effect of nitrogen from tomato hydrolysates on enhancing the lipid accumulation and the glucose uptake using *Cunninghamella echinulata* species. Recent studies showed that shrimp hydrolysates rich in N-acetylglucosamine have been used to accumulate SCO (Wu et al. 2010; Zhang et al. 2011). Finally, sorghum hydrolysates were also demonstrated to be a suitable source of carbon for SCO production in *Mortierella isabellina* (Economou et al. 2011a).

9.3.3.1 Wastewaters

Agro-industry wastewaters have gained more interest for biomass utilization due to their greater availability and their harmful effect on the environment. Such wastewater treatment consumes lots of energy input which in turn greatly increases the manufacturing cost (Yang et al. 2005). As agro-industry wastewater and sewage water are enriched in polysaccharides, they may possibly be a carbon source for the SCO production. For instance, Angerbauer et al. (2008) showed accumulation of lipids using sewage water fermentation. Similarly, Yousuf et al. (2010) used olive mill wastewaters and successfully converted and accumulated the lipids. Moreover, lipids for biodiesel production were accumulated in the bacterial cells using monosodium glutamate wastewater as a feedstock by Xue et al. (2006). The researchers also investigated to improve the yield by optimizing the levels of nitrogen and glucose during lipid accumulation followed by scaling up the fermentor to a larger scale (300 L).

9.3.3.2 Glycerol

Traditionally, glycerol is produced through chemical reaction and biological fermentation; however, it can be obtained from the soap industries too (Wang et al. 2001). The substitution of soap by detergents increased the glycerol price again in the market. But alternatively, the glycerol can be produced cheaply in large amount during the biodiesel production. Biodiesel is known as fatty acid methyl ester produced when the triglycerides react with ethanol or methanol producing glycerol as a by-product. Consequently, the biodiesel industry makes glycerol as an inexpensive raw material for SCO production and lipid fermentation (Papanikolaou and Aggelis 2002). They also reported the SCO production using *Yarrowia lipolytica* species from the agro-waste and glycerol. In addition to that, citric acid and 1,3-propanediol production by *Mortierella isabellina* was also reported using glycerol as a carbon substrate (Chi et al. 2007). Pyle et al. (2008) showed DHA production using oleaginous *Schizochytrium limacinum* algae species.

9.3.3.3 Whey

Generally, whey is generated as by-product from cheese manufacturing industry. The whey is found to be used as an attractive and prominent substrate for SCO production and it's highly generated in the USA and Europe (Ahn et al. 2000; Koller et al. 2005). Moreover, many researchers reported that oleaginous microbes are capable to utilize whey as a sole carbon source for SCO production (Ykema et al. 1988, 1990; Akhtar et al. 1998; Daniel et al. 1999). Recent studies by Vamvakaki et al. (2010) revealed that growth of *Mortierella isabellina* on cheese whey produced cost-effective γ-linolenic acid.

9.3.4 SCO Production on Hydrophobic Materials

Hydrophobic materials as a feedstock offer a significant trend line for SCO production as these materials do not limit nitrogen during bacterial growth. To date, many researchers have utilized hydrophobic materials from various organic industrial wastes for the production of SCO (Papanikolaou and Aggelis 2003; Zhu et al. 2008; Huang et al. 2009). For instance, Aggelis and Sourdis (1997) utilized vegetable oils as substrate for the growth of oleaginous microorganisms to accumulate lipids.

So far, every year, several hydrophobic materials are produced during various industrial processes, and many of these materials have been used for SCO production in the past decades. The lipid catabolism and its accumulation in the yeast strain *Yarrowia lipolytica* using industrial fats as a substrate were studied by Papanikolaou and Aggelis (2003). Interestingly, in their study, *Yarrowia lipolytica* was found accumulating a little cocoa butter when it was cultured on stearin (saturated free fatty acids) and glycerol. SCO can be a possible substitute for valuable fats because the lipid metabolism and accumulation process via de novo pathway are found similar in microbial lipid pathway. Moreover, there are no notable changes in the composition and in the structure throughout the process, e.g. cocoa butter (Zhu et al. 2008; Huang et al. 2009). But lipid metabolism via ex novo pathway produces new fatty acids at intra- and extracellular levels. This process is termed as bio-modification of lipids. This pathway upgrades the tailor-made value-added lipids that were produced from the low-fat substrates (Papanikolaou and Aggelis 2011). Though it is clearly known that the availability of hydrophobic materials (mainly from food and oil wastes) is lesser than that of hydrophilic materials, the significant functions of the hydrophobic substrates drive the researchers to do experiment on the same.

9.3.5 SCO Production on Lignocellulosic Biomass

Lignocellulosic biomass, an inexpensive substrate, is more likely to reduce the SCO production cost. However, the following two main factors which would inhibit the industrialization process are as follows: first is the raw material conveyance and the second is that its resource or availability can't be constant. Therefore, both these factors could affect the steady SCO production. Hence, in recent years, lignocellulosic biomass, the renewable and easily obtainable natural feedstock for SCO, has been a centre of attention of various researchers.

In industrial-scale ethanol production and also in many other fields, the lignocellulosic biomass has been used effectively, but still the microbial oil production from this biomass is at its early stages. The potential of lignocellulosic hydrolysates for SCO production was first stated by Zhao (2004). He also stated the biomass-to-biodiesel plan, in which he mentioned the three steps of the entire process. First step is converting lignocellulosic biomass to fermentable sugars; secondly, oleaginous microorganisms convert the fermentable sugars into lipids; and finally, the lipids are converted into biodiesel (Zhao 2004). Nearly three decades ago, the hydrolysis of lignocellulosic biomass has been achieved, and those techniques are quite established now (Wyman 1994; Hamelinck et al. 2005; Kumar et al. 2008). Moreover, the bioconversion which includes biological and chemical techniques to obtain biodiesel from lipids has been intensively studied in the last few decades (Demirbas 2007; Adamczak et al. 2009). Consequently, in recent years, many researchers have concentrated on the selection of appropriate strains for the production of lipids from lignocellulosic hydrolysates. So far, the oleaginous microorganisms that are capable to utilize inexpensive substrates for SCO production are *Lipomyces starkeyi* (Zhao et al. 2008), *Yarrowia lipolytica* (Papanikolaou et al. 2006), *Rhodosporidium toruloides* (Li et al. 2007) and *Rhodotorula glutinis* (Xue et al. 2008). Nevertheless, some oleaginous microorganisms showed their capability to utilize lignocellulosic hydrolysates to accumulate lipids particularly for acid lignocellulosic hydrolysates.

The SCO production process while using lignocellulosic hydrolysates as substrate has some limitations. Only few oleaginous strains such as *Lipomyces starkeyi* (Zhao et al. 2008), *Trichosporon fermentans* (Zhu et al. 2008) and *Mortierella isabellina* (Fakas et al. 2009) have the capability to utilize xylose for SCO production. Furthermore, *T. fermentans* has shown to produce oil on bagasse hydrolysates (Huang et al. 2012b) and rice straw (Huang et al. 2009). In recent times, some oleaginous yeast strains were reported for the production of SCO on wheat straw hydrolysates, namely, *Cryptococcus curvatus*, *Rhodosporidium toruloides*, *Rhodotorula glutinis*, *Yarrowia lipolytica* and *Lipomyces starkeyi*. Among them, *Cryptococcus curvatus* exhibited the maximum lipid yield (Yu et al. 2011). Furthermore, sugarcane bagasse hydrolysates (Tsigie et al. 2011) and rice hull (Economou et al. 2011b) were consumed by *Mortierella isabellina* and *Yarrowia lipolytica*, respectively, for SCO production. Recent research studies have showed that *Trichosporon dermatis*, an oleaginous yeast, is able to provide high lipid yield and high lipid coefficient using corncob enzymatic hydrolysates (Huang et al. 2012a). The above-mentioned studies proved every possible way of SCO production from lignocellulosic biomass.

Secondly, oleaginous microorganisms have to overcome the obstacles from inhibitors existing in lignocellulosic hydrolysates to produce lipids (Almeida et al. 2007). Various modes of purification could eliminate these inhibitors (Palmqvist and Hahn-Hägerdal 2000a), or adaptation (Martín et al. 2007) or microorganism alteration (Liu et al. 2009) may possibly raise their resistance to inhibitors. Reports about the consequences of inhibitors on growth and the metabolite accumulation may possibly state the optimization of fermentation and detoxification processes

Table 9.2 Composition of SCOs, accumulated by fungi

Strain	Composition, %				Reference
	C16:0	C18:1	C18:2	C18:3	
Aspergillus sydowii	–	62.9	–	–	Azeem et al. (1999)
Fusarium oxysporum	47	52.4	–	–	Azeem et al. (1999)
Fusarium equiseti	–	46	–	–	Azeem et al. (1999)
Mucor circinelloides	–	28	–	22.5	Vicente et al. (2010)
Mortierella isabellina	–	55.5	–	–	Liu and Zhao (2007)
Aspergillus terreus MTCC 6324	–	–	–	22.5	Kumar et al. (2010)
Mangrove fungal isolates	–	–	23	–	Khot et al. (2012)

which in turn helps to improve the strain (Palmqvist and Hahn-Hägerdal 2000b). In recent times, only a small number of works have looked into the consequence of the growth of oleaginous microorganisms and their lipid accumulation (Chen et al. 2009; Huang et al. 2011, 2012c). Therefore, it is essential to carry out additional studies to discover the work of the inhibitors on oleaginous microbes.

In brief, the likelihood of the production of SCO using lignocellulosic hydrolysates has been established. Lignocellulosic hydrolysates could work as a base substrate for the commercialization of SCO production. Yet, difficulties occur in this process and it must be overcome for industrial purpose.

9.4 Lipid Composition

There is recent interest in few fungi as appropriate substitute for source of SCOs, as those have the ability to alter some of the raw materials into a series of value-added products, such as lipids rich in polyunsaturated fatty acids (PUFAs) which have higher than 18 carbon atoms. A certain count of fatty acids has been related with specific health stimulating effects, namely, γ-linolenic acid (GLNA, C18:3ω-6 all cis 6,9,12-octadecatrienoic acid), dihomo-γ-linolenic acid (DHGLNA, C20:3ω-6 all cis 8,11,14-octadecatrienoic acid), arachidonic acid (ARA, C20:4ω-6 all cis 5,8,11,14-eicosatetraenoic acid), eicosapentaenoic acid (EPA, C20:5ω-3 all cis 5,8,11,14,17) and docosahexaenoic acid (DHA, C22:6ω-3 all cis 4,10,13,16,19). Like vitamins, these essential fatty acids cannot be produced within body system; hence they should be available through daily diet and therefore have commercial importance (Aggelis et al. 1988; Lechevalier and Lechevalier 1988; Larsson et al. 2006).

Azeem et al. (1999) stated that the fatty acid profile of microbial lipids obtained from *Aspergillus sydowii*, *Fusarium oxysporum* and *F. equiseti* had high ratio of unsaturated fatty acids, especially oleic acid (62.9, 52.4 and 46%, respectively). These microbial oils match the edible oils, namely, palm oil and groundnut, which contain C16 and C18 fatty acids esterified in the form of triacylglycerol (Table 9.2). Ellis et al. (2002) identified that the most prominent fatty acids found in the mycelium of *F. oxysporum* B1 were 18:1, 16:0 and 18:0, which totals up to 47% of overall fatty acids.

Mangrove fungal isolates comprise mainly C16 and C18 series which contains high ratio of saturated and monounsaturated fatty acids. This parameter possibly helps to identify the biofuel quality of fungal-based diesel. The fraction is almost alike to the cooked vegetable oil feedstock utilized for biodiesel such as, soybean, rapeseed, palm and sunflower (Leung et al. 2010). But, the fatty acid profiles of oleaginous and cyanobacteria are different, which indicates a domination of C14 and C16 fatty acids in *Chlorella* sp. which is rich in C18 (Hu et al. 2008). It has been reported that stearic acid contents were high (48–57%) for *Aspergillus* sp. (Subhash and Mohan 2011), whereas lower content was identified in *M. circinelloides* (7%) (Vicente et al. 2010) and *M. isabellina* (1%) (Liu and Zhao 2007). On the other hand, palmitic acids were high in *M. isabellina* (28%) (Liu and Zhao 2007) and *M. circinelloides* (20.7%) (Vicente et al. 2010), while trace amounts were found in *Aspergillus* sp. (Subhash and Mohan 2011). Oleic acid (C18:1) contents have been found to be 0.1–1.6% in *Aspergillus* sp., 28% in *M. circinelloides* and 55.5% in *M. isabellina* (Liu and Zhao 2007; Vicente et al. 2010; Subhash and Mohan 2011). Fungal SCOs have abundant polyunsaturated fatty acids (PUFAs) unlike the vegetable oils, and it's mostly studied for the production of PUFAs. Yet, PUFAs containing higher than four double bonds are not preferred for biodiesel of higher quality. In recent work with mangrove fungal isolates (Khot et al. 2012), dominant PUFA member that was identified in the SCOs was linoleic acid (C18:2) with its content ranging from minimum of 8.7% to 23.3%. The linolenic acid (C18:3) content was reported to be insignificant, and PUFAs containing four or more double bonds were not identified, which is comparable to the SCOs of *M. isabellina* (2.4%) and *Aspergillus* sp. (0.09%) (Liu and Zhao 2007; Subhash and Mohan 2011). On the contrary, greater PUFA levels were reported in the biodiesel derived from *M. circinelloides* and *A. terreus* MTCC 6324, whereby PUFAs comprised of C18:3 (22.5%) and C15:4, C17:4, C32:3 and C33:4 (9%), respectively, were obtained (Kumar et al. 2010; Vicente et al. 2010).

9.5 Lipid Extraction Methods

The microorganism selection and optimization of cultural conditions have been the main focus of the development of microbial lipid production. However, almost all the extraction mechanism applied to microbial system (Fig. 9.5) has been initially defined for plant materials and animal tissues, whereas only little attention has been given for the oil isolation. Hence, to further cultivate this microbial biotechnology research, reliable methods for recovery and cleansing of microbial oils have to be employed. The origin of the microbial cells and the kind of extract desired are the choices of segregation methods. One major issue that can cause lots of problems is the failure to avoid lipolysis in lipid recovery processing. Moreover, adequate treatment should be employed to reduce auto-oxidative degradation and the existence of

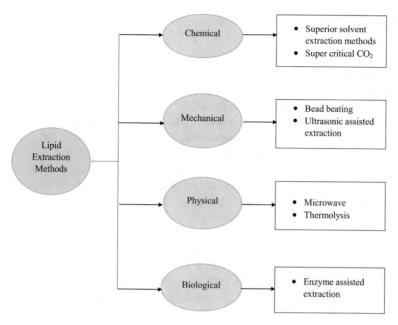

Fig. 9.5 SCO extraction methods

artefacts. SCO extraction and cleansing, in many processing phases, consume organic solvents to segregate oil away from other intracellular components. As said by Ratledge et al. (2006), wet cell biomass has to undergo drying procedure prior to extraction step, and this drying is an energy-consuming process. Hence, alternate procedures that utilize wet biomass could prove effectiveness in reducing production costs. Additionally, in case microbial oils are intended to be applied for human purposes, solvents should be acceptable in terms of toxicity, management, protection and budget.

9.5.1 Folch Method

The Folch process (Folch et al. 1957) involves treating cells with chloroform/methanol for lipid extraction from endogenous cells. In brief, the cells are blended well with 1/4th volume of salt solution. The subsequent mixture is allowed to divide into two layers where lipids remain in the upper phase. This first processing step is one of the earliest initiatives in lipid extraction, which forms the base for the current and future extraction steps. The main advantage of this method is that it's quick and simple in processing a huge number of samples. The Folch step provided the highest lipid yield for *L. starkeyi* and *R. toruloides* of about 47% w/w and 42% w/w, respectively. This procedure is less accurate when compared with other latest methods.

9.5.2   Bligh and Dyer Method

Lipid extraction and segregation are carried out concurrently in Bligh and Dyer (1959) method, in which proteins get accumulated in the interface layer between two liquid phases. As of now, this is the most used procedure for lipid extraction. Unlike Folch method, Bligh and Dyer method differs mainly in solvent/solvent and solvent/tissue ratios. The method extracts lipids using 1:2 (v/v) ratio of chloroform/methanol from homogenized cell suspension. The upper chloroform phase, containing lipids, is extracted and processed by different ways, and those details are not reported here. Many revisions have been adopted by researchers in order to develop this method. Adding 1 M NaCl instead of water is the usual modification done to avoid the acidic lipids being bound to denatured lipids. Besides that, the addition of HCl (Jensen 2008) and 0.2 M phosphoric acid (Hajra 1974) to the salt solution enhances the recovery of lipids with additional segregation time, compared to the initial extraction methods. Likewise, the recovery of acidic phospholipids was improved by addition of acetic acid (0.5% v/v) to the water phase (Weerheim et al. 2002). Moreover, recently, the most effective way for the extraction of plant sphingolipids was proved to be Hajra's described method (Markham et al. 2006).

9.5.3   Extraction of All Classes of Lipids

The current and in-depth method that was proposed by Matyash et al. (2008) is a revised method of Folch/Bligh and Dyer method. Nearly all major classes of lipids have been recovered well by the above-mentioned extraction methods. The lipidome profile was precisely generated from lipids extracted using methyl tert-butyl ether (MTBE), since it develops a less dense lipid-rich extractable organic upper phase. In short, 1.5 ml of methanol is added and mixed thoroughly (vortexing) for a 200 ml sample. Following that, 5 ml of MTBE is added to the tube, and the tube was left for incubation at room temperature for 1 h. After incubation, 1.25 ml of water was added to cool the incubator tubes. Then, to develop phase segregation, the tubes are left at room temperature for 10 min. The organic phase (top layer) was collected after the centrifugation at 1000 g for 10 min. The complete lipid recovery is obtained from other organic phase (lower phase) by extracting them with new addition of 2 ml of MTBE/methanol/water (10/3/2.5, v/v/v). The solvent phases are then vacuum dried to remove off remaining solvent from both organic phases having the lipid extract. Until further study, the extracted lipids were dissolved in 200 ml of chloroform/methanol/water (60/30/4.5, v/v/v). The aforementioned procedures are suitable for extraction of oil from all kinds of microbial cells.

9.5.4   Superior Solvent Extraction Methods

Chloroform, used in previous methods, may be a very effective solvent; yet applying it for the commercial-scale lipid extraction is prohibited due to health and environmental risks (Sheng et al. 2011). The use of particular solvent is specific to the class of lipids to be extracted. Jones et al. (2012) showed that greater lipid recovery can be obtained through the use of 2-ethoxyethanol (2-EE) than the usual extraction solvents such as chloroform, hexane and methanol. Besides that, higher lipid recovery was obtained through the accelerated solvent extraction (ASE) process operating on pressure or heat. It helps in reducing the total processing time while solvent would be recovered for reuse (Cooney et al. 2009). Worldwide, studies are being done for developing the solvent extraction techniques and combined solvent/physical extraction systems. However, when it is employed in a bigger industrial scale, there should have respective drawbacks in these extraction systems as they use organic solvents.

9.5.5   Bead Beating

Bead beating is a type of mechanical cell disruption technique, where an immediate damage to the biomass cells is created by rapid spinning motion of the microbial biomass slurry against fine bead globules (Lee et al. 1998). In this process, the microbial cells are broken due to the effect of granulating beads against the microbial cells. A wide range of microbial cells can be handled by this method. The two normal types of bead mills are agitating beads and shaking vessels. In the shaking vessel type, the microbial cells are ruptured by agitating the whole culture medium vessel. Generally, different medium vessels are shaken on an agitating stage, and this sort of method is applicable for tests which needs similar disruption treatment conditions. Thus, this set-up is solely utilized on a lab-scale experiment. Enhanced disruption and extraction efficiencies are acquired with the agitating beads. These rotating agitator vessels are lined with cooling jackets to warm up the sample and prevent the denaturation of heat-sensitive biomolecules. Certainly, the consolidated impact summing up collision, agitation and grinding of the beads delivers a more powerful disruption process (Lee et al. 2012). Beads made of titanium carbide, zirconia-silica or zirconium oxide can improve the disturbance rate and extraction proficiency of fungal cells, apparently in light of their more prominent hardness and thickness (Hopkins 1991).

9.5.6   Ultrasonic-Assisted Extraction

Ultrasonic-assisted extraction is another method to overcome the issues connected to the routine mechanical disruption techniques. The procedure is straightforward with simple set-up conditions, bestowing high-quality final product and also avoiding the further treatment of wastewater produced amid the process. Moreover, the procedure is eco-friendlier and can be finished in less time with high reproducibility. The input energy needed is less compared to traditional techniques, and ultrasonic-assisted extraction can even be worked at lower temperatures (Chemat and Khan 2011). With liquid culture mediums, there are two main mechanisms by which ultrasound can break the cells, i.e. cavitation and acoustic streaming. Cavitation is the generation of microbubbles as a consequence of the produced ultrasound waves, which pressurize the cell wall to rupture (Suslick and Flannigan 2008), and acoustic streaming encourages blending of microbial cultures (Khanal et al. 2007). The ultrasonic waves create stable and transient cavitation because of fast pressure/decompression cycles that take place amid the treatment. Transient cavitation results from unsteady motions, which will eventually implode. Along these lines, sonication breaks the cell wall because of the cavitation impact (Hosikian et al. 2010). Microstreaming and high mass transfer that occurs due to cavitation and bubble collapse are the two basic strides to decide the lipid yield extraction efficiency (Adam et al. 2012). The two fundamental types of sonicators are horn and bath, and both processors are normally utilized in batch operations; however, it could also be adjusted for continuous operations (Hosikian et al. 2010). Horns comprise of piezoelectric generators made of lead zirconate titanate crystals, and these generators vibrate with 10–15 mm amplitude, whereas sonicator utilize transducers, which are set at the base of reactor to create ultrasonic waves. The number of transducers and its arrangement in the reactor are decided based upon reactor's capacity and shape (Lee et al. 2012). The real standpoint of sonication procedure is that it produces generally low temperatures in comparison with microwave reactors and autoclaves, thus prompting less denaturation of biomolecules. Moreover, it doesn't need the external addition of beads or chemicals, which must be expelled afterwards in the process, and that in turn will incur additional cost (Harrison 1991). Nevertheless, very long ultrasonication prompts the generation of free radicals that might be impeding to the oil quality being extracted (Mason et al. 1994).

9.5.7 Microwave

In the past, application of microwave radiation was restricted for estimating trace metals and organic contaminants (Huffer et al. 1998; Marcato and Vianello 2000). The microwave-assisted lipid extraction was initially published in the mid-1980s (Ganzler et al. 1986). The initial publications provided details on microwave extraction techniques for separating lipids from seeds, nourishments, bolsters and soil, which proved microwave extraction to be efficient than the routine techniques. Therefore, microwave-assisted lipid extraction was found to be more economical and safe method; also it can process wet algal biomass directly without dewatering the biomass (Pare et al. 1996). A polar or dielectric material introduced into an electric field, for example, microwaves, will produce heat in view of frictional strengths emerging from intermolecular and intramolecular developments (Amarni and Kadi 2010). Water vapour formed due to intracellular heating disrupts the cells from inside. This prompts the electroporation impact, which in turn breaks the cell membrane, thus resulting in extraction of intracellular compounds (Rosenberg and Bogl 1987). Hence, fast generation of pressure and heat within the cell brings out high-quality extracts with high recovery rate (Hemwimon et al. 2007). Similarly, micro-waves could be utilized to extract and transesterify the SCO into biodiesel. Short reaction time, low cost and high efficiency are the main advantages while using microwave-assisted extraction. In addition, it was reported that the extraction of biodiesel in a microwave-assisted extraction procedure is around 15–20 min, which is quite faster than the ordinary extraction techniques which operate up to 6 h (Refaat et al. 2008). Nonetheless, maintenance cost, especially on a business scale, is the major drawback with the microwave-assisted extraction technique.

9.5.8 Isotonic Extraction Method

Ionic fluid for lipid extraction is another emerging method in this field. Studies on this topic have been reported by different scientists, and it is more bio-accustomed despite the comprehensive procedures (Li et al. 2010; Wang et al. 2011; Huang et al. 2013). The main goal of this method is to substitute toxic chemicals with ionic flu-ids, also called as green solvent. Ionic fluids are nonaqueous salt solution kept up at fluid state at temperatures around 0–140 °C. Asymmetric cation and an inorganic anion are the basic building blocks of ionic fluids. The solvent's hydrophobicity, conductivity, polarity and solubility can be determined by the combination of anion and cation present in the ionic fluids (Cooney et al. 2009). There is no record on financial feasibility of this technique, and it is too soon to anticipate that this tech-nique is one of the better strategies for lipid extraction. Nevertheless, this technique seems, by all factors, to be feasible.

9.5.9 Osmotic Pressure Method

A novel approach to contend with other extraction techniques is the osmotic pres-sure method (Adam et al. 2012). Two different kinds of osmotic stresses can create cell damage, namely, hyper-osmotic and hypo-osmotic. The microbial cells undergo hyper-osmotic stress when the salt concentration is higher in the outer region; thus, the cells shrink as liquids inside the cells diffuse outside and damaging the cell walls. On the other hand, hypo-osmotic stress takes place when the salt concentra-tion is low in outer region; liquid from outside, streams into the cells, and the cells bulge or rupture on reaching high stress. Hypo-osmotic stress is the commonly used method for extracting intracellular substances from biomass.

9.5.10 Enzyme-Assisted Extraction

A unique method of extracting lipids by facilitating cell disruption includes the use of enzymes. But, the lipid profile and type of microorganism majorly affect this method (Liang et al. 2012). Besides that, this method is rapid, characterized with high specificity/selectivity, and also operates at low temperatures (Taher et al. 2014). The low operating costs and low energy requirements are the advantages of this method since it overcomes the disadvantage of being a cost-intensive technique.

9.5.11 Supercritical CO_2

Over the recent years, supercritical CO_2 has been popular as a renewable extraction technology since CO_2 is an inert, easily available and GRAS (Hegel et al. 2011). Hegel et al. (2011) carried out experiments on lipid extraction from yeast and found that supercritical CO_2 with a 9% w/w ethanol supplementation achieved good neutral lipid separation. The burden of high energy requirements remains a major trade-off for using supercritical CO_2 extraction, which render the application of this technique unattractive for large-scale processes.

9.6 Processing of SCO to Biofuels

There are basically two ways to perform transesterification of SCO: (i) direct in situ transesterification of SCO within the cell, without the need to extract SCO from the biomass, and (ii) indirect ex situ transesterification followed after extraction of SCO from cells. High production costs and capital investments are associated with extraction of SCO from biomass by solvent extraction or other methods. Therefore, it is recommended to consider the direct transesterification of SCO without extracting it from cells for further research studies. The direct transesterification usually involves centrifugation of biomass produced from fermentation followed by cleaning the cells with water, then drying the cells until it reaches constant weight and finally mixing the dried cells with a HCL or H_2SO_4 solution and methanol (Liu and Zhao 2007). The direct acid-catalysed transesterification of SCO-rich microbial biomass from one fungal strain (*M. isabellina*) and two yeasts (*L. starkeyi* and *R. toruloides*) has been reported by Liu and Zhao (2007), which resulted in FAMEs (fatty acid methyl esters) with CN of 56.4, 59.9 and 63.5, respectively, and lipid to FAME yields higher than 90% (w/w). Maximum FAMEs were recovered when using 0.2 mol/L H_2SO_4 at 70 °C for 20 h with a biomass-to-methanol ratio of 1:20 (w/v). Vicente et al. (2009) carried out lipid extraction by three different solvent systems that include chloroform-methanol, chloroform-methanol-water and n-hexane from SCO produced by the fungal strain *Mucor circinelloides*. In addition, the authors also compared the efficiency of direct transesterification for biodiesel production with indirect transesterification. High-purity FAMEs (>99%) were produced from direct transesterification method, whereas indirect transesterified FAMEs showed the purity in the range of 91.4–98%. Moreover, the yield of FAMEs from direct transesterification significantly increased with the presence of the acid catalyst (Vicente et al. 2009). The operating conditions used by Vicente et al. (2009) were 8% (w/w relatively to the microbial oil) BF_3, H_2SO_4 or HCL for 8 h at 65 °C with a methanol-to-oil molar ratio of 60:1.

9.6.1 Economical Aspects

In SCO production, the downstream processing costs are one of the major issues to be solved for obtaining full economic efficiency of microbial lipids. In order to extract the lipids from microbial biomass, the first and most important step is cell disruption, because it influences the efficiency of subsequent operations and overall efficiencies (Ratledge et al. 1984). Moreover, the lipid synthesis from microbes is not cheaper than plant oil, because the microbes require expensive carbon sources and sophisticated extraction methods. In addition, the oil yields from microbial sources are relatively low compared with the yields from plant or animal sources (Pometto et al. 2005). But, these microbes can be genetically engineered to utilize low-cost substrates such as whey, vegetable oils, industrial wastes, etc. In an economical point of view, waste or by-product streams can be used as cheaper sources for the production of SCO. According to Steen et al. (2010), the market value for fatty alcohols, aldehydes and wax esters was ~3 billion $ in 2004 with a price of 1500 $/per ton. The microbial production of tailor-made fatty acid derivatives offers therefore a huge market and an economic alternative for the production of biodiesel. Another more economical approach is modifying these organisms to produce value-added products which cannot be obtained from other sources. This will be more beneficial than modifying plant oils into higher-value products such as cocoa butter.

9.7 Industrial Aspects and Developments

In commercial manufacturing plants, oil-rich biomass is produced by fermentation of sugar-rich media. These cells are then separated by centrifugation. Afterwards, the oils are extracted by disrupting the cell walls in the presence of a solvent such as hexane and evaporating the solvent by drying under vacuum conditions. Finally, the extracted crude oil is refined through a series of steps including neutralization, degumming, bleaching and deodorization (Pometto et al. 2005). The first commercial microbial oil was produced in 1985 from the microorganism, *Mucor circinelloides*. The oil was produced in order to use it as alternative to the oil produced from the seeds of evening primrose, *Oenothera biennis*. The oil was rich in γ-linolenic acid and sold under the trade name of 'Oil of Javanicus' (Wynn and Ratledge 2006). The commercial breakthrough in microbial oil occurred after supplementing infant formula with arachidonic acid and docosahexaenoic acid (ARASCO™ and DHASCO™). This commercial product has been available in many countries like Australia, Europe and the Far East (Ochsenreither et al. 2016). Besides that, in 1988, the Lion Corp of Japan developed and patented the production process of ARA-rich oil. Moreover, before microbial single cell oil was used for infant formula, it was first used in the 1980s as a substitute for cocoa butter (Varma and Podila 2005). But, the process became uncompetitive when the world price of cocoa butter dropped from \$8000/ton to <\$2500/ton (Wynn et al. 2001; Wynn and Ratledge 2006; Ochsenreither et al. 2016). Furthermore, many polyunsaturated fatty acids (PUFA), which have gained interest as dietary supplements and nutraceuticals, are being synthesized using microorganisms, mostly fungal and algal species. In a commercial point of view, single cell oil became successful in the twenty-first century although the first SCO was launched on the market in 1985. However, major obstacles are still the extraction of the intracellular SCOs being difficult and cost-intensive. Therefore, commercialization has only occurred for high-value products. A general transfer of established oil extraction methods for plant oils is usually not applicable for oleaginous microorganisms due to the small size and the risk of emulsification. A step towards better profitability might be the realization of high cell densities achieved in short time, product purity, i.e. the enrichment of single desired fatty acids in SCO achieved by metabolic engineering, and the development of an universal and cost-efficient method for SCO recovery.

9.8 Conclusion

A remarkable number of research works have been devoted around the world to establish the suitability of SCO as substitute of plant oils for biofuel production. In terms of the energy and economic cost, yet it is not feasible for industrial production. Recent research is increasingly concerned with genetic modification of lipid metabolism in the cell. Moreover, overexpression of relevant genes and of enzymes can increase the productivity of SCO and can help to control their structure and properties for future use. However, future study should address the following issues:

(i) The enzymatic hydrolysis of biomass appears very promising, though improvements are required as regards different aspects of the process, in particular the enzyme separation and reutilization after the biomass treatment.

(ii) Currently, second-generation biofuels are sought, to produce with no utilization of fertile soils. Consequently, lignocellulosic materials should be obtained from plants able to grow in the presence of semi-fertile soils, usually not exploited for agriculture.

(iii) As the microbial oils contain high levels of free fatty acids (FFA), suitable catalyst is required, in that the traditional alkaline catalyst (NaOH) may interact with FFA, leading to soap formation and reducing the biodiesel yield, as well as the quality of the co-produced glycerol. The enzymatic synthesis of the biodiesel, based on the use of microbial lipases, offers significant improvements in the process efficiency.

(iv) Direct transesterification becomes very promising, but one needs to develop in situ microbial cell disruption technique to maximize the oil extraction.

References

Adam F, Abert-Vian M, Peltier G, Chemat F (2012) "Solvent-free" ultrasound-assisted extraction of lipids from fresh microalgae cells: a green, clean and scalable process. Bioresour Technol 114:457–465

Adamczak M, Bornscheuer UT, Bednarski W (2009) The application of biotechnological methods for the synthesis of biodiesel. Eur J Lipid Sci Technol 111(8):800–813

Aggelis G, Sourdis J (1997) Prediction of lipid accumulation-degradation in oleaginous microorganisms growing on vegetable oils. Antonie Van Leeuwenhoek 72(2):159–165

Aggelis G, Ratomahenina R, Arnaud A, Galzy P, Martin-Privat P, Perraud J, Pina M, Graille J (1988) Etude de l'influence des conditions de culture sur la teneur en acide gamma limolénique de souches de Mucor. Oleagineux 43(7):311–317

Ahn WS, Park SJ, Lee SY (2000) Production of Poly (3-hydroxybutyrate) by fed-batch culture of recombinant Escherichia coliwith a highly concentrated whey solution. Appl Environ Microbiol 66(8):3624–3627

Akhtar P, Gray J, Asghar A (1998) Synthesis of lipids by certain yeast strains grown on whey permeate. J Food Lipids 5(4):283–297

Almeida JR, Modig T, Petersson A, Hähn-Hägerdal B, Lidén G, Gorwa-Grauslund MF (2007) Increased tolerance and conversion of inhibitors in lignocellulosic hydrolysates by Saccharomyces cerevisiae. J Chem Technol Biotechnol 82(4):340–349

Alvarez H, Steinbüchel A (2002) Triacylglycerols in prokaryotic microorganisms. Appl Microbiol Biotechnol 60(4):367–376

Amaretti A, Raimondi S, Sala M, Roncaglia L, De Lucia M, Leonardi A, Rossi M (2010) Single cell oils of the cold-adapted oleaginous yeast Rhodotorula glacialis DBVPG 4785. Microb Cell Factories 9(1):1

Amarni F, Kadi H (2010) Kinetics study of microwave-assisted solvent extraction of oil from olive cake using hexane: comparison with the conventional extraction. Innovative Food Sci Emerg Technol 11(2):322–327

Angerbauer C, Siebenhofer M, Mittelbach M, Guebitz G (2008) Conversion of sewage sludge into lipids by Lipomyces starkeyi for biodiesel production. Bioresour Technol 99(8):3051–3056

Azeem A, Neelagund Y, Rathod V (1999) Biotechnological production of oil: fatty acid composition of microbial oil. Plant Foods Hum Nutr 53(4):381–386

Bednarski W, Leman J, Tomasik J (1986) Utilization of beet molasses and whey for fat biosynthesis by a yeast. Agric Wastes 18(1):19–26

Beopoulos A, Chardot T, Nicaud J-M (2009) Yarrowia lipolytica: a model and a tool to understand the mechanisms implicated in lipid accumulation. Biochimie 91(6):692–696

Bligh EG, Dyer WJ (1959) A rapid method of total lipid extraction and purification. Can J Biochem Physiol 37(8):911–917

Chatzifragkou A, Fakas S, Galiotou-Panayotou M, Komaitis M, Aggelis G, Papanikolaou S (2010) Commercial sugars as substrates for lipid accumulation in Cunninghamella echinulata and Mortierella isabellina fungi. Eur J Lipid Sci Technol 112(9):1048–1057

Chemat F, Khan MK (2011) Applications of ultrasound in food technology: processing, preservation and extraction. Ultrason Sonochem 18(4):813–835

Chen X, Li Z, Zhang X, Hu F, Ryu DD, Bao J (2009) Screening of oleaginous yeast strains tolerant to lignocellulose degradation compounds. Appl Biochem Biotechnol 159(3):591–604

Cherubini F (2010) The biorefinery concept: using biomass instead of oil for producing energy and chemicals. Energy Convers Manag 51(7):1412–1421

Cherubini F, Bird ND, Cowie A, Jungmeier G, Schlamadinger B, Woess-Gallasch S (2009) Energy-and greenhouse gas-based LCA of biofuel and bioenergy systems: key issues, ranges and recommendations. Resour Conserv Recycl 53(8):434–447

Chi Z, Pyle D, Wen Z, Frear C, Chen S (2007) A laboratory study of producing docosahexaenoic acid from biodiesel-waste glycerol by microalgal fermentation. Process Biochem 42(11):1537–1545

Cooney M, Young G, Nagle N (2009) Extraction of bio-oils from microalgae. Sep Purif Rev 38(4):291–325

Daniel H-J, Otto R, Binder M, Reuss M, Syldatk C (1999) Production of sophorolipids from whey: development of a two-stage process with Cryptococcus curvatus ATCC 20509 and Candida bombicola ATCC 22214 using deproteinized whey concentrates as substrates. Appl Microbiol Biotechnol 51(1):40–45

De Gorter H, Just DR (2010) The social costs and benefits of biofuels: the intersection of environmental, energy and agricultural policy. Appl Econ Perspect Policy 32(1):4–32

Demirbas A (2007) Progress and recent trends in biofuels. Prog Energy Combust Sci 33(1):1–18

Doelle MB, Doelle HW (1990) Sugar-cane molasses fermentation by Zymomonas mobilis. Appl Microbiol Biotechnol 33(1):31–35

Economou CN, Aggelis G, Pavlou S, Vayenas D (2011a) Modeling of single-cell oil production under nitrogen-limited and substrate inhibition conditions. Biotechnol Bioeng 108(5):1049–1055

Economou CN, Aggelis G, Pavlou S, Vayenas D (2011b) Single cell oil production from rice hulls hydrolysate. Bioresour Technol 102(20):9737–9742

Ellis RJ, Geuns J, Zarnowski R (2002) Fatty acid composition from an epiphytic strain of Fusarium oxysporum associated with algal crusts. Acta Microbiol Pol 51(4):391–394

Evans CT, Ratledge C (1983) Biochemical activities during lipid accumulation in Candida curvata. Lipids 18(9):630–635

Fakas S, Galiotou-Panayotou M, Papanikolaou S, Komaitis M, Aggelis G (2007) Compositional shifts in lipid fractions during lipid turnover in Cunninghamella echinulata. Enzym Microb Technol 40(5):1321–1327

Fakas S, Papanikolaou S, Batsos A, Galiotou-Panayotou M, Mallouchos A, Aggelis G (2009) Evaluating renewable carbon sources as substrates for single cell oil production by Cunninghamella echinulata and Mortierella isabellina. Biomass Bioenergy 33(4):573–580

Folch J, Lees M, Sloane-Stanley G (1957) A simple method for the isolation and purification of total lipids from animal tissues. J Biol Chem 226(1):497–509

Ganzler K, Salgó A, Valkó K (1986) Microwave extraction: a novel sample preparation method for chromatography. J Chromatogr A 371:299–306

Gnansounou E, Dauriat A, Villegas J, Panichelli L (2009) Life cycle assessment of biofuels: energy and greenhouse gas balances. Bioresour Technol 100(21):4919–4930

Hajra AK (1974) On extraction of acyl and alkyl dihydroxyacetone phosphate from incubation mixtures. Lipids 9(8):502–505

Hamelinck CN, Van Hooijdonk G, Faaij AP (2005) Ethanol from lignocellulosic biomass: techno-economic performance in short-, middle-and long-term. Biomass Bioenergy 28(4):384–410

Harrison ST (1991) Bacterial cell disruption: a key unit operation in the recovery of intracellular products. Biotechnol Adv 9(2):217–240

Hegel PE, Camy S, Destrac P, Condoret J-S (2011) Influence of pretreatments for extraction of lipids from yeast by using supercritical carbon dioxide and ethanol as cosolvent. J Supercrit Fluids 58(1):68–78

Hemwimon S, Pavasant P, Shotipruk A (2007) Microwave-assisted extraction of antioxidative anthraquinones from roots of Morinda citrifolia. Sep Purif Technol 54(1):44–50

Holdsworth JE, Ratledge C (1991) Triacylglycerol synthesis in the oleaginous yeastCandida curvata D. Lipids 26(2):111–118

Hopkins T (1991) Physical and chemical cell disruption for the recovery of intracellular proteins. Bioprocess Technol 12:57–83

Hosikian A, Lim S, Halim R, Danquah MK (2010) Chlorophyll extraction from microalgae: a review on the process engineering aspects. Int J Chem Eng 2010:1–11

Hu Q, Sommerfeld M, Jarvis E, Ghirardi M, Posewitz M, Seibert M, Darzins A (2008) Microalgal triacylglycerols as feedstocks for biofuel production: perspectives and advances. Plant J 54(4):621–639

Huang C, Zong M-H, Wu H, Liu Q-P (2009) Microbial oil production from rice straw hydrolysate by Trichosporon fermentans. Bioresour Technol 100(19):4535–4538

Huang C, Wu H, Liu Q-P, Li Y-Y, Zong M-H (2011) Effects of aldehydes on the growth and lipid accumulation of oleaginous yeast Trichosporon fermentans. J Agric Food Chem 59(9):4606–4613

Huang C, Chen X-F, Xiong L, Ma L-L (2012a) Oil production by the yeast Trichosporon dermatis cultured in enzymatic hydrolysates of corncobs. Bioresour Technol 110:711–714

Huang C, Wu H, Li R-F, Zong M-H (2012b) Improving lipid production from bagasse hydrolysate with Trichosporon fermentans by response surface methodology. New Biotechnol 29(3):372–378

Huang C, Wu H, Liu Z-J, Cai J, Lou W-Y, Zong M-H (2012c) Effect of organic acids on the growth and lipid accumulation of oleaginous yeast Trichosporon fermentans. Biotechnol Biofuels 5(1):1

Huang Q, Wang Q, Gong Z, Jin G, Shen H, Xiao S, Xie H, Ye S, Wang J, Zhao ZK (2013) Effects of selected ionic liquids on lipid production by the oleaginous yeast Rhodosporidium toruloides. Bioresour Technol 130:339–344

Huffer JW, Westcott JE, Miller LV, Krebs NF (1998) Microwave method for preparing erythrocytes for measurement of zinc concentration and zinc stable isotope enrichment. Anal Chem 70(11):2218–2220

Jensen SK (2008) Improved Bligh and Dyer extraction procedure. Lipid Technol 20(12):280–281

Jones J, Manning S, Montoya M, Keller K, Poenie M (2012) Extraction of algal lipids and their analysis by HPLC and mass spectrometry. J Am Oil Chem Soc 89(8):1371–1381

Kamm B, Kamm M, Schmidt M, Hirth T, Schulze M (2006a) Lignocellulose- based chemical products and product family trees. In: Kamm B, Kamm M, Gruber P (eds) Biorefineries-industrial processes and products, Status quo and future directions, vol 2. Wiley-VCH, Weinheim, pp 97–149

Kamm B, Kamm M, Gruber P (2006b) Biorefinery systems –an overview. In: Kamm B, Kamm M, Gruber P (eds) Biorefineries – industrial processes and products. Status quo and future directions, vol 1. Wiley-VCH, Weinheim, pp 3–40

Khanal SK, Grewell D, Sung S, Van Leeuwen J (2007) Ultrasound applications in wastewater sludge pretreatment: a review. Crit Rev Environ Sci Technol 37(4):277–313

Khot M, Kamat S, Zinjarde S, Pant A, Chopade B, RaviKumar A (2012) Single cell oil of oleaginous fungi from the tropical mangrove wetlands as a potential feedstock for biodiesel. Microb Cell Factories 11(1):1

Koutinas AA, Wang R, Campbell GM, Webb C (2006) A whole crop biorefinery system: a closed system for the manufacture of non-food-products from cereal. In: Kamm B, Kamm M, Gruber P (eds) Biorefineries-industrial processes and products, status quo and future directions, vol 1. Wiley-VCH, Weinheim, pp 165–191

Koller M, Bona R, Braunegg G, Hermann C, Horvat P, Kroutil M, Martinz J, Neto J, Pereira L, Varila P (2005) Production of polyhydroxyalkanoates from agricultural waste and surplus materials. Biomacromolecules 6(2):561–565

Kromus S, Kamm B, Kamm M, Fowler P, Narodoslawsky M (2006) The green biorefinery concept – fundamentals and potentials. In: Kamm B, Kamm M, Gruber P (eds) Biorefineries –biobased industrial processes and products. Status quo and future directions, vol 1. Wiley-VCH, Weinheim, pp 253–294

Kumar R, Singh S, Singh OV (2008) Bioconversion of lignocellulosic biomass: biochemical and molecular perspectives. J Ind Microbiol Biotechnol 35(5):377–391

Kumar AK, Vatsyayan P, Goswami P (2010) Production of lipid and fatty acids during growth of Aspergillus terreus on hydrocarbon substrates. Appl Biochem Biotechnol 160(5):1293–1300

Larsson K, Quinn P, Sato K, Tiberg F (2006) Lipids: structure, physical properties and functionality. Oily Press Bridgwater, Bridgwater

Lechevalier H, Lechevalier M (1988) Chemotaxonomic use of lipids—an overview. Microbial Lipids 1:869–902

Lee SJ, Yoon B-D, Oh H-M (1998) Rapid method for the determination of lipid from the green alga Botryococcus braunii. Biotechnol Tech 12(7):553–556

Lee AK, Lewis DM, Ashman PJ (2012) Disruption of microalgal cells for the extraction of lipids for biofuels: processes and specific energy requirements. Biomass Bioenergy 46:89–101

Leung DY, Wu X, Leung M (2010) A review on biodiesel production using catalyzed transesterification. Appl Energy 87(4):1083–1095

Li Y, Zhao ZK, Bai F (2007) High-density cultivation of oleaginous yeast Rhodosporidium toruloides Y4 in fed-batch culture. Enzym Microb Technol 41(3):312–317

Li Q, Du W, Liu D (2008) Perspectives of microbial oils for biodiesel production. Appl Microbiol Biotechnol 80(5):749–756

Li Q, Jiang X, He Y, Li L, Xian M, Yang J (2010) Evaluation of the biocompatibile ionic liquid 1-methyl-3-methylimidazolium dimethylphosphite pretreatment of corn cob for improved saccharification. Appl Microbiol Biotechnol 87(1):117–126

Liang K, Zhang Q, Cong W (2012) Enzyme-assisted aqueous extraction of lipid from microalgae. J Agric Food Chem 60(47):11771–11776

Liu B, Zhao ZK (2007) Biodiesel production by direct methanolysis of oleaginous microbial biomass. J Chem Technol Biotechnol 82(8):775–780

Liu ZL, Ma M, Song M (2009) Evolutionarily engineered ethanologenic yeast detoxifies lignocellulosic biomass conversion inhibitors by reprogrammed pathways. Mol Gen Genomics 282(3):233–244

Marcato B, Vianello M (2000) Microwave-assisted extraction by fast sample preparation for the systematic analysis of additives in polyolefins by high-performance liquid chromatography. J Chromatogr A 869(1):285–300

Markham BL, Ong L, Barsi JA, Mendenhall JA, Lencioni DE, Helder DL, Hollaren DM, Morfitt R (2006) Radiometric calibration stability of the EO-1 advanced land imager: 5 years on-orbit, Remote Sensing, International Society for Optics and Photonics

Martín C, Marcet M, Almazán O, Jönsson LJ (2007) Adaptation of a recombinant xylose-utilizing Saccharomyces cerevisiae strain to a sugarcane bagasse hydrolysate with high content of fermentation inhibitors. Bioresour Technol 98(9):1767–1773

Mason T, Lorimer J, Bates D, Zhao Y (1994) Dosimetry in sonochemistry: the use of aqueous terephthalate ion as a fluorescence monitor. Ultrason Sonochem 1(2):S91–S95

Matyash V, Liebisch G, Kurzchalia TV, Shevchenko A, Schwudke D (2008) Lipid extraction by methyl-tert-butyl ether for high-throughput lipidomics. J Lipid Res 49(5):1137–1146

Meng X, Yang J, Xu X, Zhang L, Nie Q, Xian M (2009) Biodiesel production from oleaginous microorganisms. Renew Energy 34(1):1–5

Muller EE, Sheik AR, Wilmes P (2014) Lipid-based biofuel production from wastewater. Curr Opin Biotechnol 30:9–16

Naik SN, Goud VV, Rout PK, Dalai AK (2010) Production of first and second generation biofuels: a comprehensive review. Renew Sust Energ Rev 14(2):578–597

Ochsenreither K, Glück C, Stressler T, Fischer L, Syldatk C (2016) Production strategies and applications of microbial single cell oils. Front Microbiol 7:1539

Palmqvist E, Hahn-Hägerdal B (2000a) Fermentation of lignocellulosic hydrolysates. I: inhibition and detoxification. Bioresour Technol 74(1):17–24

Palmqvist E, Hahn-Hägerdal B (2000b) Fermentation of lignocellulosic hydrolysates. II: inhibitors and mechanisms of inhibition. Bioresour Technol 74(1):25–33

Papanikolaou S, Aggelis G (2002) Lipid production by Yarrowia lipolytica growing on industrial glycerol in a single-stage continuous culture. Bioresour Technol 82(1):43–49

Papanikolaou S, Aggelis G (2003) Modeling lipid accumulation and degradation in Yarrowia lipolytica cultivated on industrial fats. Curr Microbiol 46(6):0398–0402

Papanikolaou S, Aggelis G (2011) Lipids of oleaginous yeasts. Part I: biochemistry of single cell oil production. Eur J Lipid Sci Technol 113(8):1031–1051

Papanikolaou S, Komaitis M, Aggelis G (2004) Single cell oil (SCO) production by Mortierella isabellina grown on high-sugar content media. Bioresour Technol 95(3):287–291

Papanikolaou S, Galiotou-Panayotou M, Chevalot I, Komaitis M, Marc I, Aggelis G (2006) Influence of glucose and saturated free-fatty acid mixtures on citric acid and lipid production by Yarrowia lipolytica. Curr Microbiol 52(2):134–142

Pare J, Matni G, Belanger J, Li K, Rule C, Thibert B, Yaylayan V, Liu Z, Mathé D, Jacquault P (1996) Use of the microwave-assisted process in extraction of fat from meat, dairy, and egg products under atmospheric pressure conditions. J AOAC Int 80(4):928–933

Polakowski T, Bastl R, Stahl U, Lang C (1999) Enhanced sterol-acyl transferase activity promotes sterol accumulation in Saccharomyces cerevisiae. Appl Microbiol Biotechnol 53(1):30–35

Pometto A, Shetty K, Paliyath G, Levin RE (2005) Food biotechnology. CRC Press, Florida, USA

Pyle DJ, Garcia RA, Wen Z (2008) Producing docosahexaenoic acid (DHA)-rich algae from biodiesel-derived crude glycerol: effects of impurities on DHA production and algal biomass composition. J Agric Food Chem 56(11):3933–3939

Ratledge C (1991) Microorganisms for lipids. Acta Biotechnol 11(5):429–438

Ratledge C (1993) Single cell oils—have they a biotechnological future? Trends Biotechnol 11(7):278–284

Ratledge C (2004) Fatty acid biosynthesis in microorganisms being used for single cell oil production. Biochimie 86(11):807–815

Ratledge C, Dawson PSS, Rattray J (1984) Biotechnology for the oils and fats industry. The American Oil Chemists Society, USA

Ratledge C, Hopkins S, Gunstone F (2006) Lipids from microbial sources. Modifying lipids for use in food:80–113

Refaat A, El Sheltawy S, Sadek K (2008) Optimum reaction time, performance and exhaust emissions of biodiesel produced by microwave irradiation. Int J Environ Sci Technol 5(3):315–322

Rosenberg U, Bogl W (1987) Microwave thawing, drying, and baking in the food industry. Food Technol 41:85

Rumbold K, van Buijsen HJ, Overkamp KM, van Groenestijn JW, Punt PJ, van der Werf MJ (2009) Microbial production host selection for converting second-generation feedstocks into bioproducts. Microb Cell Factories 8(1):1

Santala S, Efimova E, Kivinen V, Larjo A, Aho T, Karp M, Santala V (2011) Improved triacylglycerol production in Acinetobacter baylyi ADP1 by metabolic engineering. Microb Cell Factories 10(1):1

Sheng J, Vannela R, Rittmann BE (2011) Evaluation of methods to extract and quantify lipids from Synechocystis PCC 6803. Bioresour Technol 102(2):1697–1703

Steen EJ, Kang Y, Bokinsky G, Hu Z, Schirmer A, McClure A, Del Cardayre SB, Keasling JD (2010) Microbial production of fatty-acid-derived fuels and chemicals from plant biomass. Nature 463(7280):559

Subhash GV, Mohan SV (2011) Biodiesel production from isolated oleaginous fungi Aspergillus sp. using corncob waste liquor as a substrate. Bioresour Technol 102(19):9286–9290

Subramaniam R, Dufreche S, Zappi M, Bajpai R (2010) Microbial lipids from renewable resources: production and characterization. J Ind Microbiol Biotechnol 37(12):1271–1287

Suslick KS, Flannigan DJ (2008) Inside a collapsing bubble: sonoluminescence and the conditions during cavitation. Annu Rev Phys Chem 59:659–683

Taher H, Al-Zuhair S, Al-Marzouqi AH, Haik Y, Farid M (2014) Effective extraction of microalgae lipids from wet biomass for biodiesel production. Biomass Bioenergy 66:159–167

Takeno S, Sakuradani E, Tomi A, Inohara-Ochiai M, Kawashima H, Shimizu S (2005) Transformation of oil-producing fungus, Mortierella alpina 1S-4, using Zeocin, and application to arachidonic acid production. J Biosci Bioeng 100(6):617–622

Tanisho S, Ishiwata Y (1995) Continuous hydrogen production from molasses by fermentation using urethane foam as a support of flocks. Int J Hydrog Energy 20(7):541–545

Tsigie YA, Wang C-Y, Truong C-T, Ju Y-H (2011) Lipid production from Yarrowia lipolytica Po1g grown in sugarcane bagasse hydrolysate. Bioresour Technol 102(19):9216–9222

Vamvakaki AN, Kandarakis I, Kaminarides S, Komaitis M, Papanikolaou S (2010) Cheese whey as a renewable substrate for microbial lipid and biomass production by Zygomycetes. Eng Life Sci 10(4):348–360

Varma A, Podila GK (2005) Biotechnological applications of microbes. IK International Pvt Ltd, New Delhi

Vega EZ, Glatz BA, Hammond EG (1988) Optimization of banana juice fermentation for the production of microbial oil. Appl Environ Microbiol 54(3):748–752

Vicente G, Bautista LF, Rodríguez R, Gutiérrez FJ, Sádaba I, Ruiz-Vázquez RM, Torres-Martínez S, Garre V (2009) Biodiesel production from biomass of an oleaginous fungus. Biochem Eng J 48(1):22–27

Vicente G, Bautista LF, Gutiérrez FJ, Rodríguez R, Martínez V, Rodríguez-Frómeta RA, Ruiz-Vázquez RM, Torres-Martínez S, Garre V (2010) Direct transformation of fungal biomass from submerged cultures into biodiesel. Energy Fuel 24(5):3173–3178

Wang Z, Zhuge J, Fang H, Prior BA (2001) Glycerol production by microbial fermentation: a review. Biotechnol Adv 19(3):201–223

Wang X, Li H, Cao Y, Tang Q (2011) Cellulose extraction from wood chip in an ionic liquid 1-allyl-3-methylimidazolium chloride (AmimCl). Bioresour Technol 102(17):7959–7965

Ward OP, Singh A (2005) Omega-3/6 fatty acids: alternative sources of production. Process Biochem 40(12):3627–3652

Wee Y-J, Kim J-N, Yun J-S, Ryu H-W (2004) Utilization of sugar molasses for economical L (+)-lactic acid production by batch fermentation of Enterococcus faecalis. Enzym Microb Technol 35(6):568–573

Weerheim AM, Kolb AM, Sturk A, Nieuwland R (2002) Phospholipid composition of cell-derived microparticles determined by one-dimensional high-performance thin-layer chromatography. Anal Biochem 302(2):191–198

Werpy T, Petersen G (2004) Top Value Added Chemicals from biomass, U.S. Department of Energy, Office of scientific and technical information, No.: DOE/GO-102004-1992, www.osti. gov/bridge.

Wu S, Hu C, Zhao X, Zhao ZK (2010) Production of lipid from N-acetylglucosamine by Cryptococcus curvatus. Eur J Lipid Sci Technol 112(7):727–733

Wyman CE (1994) Ethanol from lignocellulosic biomass: technology, economics, and opportunities. Bioresour Technol 50(1):3–15

Wynn J, Ratledge C (2006) Microbial production of oils and fats. Food Sci Technol-New York-Marcel Dekker 148:443

Wynn JP, Kendrick A, Hamid AA, Ratledge C (1997) 141 Malic enzyme: A lipogenic enzyme in fungi. Biochem Soc Trans 25(4):S669–S669

Wynn JP, Hamid AA, Li Y, Ratledge C (2001) Biochemical events leading to the diversion of carbon into storage lipids in the oleaginous fungi Mucor circinelloides and Mortierella alpina. Microbiology 147(10):2857–2864

Xue F, Zhang X, Luo H, Tan T (2006) A new method for preparing raw material for biodiesel production. Process Biochem 41(7):1699–1702

Xue F, Miao J, Zhang X, Luo H, Tan T (2008) Studies on lipid production by Rhodotorula glutinis fermentation using monosodium glutamate wastewater as culture medium. Bioresour Technol 99(13):5923–5927

Yang Q, Yang M, Zhang S, Lv W (2005) Treatment of wastewater from a monosodium glutamate manufacturing plant using successive yeast and activated sludge systems. Process Biochem 40(7):2483–2488

Ykema A, Verbree EC, Kater MM, Smit H (1988) Optimization of lipid production in the oleaginous yeastApiotrichum curvatum in wheypermeate. Appl Microbiol Biotechnol 29(2–3):211–218

Ykema A, Verbree EC, Verwoert II, van der Linden KH, Nijkamp HJJ, Smit H (1990) Lipid production of revertants of Ufa mutants from the oleaginous yeast Apiotrichum curvatum. Appl Microbiol Biotechnol 33(2):176–182

Yousuf A, Sannino F, Addorisio V, Pirozzi D (2010) Microbial conversion of olive oil mill wastewaters into lipids suitable for biodiesel production. J Agric Food Chem 58(15):8630–8635

Yousuf A, Hoque M, Jahan MA, Pirozzi D (2014) Technology and engineering of biodiesel production: a comparative study between microalgae and other non-photosynthetic oleaginous microbes. Int Rev Biophys Chem (IREBIC) 5(5):125–129

Yu X, Zheng Y, Dorgan KM, Chen S (2011) Oil production by oleaginous yeasts using the hydrolysate from pretreatment of wheat straw with dilute sulfuric acid. Bioresour Technol 102(10):6134–6140

Zhang Y, Adams IP, Ratledge C (2007) Malic enzyme: the controlling activity for lipid production? Overexpression of malic enzyme in Mucor circinelloides leads to a 2.5-fold increase in lipid accumulation. Microbiology 153(7):2013–2025

Zhang G, French WT, Hernandez R, Hall J, Sparks D, Holmes WE (2011) Microbial lipid production as biodiesel feedstock from N-acetylglucosamine by oleaginous microorganisms. J Chem Technol Biotechnol 86(5):642–650

Zhao Z (2004) Toward cheaper microbial oil for biodiesel oil. J Chin Biotechnol 25(2):8–11

Zhao X, Kong X, Hua Y, Feng B, Zhao ZK (2008) Medium optimization for lipid production through co-fermentation of glucose and xylose by the oleaginous yeast Lipomyces starkeyi. Eur J Lipid Sci Technol 110(5):405–412

Zheng Y, Yu X, Zeng J, Chen S (2012) Feasibility of filamentous fungi for biofuel production using hydrolysate from dilute sulfuric acid pretreatment of wheat straw. Biotechnol Biofuels 5(1):50

Zhu M, Yu L-J, Wu Y-X (2003) An inexpensive medium for production of arachidonic acid by Mortierella alpina. J Ind Microbiol Biotechnol 30(1):75–79

Zhu L, Zong M, Wu H (2008) Efficient lipid production with Trichosporonfermentans and its use for biodiesel preparation. Bioresour Technol 99(16):7881–7885

Zoebelin H (1996) (Ed) Dictionary of Renewable Resources, WILEY-VCH, Weinheim, Germany

Chapter 10
Production of Single Cell Protein (SCP) from Vinasse

Ernesto Acosta Martínez, Jéssica Ferreira dos Santos, Geiza Suzart Araujo, Sílvia Maria Almeida de Souza, Rita de Cássia Lacerda Brambilla Rodrigues, and Eliana Vieira Canettieri

Abstract Sustainable development has been a major focus of twenty-first-century research, and the world economy is undergoing profound changes, including the minimization and use of waste as well as the search for new materials to replace traditional sources derived from fossil fuels.

Research on biofuels has made Brazil a pioneer in the production of ethanol and cachaça from sugarcane. After fermentation and distillation of the wort, vinasse is generated as a by-product.

Vinasse is a toxic effluent that poses a potential hazard to surface and groundwater. This chapter discusses the treatment and application of vinasse, food industry, and waste management: biotechnological production and single cell protein (SCP) production.

10.1 Introduction

The worldwide interest in the development of the biofuels has increased due to a greater concern for the production of cleaner and renewable energy sources.

Brazil is the largest sugarcane producer in the world, followed by India, China, Paquistão, Thailand, and Mexico (Christofoletti et al. 2013).

E. A. Martínez (✉) · J. F. dos Santos · G. S. Araujo · S. M. A. de Souza
Department of Technology, Food Engineering, State University of Feira de Santana,
Feira de Santana, BA, Brazil
e-mail: ernesto.amartinez@uefs.br

R. de C. L. B. Rodrigues
Department of Biotechnology, Engineering School of Lorena, São Paulo University,
Lorena, SP, Brazil

E. V. Canettieri
Department of Energy, São Paulo State University, Guaratinguetá, SP, Brazil

© Springer International Publishing AG, part of Springer Nature 2018
S. Kumar et al. (eds.), *Fungal Biorefineries*, Fungal Biology,
https://doi.org/10.1007/978-3-319-90379-8_10

The sucroalcohol sector is in constant transformation; besides sugar, the processing plants have focused on the production of ethanol and, more recently, have turned their attention to bioelectricity, alcohol-chemical, and marketing of carbon credits (Souza and Macedo 2010).

The biorefinery concept requires the production of nonenergetic and energetic outlets, which apply to processes that primarily generate biobased products from the integral biomass use, such as biomaterials, lubricants, chemicals, food, feed, etc., from process wastes used to produce heat and power (Jong and Jungmeier 2015). The biorefinery concept has been practiced widely with the aim of obtaining value-added products to improve process economics and environmental sustainability (Cherubini 2010; Martinez-Hernandez et al. 2013; Jong and Jungmeier 2015).

Jong and Jungmeier (2015) stated that all biorefineries should be accessed for the entire value chain in their environmental, economic, and social sustainability covering their entire life cycle. This evaluation should also consider possible consequences such as the competition between food and biomass resources, the impact on water use and quality, changes in land use, balance of soil carbon and fertility, net greenhouse gas emissions, impact on biodiversity, potential toxicological dangers, and energy efficiency.

The Brazilian National Alcohol Program was created to promote partial substitution of gasoline in light vehicles with hydrated ethanol, as part of the actions adopted by the federal government in reducing the impact of increased oil prices in the 1970s. According to CONAB (2016), sugarcane production was estimated for the 2016/2017 harvest to be 694.5 million tons, and ethanol production should remain above 27.9 billion liters (11.3 billion liters of anhydrous and 16.5 billion liters of hydrated ethanol). Furthermore, 40,000 producers and 4000 brands of cachaça are on the national market, which is estimated to have the installed capacity in Brazil of 1.2 billion of liters per year (Schoeninger et al. 2014). From the data, it is possible to infer that the generated volume of vinasse is in the range from 3.01 to 9.6 billion of liters per year.

On average, production of 1 m^3 ethanol yields 8–18 m^3 of vinasse (Vadivel et al. 2014; Nair and Taherzadeh 2016; Leme and Seabra 2017). Currently, the largest ethanol producers in the world are the United States and Brazil, which produce about 58.59 and 26.12 billion liters per year, respectively, corresponding to more than 83% of the ethanol supplied globally (Ferreira et al. 2010; Nitayavardhana and Khanal 2010; Sydney 2014; Sartori et al. 2015; CONAB 2017; Mohsenzadeh et al. 2017). Considering only the annual projection of vinasse in the largest ethanol producers on the world, United States and Brazil, it is possible to infer that the volume of vinasse is between 469 and 1054 billion liters and 209–470 billion liters, respectively.

In addition to the problems related to the volume of vinasse generated, there is a high potential for environmental contamination represented by this effluent, which has noteworthy characteristics, such as dark color, strong odor, low pH, high potassium content, high organic content, and high chemical oxygen demand (Silva et al. 2011, 2013, 2014; De Oliveira et al. 2013a; Nascimento 2017). Vinasse is a waste with a high content of salt and organic material, and its elimination into soils and water bodies causes serious problems, mainly salinization with increased salt concentration and alteration of pH levels (Bermúdez-Savón et al. 2000; Moraes et al. 2015).

The first application of the vinasse was for microbial growth to generate a commercial product, such as animal feed. Single cell protein appears as an economically and environmentally friendly alternative. These procedures could reduce the accumulation of vinasse and its direct disposal into water or soil.

Many studies have been published dealing with the use of vinasse from different raw material such as cane sugar (Nitayavardhana and Khanal 2010; Nair and Taherzadeh 2016; Dos Santos et al. 2016), cachaça from sugarcane juice (Pires et al. 2016), beet (Coca et al. 2015), rum, molasses (Nitayavardhana and Khanal 2010), and corn alcohol (Ielchishcheva et al. 2016) for single cell protein (SCP) production. This work presents an updated application data and treatment of vinasse for protein production. In this scenario, the development of research is essential to reduce environmental impacts and use this resource for human and animal consumption.

10.2 Treatment and Application of the Vinasse

The main constituent of the vinasse, stillage, restil, or distiller syrup is organic matter, primarily in the form of organic acids, and a smaller amount of positive ions such as potassium (K), calcium (Ca), and magnesium (Mg). Its nutritional value is connected to the origin of the wort. In general, vinasse is composed of 93% water and 7% solids, of that 75% is organic matter (Cortez et al. 1992; Parajó et al. 1995; Smith 2009). Due to all these characteristics, vinasse is an attractive source of nutrients that can be applied for irrigation or used as a microbiological culture medium (Moraes et al. 2015; Pires et al. 2016).

Currently, vinasse has been mostly used for fertirrigation practices, during sugarcane cultivation, to replace mineral fertilizers such as phosphorus (Christofoletti et al. 2013; Reis and Hu 2017). Other proposals for use of vinasse include recycling of vinasse in fermentation, concentration by evaporation, yeast production, energy and heat production, and as an additive for animal feed (Christofoletti et al. 2013; Nair and Taherzadeh 2016). Recent studies have proposed using vinasse for anaerobic digestion, which allows the recovery of part of its energy content through the production of biogas (Moraes et al. 2015; Janke et al. 2016; Utami et al. 2016; Moraes et al. 2017).

Figure 10.1 illustrates the flow of sugarcane processing for the production of sugar and cane bioethanol, highlighting the by-products and wastes from the process. Table 10.1 shows the characteristics of vinasse from molasses wort, cane juice wort, and mixed wort at a constant temperature range (81.1–100 °C). There is significant variation in values of BOD and COD (minimum range of the juice wort and maximum range of molasses wort) and composition of the effluent related to the presence of solids and minerals.

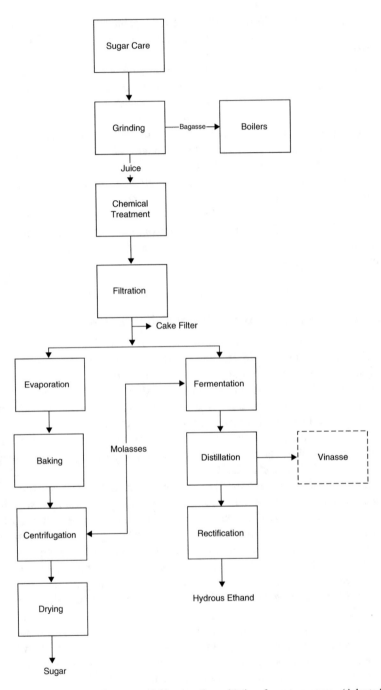

Fig. 10.1 Flow diagram of sugar and bioethanol production from sugarcane. (Adapted from BNDES (2008))

Table 10.1 Characteristics of the vinasse from molasses wort, cane juice, and mixed worts in the corresponding temperature range (81.1–100 °C)

Parameter	Molasses wort	Juice wort	Mixed wort
pH	4.20–5.00	3.70–4.60	4.40–4.60
BOD (mg/LO$_2$)	25,000	6000–16,500	19,800
COD (mg/LO$_2$)	65,000	15,000–33,000	45,000
Ratio DBO/DQO (biodegradability, F$_b$)[a]	0.59	0.61–0.77	0.68
Ratio DQO/DBO (treatability)[b]	2.5	2.5–2.0	2.3
Total solids (mg/L)	81,500	23,700	52,700
Volatile solids (mg/L)	60,000	20,000	40,000
Fixed solids (mg/L)	21,500	3700	12,700
Nitrogen (mg/L N)	450–1600	150–700	480–710
Phosphorus (mg/L P$_2$O$_5$)	100–290	10–210	9–200
Potassium (mg/L K$_2$O)	3740–7830	1200–2100	3340–4600
Calcium (mg/L CaO)	450–5180	130–1540	1330–4570
Magnesium (mg/L MgO)	420–1520	200–490	580–700
Sulfate (mg/L RO$_4^-$)	6400	600–760	3700–3730
Carbon (mg/L C)	11,200–22,900	5700–13,400	8700–12,100
Ratio C/N	16.00–16.27	19.70–21.07	16.40–16.43
Organic matter (mg/L)	63,400	19,500	38,000
Reducing substances (mg/L)	9500	7900	8300

Adapted from Society of Sugar and Alcohol Producers – SOPRAL (1986)
Where: [a]Biodegradability of vinasse (F$_b$), determined from the ration – BOD/(0.65*COD). [b]Treatability of the vinasse, determined from the ration – COD/BOD

The biodegradability of a compound and its environmental impact in a receiving body can be estimated through the relationship between the biochemical oxygen demand (BOD) and chemical oxygen demand (COD). If the COD/BOD ratio is less than 2.5, the effluent is considered readily biodegradable. If that ratio is greater than/equal to 2.5 and less than 5.0, a treatment with higher complexity is required; if it is greater than 5.0, the biological process of the application is not interesting. The biological treatability of the effluent can also be estimated from the BOD. Thus, the higher the BOD value, the greater the biological treatability of organic compounds present in the effluent (Jardim and Canela 2004; Braga et al. 2012).

From the results obtained for the BOD/COD and COD/BOD ratios, shown in Table 10.1, despite having a high organic load and representing an imminent danger to the environment, depending on the type of process, vinasse can be considered an effluent with significant biodegradability and treatability, and therefore, it can easily suffer from degradation.

Despite its liquid consistency, vinasse is considered a solid waste by NBR 10004/2004, because there is no technical and economical solution for efficient conventional treatment, which allows its release into waterways, according to the standards required by law (ABNT 2004). The indiscriminate application of the vinasse in the soil can cause an imbalance of nutrients and increase the risk of soil salinization and leaching, consequently, contaminating the groundwater. In water

Table 10.2 Developments in regulatory activities for the disposal, use, and application of vinasse in Brazil

Law	Description
MINTER Ordinance n° 323, of 11/29/1978	Prohibits the release of vinasse into surface waters
MINTER Ordinance n° 323, de 11/03/1980	Prohibits the release of vinasse into surface waters
CONAMA Resolution n° 0002, of 06/05/1984	Determination of studies and project resolution presentation containing standards to control pollution by effluents from alcohol distilleries and by water from the cane wash
CONAMA Resolution n° 0001, of 01/23/1986	Mandates Environmental Impact Assessment (EIA) and Environmental Impact Report (EIR) for newly installed industries or any extension made to the existing ones
Law n° 6134, de 06/02/1988, article 5, from state of São Paulo	The liquid, solid, or gaseous waste and solids from agricultural, industrial, commercial, or any other natural activities can only be conducted or released in a way that does not to pollute groundwater
CETESB, 2005, Stillage – Criteria and procedures for agricultural soil application. 1st edition	Establishes the criteria and procedures for storage, transportation, and application of vinasse generated by sugarcane activity in sugarcane processing
CETESB, 2006, Stillage – Criteria and procedures for agricultural soil application. 2nd edition	New version with some changes
CETESB, 2015, Stillage – Criteria and procedures for agricultural soil application. 3rd edition	Establishes criteria and procedures for the storage, transport, and application of vinasse generated by sugarcane activity in sugarcane processing, in the soil of the state of São Paulo

From Hassuda (1989) and CETESB (2015)

bodies, the organic load of vinasse causes the proliferation of microorganisms that consume the dissolved oxygen in the water making it difficult to use the contaminated water, for example, as a source of drinking water supply. Furthermore, its disposal into waterways causes bad smell and contributes to the aggravation of endemic diseases such as malaria, amoebiasis, and schistosomiasis (Corazza 1998; ANA 2009).

Table 10.2 presents the evolution of the regulation on disposal, use, and application of vinasse in Brazil between 1978 and 2015.

The diversification initiatives in the second half of the 1990s only aimed to overcome the market crisis that the agricultural sugarcane industry was facing. Today, sugar-alcohol enterprises are capable of producing according to environmental standard, because the possible environmental impacts have been solved through technological advances, such as the use of waste as by-products or raw materials, development of new equipment, and genetic improvement of sugarcane varieties. These changes, in addition to contributing to production sustainability, bring energy and economic advantages, describing Brazil as a pioneer in the technological

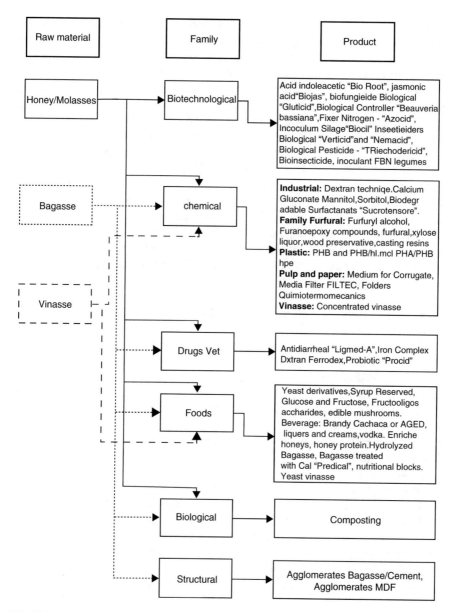

Fig. 10.2 Family of products. (From SEBRAE (2005))

innovation of ethanol. Thus, the benefits have transcended the social, political, and environmental character and gained economic support (SEBRAE 2005).

The raw materials from the sugar and alcohol industry and the functional framework of the products related to each technological application are shown in Fig. 10.2. There are various techniques and ways to use vinasse. Various vinasse utilization and treatment methods will be discussed below.

Direct soil application or fertirrigation (fertilization and irrigation) has been a widespread technique since the beginning of studies about the composition and pollution potential of vinasse. The high concentrations of water and organic and mineral matter are an attraction for the partial or full replacement of mineral fertilizer. Its use can alter the soil characteristics by promoting modifications in its chemical properties, helping to increase the availability of some elements for the sugarcane plants (Silva et al. 2007). However, the application without dosage criteria can cause a nutritional imbalance resulting in undesirable effects such as compromising of the quality of the cane for sugar production, pollution of the water table, and soil salinization (Cortez et al. 1992; Leme and Seabra 2017).

Concentration of vinasse: The water is removed without loss of solids, to decrease the initial volume, which makes the fertigation of many remote sugarcane plantations viable. The organic characteristics of fresh vinasse are maintained and can be used as animal feed and as biomass. This alternative investment helps to eliminate contamination and avoid fines and taxes due to noncompliance with environmental standards. It also adds value to the production of sugar and alcohol. However, the real applicability of the technology production must be assessed, because the concentrated product is more difficult for the soil to absorb; fluid transport is more complex, since product is corrosive to the equipment (SEBRAE 2005; ANA 2009).

Recycling of vinasse: The recycling aims to reduce the disposal volume. It can remove a part of the solid organic and inorganic constituents, consequently, reducing the polluting load. A technical limit on the use of vinasse for this purpose prevents the reduction of vinasse disposal; though effective, it can be very significant. The vinasse can also be used to form a portion of the water for molasses dilution, with an employed mechanism to replace the water as diluent (ratio 1:3 between vinasse and water) during a certain number of cycles, which results in less investment in nutrients and, consequently, reduces the effluent flow (CETESB 2015; Laime et al. 2011).

Physicochemical treatment: The physicochemical treatments of the vinasse result in the removal of BOD and COD, reducing the organic load. The various treatments that have been explored include adsorption, coagulation/flocculation, oxidation processes, treatments with membranes, and evaporation/combustion. Some treatments have shown little success. Sedimentation, for example, has been shown as unsatisfactory even with the addition of coagulants and other additives, such as alum, lime, ferric chloride (Satyawali and Balakrishnan 2008).

Souza et al. (2016) used the process of coagulation/flocculation for pretreatment of vinasse, which achieved a reduction of 67% in chemical oxygen demand (COD). Its toxicity was reduced up to ten times after the photocatalytic treatment with TiO_2–34 and TiO_2–87.

Use of vinasse in construction: The manufacture of building materials, especially blocs, from the vinasse has been studied. In concentrated form, the vinasse can be used as a soil binding agent, with positive effects on mechanical resistance, and can be added to the cement mass. However, the possibility of reducing the vinasse disposal is limited, because the economic viability of this alternative would be restricted to buildings near the origin of vinasse, due to transport costs (Freire and Cortez 2000).

Combustion of vinasse: The combustion is the alternative where the waste is concentrated and burned in the boiler. The high consumption of energy to evaporate the vinasse water, however, does not compensate the energy saved at the distillery. Research, into this alternative, should aim to improve the energy balance (Corazza 1998; Laime et al. 2011).

Production of methane (anaerobic digestion): Technical problems, such as long retention time and the granulation of microorganisms in mud, were overcome, and the anaerobic digestion of vinasse is now considered technically viable. Although the BOD is significantly reduced, an additional or end-of-pipe treatment is required. End-of-pipe technology is used to treat and control waste at the end of the production process. The economic viability of this technology, however, is hampered by several factors, including the lack of valorization of biogas as an alternative fuel, the adoption of appropriate methodologies for the use of fertigation, and the decline of the National Alcohol Program (Laime et al. 2011). Leme and Seabra (2017) state that there is a lack of studies exploring the use of vinasse for the production of biomethane.

Bermúdez-Savón et al. (2000) studied the biodegradation of wastewaters from a distillery. The tests were performed in batch anaerobic digesters for methane production, and the results showed good degradability using neutralizers and fertilizers. These authors stated that the anaerobic treatment can reduce more than half of the polluting organic load.

Feeding animals: Vinasse has been used in liquid (LV) and concentrated (CV) form as an additive in animal feed due to the presence of the organic acids, which improve nutrient use, digestion, vitamin D synthesis, and vitamin C and mineral absorption, which facilitate food metabolism. Liquid vinasse has a good amino acid and mineral profile that may work together with cell proliferation and integrity of the intestinal mucosa (Hidalgo et al. 2009). Chará and Suárez (1993) evaluated the effect of vinasse mixture in cane juice in the feed of Beijing duck and found good results with a mixture of up to 40% of vinasse. According to De Oliveira et al. (2013b), the better development of the intestinal mucosa tissues in rabbits fed with LV could be due to the low pH value of the vinasse and diets with LV levels over 75 g/kg (5.98 and 5.89). On the other hand, CV use could increase the complete consumption of rations, allowing greater use of bulks due to the presence of vitamins, minerals, proteins, and energy (Crochet 1967).The vinasse concentration process consists of a multi-effect evaporation system until it becomes concentrated to 55–60 °Brix (Ferreira 2012; Cruz et al. 2013), which is recommended due to its greater ease of handling and reduced transport costs and conservation problems. Sarria and Preston (1992) used CV for feeding pigs and observed mean rates of growth (kg/d): 590, 679, and 810 for soybean grain and 631, 651, and 650 for soybean when sugarcane juice was replaced with 0, 10, and 20% vinasse (48 °Brix). Gorni et al. (1987) found that the use of CV (0–12%) in corn and soybean meal rations did not alter the performance and characteristics of pig carcasses in the growth and finishing phases. Segundo Lâcorte et al. (1989) carried out a feed lot trial with rations formulated to provide 1.0 kg weight gain per day, which were composed by 50% auto-hydrolyzed sugarcane, 12.7% protein source (CV), 17.8% corn, and 12.5% sugarcane stalks as roughage complement. The mean weight gain was 580 g/animal-day for the ration containing CV. The vinasse palatability seems to be the preponderant factor that limits substitution, and its high hygroscopicity could cause formation of lumps in the whole ration, hindering their grip and chewing (Piekarski 1983; Lâcorte et al. 1989). The characteristics and nutritional properties of vinasse on rumen dynamics and performance of animals were reviewed by Da Costa et al. (2015).

10.3 Food Industry and Waste Management: Biotechnological Production

The food industry (agribusiness) is considered one of the most important manufacturing industries, and the cost of food is responsible for at least 20–30% of household budgets. While there is an excess of food production in some regions, the scarcity and insufficient production in Central Africa, China, and a significant part of South America have been a recurring problem (Gava 1998; Smith 2009). The production of adequate food, which supplies a good portion of the world population, is the great challenge for a sustainable future. It is estimated that the size of the urban population will be at least twice the rural population in 2030 (Smith 2009).

Agriculture represents a significant portion of food production in Brazil, reaching about one billion tons per year. Table 10.3 specifies the main cultivation processed in Brazil and its primary wastes (Pastore et al. 2013).

Table 10.3 Main crops processed in Brazil and its main wastes

Crop	Annual production (millions of tons)[a]	Wastes from processing[b]	Waste production (million of tons)[c]
Sugarcane	693.49	Vinasse	406.25
		Bagasse	138.7
Soy	62.49	Rind	3.12
		Molasses	11.92
		Vinasse	33.34
Corn	52.36	Bran	9.43
Cassava	26.04	Bagasse	26.04
		Cassava	15.63
Orange	18.66	Citric pulp	9.14
Rice	11.95	Rind	1.55
		Bran	9.56
Total	864.99		664.68
Dry equivalent biomass	208.75[c]		121.401[c]
Other crops over 1000 ton/years	44.1	Variable	

From Pastore et al. (2013)
Where: [a]Average 2008–2009; [b] wastes with variable content of solids; [c] considering the respective content of solids

The by-products from food processing are rich sources. In orange juice processing, for example, the essential oil has wide application for the cosmetics industry; the pectin is useful to produce gel; and the dried wastes can be used as fertilizer and animal feed (Gava 1998; SEBRAE 2005; Smith 2009).

The food industry is characterized by increasing fragmentation of its production process, which promotes intensification in creation of new products, attending different segments of market, with specific patterns of consumption. The aim of food biotechnology is the integration of modern biological knowledge, with the techniques and principles of bioengineering for processing and preserving food (SEBRAE 2005; Smith 2009).

The use of new techniques will not only enable the industry to improve its products but also directly influence the volume and quality of production. Therefore, the study of submerged fermentation and its peculiarities is very important, as it consists in a widespread alternative.

10.4 Single Cell Protein (SCP) Production

The term single cell protein (SCP) has been used to name the dried cells of microorganisms used as functional ingredient in human food as well as animal feed, giving the foods desirable characteristics such as taste, foaming, retention of water and fat, texture, etc. Technically, SCP is cell mass manufacturing by cultures of microorganisms, using agricultural and industrial wastes, which are available in abundance. After fermentation, the biomass is collected and can be subjected to downstream processing steps, such as washing, cell disruption, extraction, and purification of proteins (Tuse 1984; Gava 1998; Anupama and Ravindra 2000).

SCP has very attractive features as a nutritional supplement for humans. It is basically comprised of protein, carbohydrate, fats, water, and elements such as phosphorus and potassium. One benefit over animal and plant proteins is in that its growth is neither seasonally nor climatically dependent; therefore, it can be produced year round (Suman et al. 2015).

Single cell proteins have application in animal nutrition for fattening calves, poultry, and pigs as well as fish breeding; in the foodstuff area as aroma carriers, vitamin carriers, and emulsifying aids and for improving the nutritive value of baked products, soups, ready-to-serve meals, and diet recipes; and in the technical field for paper processing, leather processing, and foam stabilizers. The production of single cell proteins occurs in a fermentation process. This is done by a selected process for the growth of the culture and the cell mass, followed by separation processes (Nasseri et al. 2011).

Submerged fermentations have been the main industrial process for production of biotechnology (Robinson et al. 2001). It represents a technology because the microorganisms are grown in large stirred tanks and covered with liquid fermentation medium, which consists of carbohydrates, vitamins, minerals, and other nutrients (Mosier and Ladisch 2009).

Cane vinasse in cultures of microorganisms to produce single cell protein has been widely used. Yeast mass cultivation for food is recommended especially in regions where human malnutrition is chronic and is also used when there is interest in reducing BOD. According to Campos et al. (2014), vinasse has a high organic content, biochemical oxygen demand (BOD >5000 mg/L), and chemical oxygen demand (COD $\geq$12,000 mg/L). It is a way of converting this waste into high value-added products while avoiding environmental pollution.

Dos Santos et al. (2016) studied the feasibility of using sugarcane vinasse as a supplement to grow *Spirulina maxima*, focusing on the most suitable cultivation strategy and biochemical composition of the biomass obtained. Pires et al. (2016) investigated the production and quality of microbial biomass from a mixed culture of *Bacillus subtilis* CCMA 0087 and two yeast strains (*Saccharomyces cerevisiae* CCMA 0137 and *S. cerevisiae* CCMA 0188). These authors reported that the yeast *S. cerevisiae* showed promising results for microbial biomass production.

The nutritional characteristics of biomass are suitable for use as a protein supplement for animal feed, particularly for some species of fish. However, pretreatment of the residue is necessary, in order to reduce the level of potassium, so it can be used as animal feed for cattle, pigs, and poultry.

Along with potassium, vinasse contains valuable components, such as betaine, glycerol, monosaccharides, amino acids, and succinic acid, which could be recovered. However, by recovering these valuable components from the vinasse, the potassium concentration significantly increases, which limits its use in animal feed The percentage of organic compounds in the product is also high, thus limiting its use in fertilizer applications (Paananen et al. 2000). To overcome this problem, it is possible to process by fractioning vinasse, which forms potassium salts/crystals (30–42%) containing most of the inorganic fraction. The pH of the vinasse is lowered with sulfuric acid. The inorganic fraction containing potassium can be used as a fertilizer, for example. The amount of potassium in the remaining organic fraction will be brought to a desired level by adjusting the amount of sulfuric acid added (Paananen et al. 2000).

Hamstra and Schoppink (1998) relate a process to isolate valuable products from vinasse. The process involves concentrating the vinasse to a dry matter content of 50–80% dry solids and then separating potassium containing crystals formed in a concentration step from a resulting supernatant, which is substantially free of potassium salts. Ammonium sulfate is added to the vinasse before or during concentration. However, various techniques proposed for potassium removal from molasses vinasse, such as chemical precipitation, electrodialysis, and reverse osmosis, are expensive and require complex and strict control of the operating conditions (Zhang et al. 2012).

Zhang et al. (2012) mentioned that ion exchange is an easier technique to remove potassium from molasses vinasse, allowing the potassium to be recycled. These authors extracted potassium ions from molasses vinasses using a strong acid-cation exchange resin. Then it was desorbed from the resin and dissolved into the eluate with a H_2SO_4 solution. The ion exchange process using ZGC108 resin is potentially an effective method to remove potassium ions from the molasses vinasse because of its simplicity and lower raw material consumption. They also evaluated the recycle and reuse of the mother liquor (once-used eluant), after extracting K_2SO_4 crystals from the eluant, and found that using the mother liquor to elute the resin column by absorbing K^+ from molasses vinasse is a potential method for industrial production of potassium salt. Meanwhile, the molasses vinasse with most potassium ions removed might be used as animal feed (Zhang et al. 2012).

However, a dosage limitation must be obeyed. In ruminants, for example, the animal feed made from vinasse must not exceed 10% of the daily diet; in swine, it should not exceed 2–3%. Some studies have shown that it increased milk production; however, it has a laxative effect in cattle (Corazza 1998).

Nitayavardhana and Khanal (2010) examined the potential of protein-rich fungal biomass production obtained from different sugar-based feedstocks such as rum vinasse, molasses vinasse, and cane vinasse for aquaculture application. Molasses vinasse was the best substrate for fungal growth with high protein production.

The economic viability to dry yeast production depends on two main factors, (1) the requirement to add ammonia and magnesium salts to vinasse and (2) the high energy consumption needed to evaporate water from vinasse (Corazza 1998). Thus, it is necessary to design fermenters to ensure greater efficiency in transferring oxygen to the wort.

Vinasse, like any agro-industrial waste, generally has varied functions in bioprocesses. Basically, it can serve as an alternative source of nutrients for microbial growth and development, generating new compounds of interest as a product of metabolism (Pastore et al. 2013).

The biological value of a protein is related to its digestibility, toxicity, and ability to provide essential amino acids, as well as to the presence of antinutritional substances. For example, proteins of vegetable origin are considered poor or incomplete because they do not have adequate levels of some essential amino acids compared to animal proteins (Cheftel et al. 1989; Sgarbieri 1996).

The yeast *Candida utilis* (Henneberg) or *Cyberlindnera jadinii* synonymous to *Pichia jadinii*, also known as torula, belongs to the class of *Ascomycetes*. Its colonies are yellowish and have a subtle odor of ester. It can show cell dimorphism and aggregate cell, depending on environmental conditions. It is isolated from vegetable and animal sources, has large habitat, and may be added directly to food without limitation (Yang et al. 1979; Hawksworth et al. 1983; Bourdichon et al. 2012).

The yeast *C. utilis* is an excellent source of protein biomass and vitamins. It has been widely used to obtain single cell protein from the culture in agro-industrial by-products, with the advantage of having toxicological characteristics considered safe for human consumption. Their cells contain high protein content (about 50% dry weight) and a diverse essential amino acid profile; thus, their biomass is widely used as a protein supplement in animal feed and in the food industry as an additive (Serzedello 1986). Table 10.4 shows the typical composition of *C. utilis* yeast.

The cells of *C. utilis* are typically cultured in batch or continuous processes, with temperatures of about 30 °C and pH around 4.5. Then, they are recovered by centrifugation and utilized in food. This yeast does not have Crabtree effect, and high concentrations of sugars do not inhibit the activity of its respiratory enzymes, so

Table 10.4 Typical composition of the yeast *Candida utilis*

Composition	(%)	Vitamins	(mg/100 g)	Minerals	(mg/100 g)
Protein	52	Thiamine	0.80	Phosphorus	2100
Carbohydrate	22	Riboflavin	4.50	Potassium	2000
Minerals	08	Niacin	55.00	Magnesium	300
Fat	07	Folic acid	0.40	Sulfur	200
Humidity	06	Pyridoxine	8.30	Sodium	100
Crude fiber	05	Pantothenic acid	9.40	Calcium	15
		Biotin	0.08	Iron	9.5
		Para-aminobenzoic acid (PABA)	1.40	Zinc	9.3
		Choline	780	Fluoride	1.2
		Inositol	460	Manganese	0.7
		Vitamin B12	0.0004	–	

From Serzedello (1986)

there is no ethanol production; instead biomass is produced (Kondo et al. 1995; Christen et al. 1999; Borzani et al. 2001).

The food and feed industries consider *C. utilis* a safe and highly nutritious source of single cell protein. Its growth from a low-value commercial substrate is a promising option for the production of protein supplements for animal feed (Villas-Boas et al. 2003). Currently, single cell protein has been produced from various microbial species such as algae, fungi, and bacteria (Anupama and Ravindra 2000). However, as discussed earlier, because of their rapid growth in agro-industrial wastes, their high protein, and their secure toxicology for public health, some species of yeasts have been widely used to produce single cell protein. The conventional production of *Candida utilis* (torula yeast) is made from molasses. The yeast growth process is continuous, aerobic, and exergonic. The composition of the vinasse produced from the yeast is very similar to that obtained with cane molasses (ICIDCA 2000).

Martelli and Souza (1978) studied the bioconversion of organic matter of the vinasse in yeast biomass to reduce the polluting power of the waste. They suggest the use of biomass to replace part of the input used in animal feed and propose a technology capable of reducing the costs of waste treatment. The culture was performed with 0.4 g/L of inoculum at 30 °C for 16 h, in conventional batch, with pH 4.5, agitated at 600 rpm, and aerated with 1.0 L $_{air}$/L.min. The H/D relation was 1.0, the generation time of the culture was 1.5 h, and the specific growth rate in exponential phase (μ_x) value was calculated as 0.461/h. The amount of reducing matter consumed was 0.34 g/L giving a cell yield factor of 0.19. The growth of *C. utilis* on vinasse without previous treatment resulted in only 50% conversion of the reducing matter into cells. The decontamination of vinasse was limited by the culture ability to use the reducing matter in the vinasse. It was concluded that the use of fresh vinasse is not recommended as raw material for the production of food biomass of *C. utilis*.

SivaRaman et al. (1984) investigated the adaptation and growth of *C. utilis* in distillery effluent produced from sugarcane molasses, for single cell protein production. The aqueous waste of the first distillation, containing 2% of total unfermented sugars, was sterilized by autoclaving (15 psi, 20 min) and supplemented with 0.2% urea. The pH of the medium was 4.5. The adaptation of *C. utilis* was conducted by weekly transfers, for 4–6 weeks, to the fresh medium supplemented with urea and incubated at 30 °C in 250 ml Erlenmeyer flask and rotary shaker at 200 rpm. The unfermented sugars in the effluent were completely used during growth of the yeast, and the BOD value of the effluent was reduced. The growth curve of *C. utilis* in medium supplemented with urea lasted for 168 h; the lag phase (adaptation) started after one day and a half. The maximum growth concentration corresponded to 32 g/L and was reached after 144 h of cultivation.

Pedraza (1989) investigated the submerged culture of *Spirulina maxima* in water-diluted vinasse in the following vinasse/water proportions: 0.5:0.5; 0.6:0.4, and 0.7:0.3, average temperature of 20 °C in rectangular tanks (20 cm × 25 cm × 30 cm), shaken glass (2.500 cm^3_{air}/min). Dilution of the vinasse in 50% water was the best condition. After 720 h of culture, the cell concentration reached only 9×10^3 cells/mL. The objective was to develop a substrate system for the growth of *Spirulina*; however, the obtained cellular concentration was insufficient, and the culture time was very long, making it impracticable on an industrial scale.

Diaz et al. (2003) studied the single cell protein production of *C. utilis* from a mixture enriched with nutrients (ammonium sulfate, urea, and malt extract) at different concentrations (0, 0.5, 1.0, 1.5 g/L), in 1 L of culture medium, room temperature, pH 4.5, and experimental driving capacity of 2 L fermenter under aeration for 24 h, and the optimal concentration of the mixture corresponded to 1 g/L, resulting in a cell concentration of 2.2×10^8 cells/mL after 5 h of fermentation.

Studies were performed with *C. utilis* in a medium composed of vinasse supplemented with nutrient salts (ammonium phosphate and sulfate) in continuous cultivation, without previous treatment. The inoculum of *C. utilis* was prepared from cultures stored on agar slants-malt overnight in a pH 4.5 medium containing cane molasses to 20 mg/mL of total reducing sugars and nutrient salts concentration (diammonium phosphate and sulfate), using a rotary shaker at 32 °C, of 2.5 L Marubishi fermenter MD5. The cells were propagated in increasing concentrations of vinasse, corresponding to a variation of 2.5 g/L to 25 g/L of K_2O. The specific growth rate (μ_{max}) ranged from 0.32/h (2.5 g/L) to 0.28/h (25 g/L), and the yield coefficient biomass conversion of substrate was 0.23 (concentration of 2.5 g/L of K_2O) and 0.18 (concentration of 25 g/L of K_2O). The results suggested that the propagation of *C. utilis* in vinasse can significantly reduce the potassium content of this waste, making it suitable for use in irrigation. However, in terms of nutritional evaluation, the cumulative potassium would have a deleterious effect on animal health, so this method is not interesting for the production of SCP (Otero-Rambla et al. 2010).

Due to the nutrient richness of vinasse, a very important application of this residue is as a growth medium for microorganisms that can transform nutrients into biomass, thus altering the chemical composition of vinasse. According to Chia et al. (2013), besides the treatment of vinasse, we obtain products that can be used as nutritional supplements. This application makes it possible to reduce the eutrophic power of the vinasse, because it decreases, for example, the concentration of nitrogen and phosphorus. The biomass retains lipids, proteins, carbohydrates, and antioxidant compounds, among others.

Rhizopus microsporus (var. *oligosporus*) was successfully cultivated on thin stillage from corn-ethanol plants under aseptic conditions (Rasmussen et al. 2007). The fungal biomass cultivated on the settled thin stillage supernatant had a relatively high protein content of 43% (dry weight) with various essential amino acids, including 1.8% lysine, 1.8% methionine, 1.5% threonine, and 0.3% tryptophan. Moreover, a significant reduction in organic matters (up to 46%) and organic acids (71% lactic acid and 100% acetic acid) was achieved within 5 days of cultivation.

Ferreira (2009) observed that the incubation of *P. sajor*-caju in vinasse resulted in the reduction of chemical oxygen demand (82.76%), biochemical oxygen demand (75.29%), phenols (98.18%), total suspended solids (97.58%), total dissolved solids (9.03%), phosphate (85.50%), sulfate (19.32%), calcium (69.56%), magnesium (24.14%), potassium (7.87%), and reducing sugars (34.19%).

Nitayavardhana and Khanal (2010) reported that organic and inorganic matters were reduced 42.02% SCOD, 24.14% nitrogen (as TKN), 25.75% phosphorus (as PO_4^{3-}), 34.29% potassium (K), and 24.41% total dissolved solids, as well as a significant reduction of 43.30% glucose in a study about the production of *Rhizopus microsporus* (var. oligosporus) on vinasse generated during sugar-based ethanol fermentation.

A study investigated the influence of cultivation in vinasse on the composition of minerals present in torula yeast by neutron activation analysis. The presence of 23 elements of the periodic table was observed. Potassium (K), calcium (Ca), chlorine (Cl), magnesium (Mg), iron (Fe), and sodium (Na) were found in high concentrations. To perform the neutron activation analysis, 500 g of yeast were stored at room temperature bombarded by radiation, and the resulting radioactivity was measured by gamma spectrometry emitted by each radioisotope. The sample (in triplicate) and the synthetic standards of the analyzed elements were subjected to long and short irradiations in an IEAR1m reactor. In short radiation, samples and standards were irradiated for 2 min in a stream thermal neutron 1011 n/cm²s¹ and particular chlorine (Cl), potassium (K), magnesium (Mg), manganese (Mn), and sodium (Na). In the long radiation, samples and standards were irradiated for 8 h in a flow of thermal neutrons 1012 n/cm²s, and after about 20 days of stabilization, radioisotopes of calcium (Ca), barium (Ba), cerium (Ce), cobalt (Co), chromium (Cr), cesium (Cs), iron (Fe), rubidium (Rb), antimony (Sb), scandium (Sc), selenium (Se), and zinc (Zn) were determined. The radiation emitted by different range radioisotopes produced in the irradiation of samples and standards was measured in a gamma ray spectrometer and then transformed into digital information by the computer program (Solcoy). The vinasse in culture resulted in a high content of mineral matter (7.15%), high potassium content (2.12%), and total phosphorous (1.61%) in yeast (Rodriguez et al. 2011).

Nitayavardhana et al. (2013) reported cultivation of the fungus *Rhizopus oligosporus* on 75% (v/v) vinasse with nutrient supplementation (nitrogen and phosphorus) in an airlift bioreactor (2.5-L working volume) at 37 °C and pH 5.0 and aeration rates equal to 0.5, 1.0, 1.5, and 2.0 volume air/volume liquid/min (vvm). The highest fungal biomass production (8.04 ± 0.80 g/g) resulted in an 80% reduction in organic content (as soluble chemical oxygen demand) at 1.5 vvm aeration rate.

Nair and Taherzadeh (2016) investigated integration of edible ascomycetes fila-mentous fungi in the existing sugar- or molasses-to-ethanol processes, grown on dilute vinasse (10% v/v), and produced protein-rich fungal biomass at 35 °C for 72 h. These authors achieved 74% glycerol utilization, a 34% decrease in chemical oxygen demand (QOD), and no significant ($p \geq 0.05$) heavy metals reduction using *Aspergillus oryzae*. On the other hand, reduction in the amount of certain heavy metals such as zinc (57%), manganese (91%), and aluminum (60%) and 19% decrease in chemical oxygen demand (QOD) were observed using *Neurospora intermedia*.

The nucleic acids at different SCPs must be reduced to acceptable limits when used in food production. In terms of food safety, fungi have lower nucleic acid con-tent than bacteria. Furthermore, fungal SCP is rich in methionine and lysine. The fungal lysine content is higher than in bacteria or algae (Calloway 1974; Parajó et al. 1995; Anupama and Ravindra 2000).

The process of obtaining of the microbial biomass can be divided into the steps of preparation of the culture: fermentation, separation, and drying of the cells. After drying, the cells can be treated to improve their nutritional and organoleptic charac-teristics. The culture conditions, previous treatment of substrates, nutritional sup-plementation, and the performance of the fermentation process can improve the final composition of SCP (Anupama and Ravindra 2000).

Figure 10.3 describes the technology for obtaining single cell protein on an industrial scale. From the environmental viewpoint, this technology consists in an interesting waste treatment system, because it can reduce the contaminant content of 70–80 kg COD/m^3 to 25–35 kg COD/m^3. It is best that the plant be located near the distillery that provides the raw material (SEBRAE 2005).

The reuse of industrial by-products should be considered part of decreasing envi-ronmental impacts and poverty reduction, as it also provides a profit opportunity in different areas. The development of a technology that allows use of the vinasse with maximum efficiency to produce functional foods would provide opportunity to develop products in different niche markets, meet the nutritional needs of less afflu-ent persons, and enhance the diet of individuals, who seek to obtain foods with dif-ferent nutritional profile.

The use of microorganisms as a food source may appear to be unacceptable to some people, but the idea of microorganism consumption as food for humans and animals is certainly an innovative solution to the global food problem (Anupama and Ravindra 2000). The use of agro-industrial wastes in the form of SCP produced from yeasts for production and enrichment of foods contributes not only to mini-mize the problem of the hunger, but it would be an important tool to the food industry.

There are no papers dealing with economic aspects related to SCP produced from vinasse or about a plant running on this concept. Economic studies exist. The use of microorganisms as a food source may appear to be unacceptable to some people, but the idea of microorganism consumption as food for humans and animals is certainly an innovative solution to the global food problem related to the anaerobic digestion of vinasse for energy purposes (Poveda 2004; Bernal and Santos 2015).

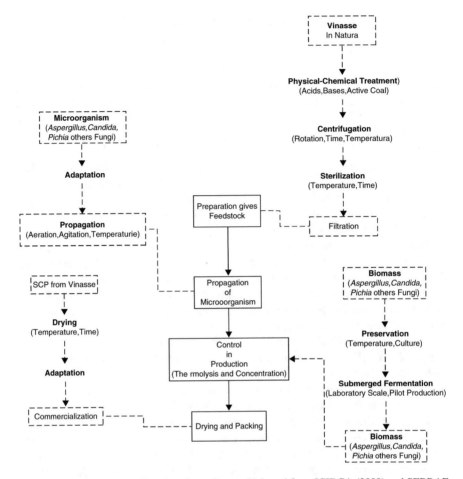

Fig. 10.3 SCP production flowchart from vinasse. (Adapted from ICIDCA (2000) and SEBRAE (2005))

The bioconversion of wastes into SCP is a cheaper and economically feasible way to overcome protein scarcity (Anupama and Ravindra 2000). Increasing concern about pollution caused by agri-wastes such as bagasse, citrus wastes, sulfite waste liquor, molasses, animal manure, starch, sewage, and wastewater from the cane and alcohol industry has stimulated interest for its conversion into commercially valuable product such as SCP (Bacha et al. 2011; Suman et al. 2015).

Some companies in the world are producing SCP including BP (United Kingdom), Kanegafuchi (Japan), and Liquichimica (Italy). In the United States, less than 15% of the plants making SCP were said to rely on hydrocarbons as the source of carbons and energy for the microorganisms (Suman et al. 2015).

According to Nitayavardhana and Khanal (2010), a typical 25 million gallons per year ethanol, plant could produce approximately 30,596 dry ton of fungal *Rhizopus microsporus* (var. oligosporus) biomass annually. If the fungal biomass provides market value similar to the soybean meal, this could generate additional revenue of at least US\$ 9.5 million annually for the bioethanol plants. This, however, does not include the costs associated with bioreactor, control systems (pH and temperature), air and nutrients supplies, and fungal biomass separation and drying.

The future of SCP will be heavily dependent on reducing production costs and improving quality by fermentation, downstream processing, and producer organisms, as a result of conventional applied genetics together with recombinant DNA technology (Omar and Sabry 1991; Srividya et al. 2015). The availability of the substrate and its proximity to the production plant are the major factors that determine the design and strategy of an SCP production process (Suman et al. 2015).

The development of fortified foods from agro-industrial by-products is a rich opportunity for profit maximization, incorporation of nutrients, and diversification of the production chain. The inclusion of protein products from the rational use of agro-processing, particularly from the activity of plants producing liquor ("cachaçarias"), will result in significant reduction in environmental impacts from this activity, as well as add value to sugarcane production.

10.5   Conclusions

This discussion, although initiated in the last century, is a contemporary subject, of an emergency and salutary character. Vinasse still remains as an effluent with high polluting potential but is also a by-product rich in organic matter minimally exploited for the production of food for commercialization on a large scale.

The success of commercialization of SCP as a food will become possible if the product is tasty, attractive, and nutritious and has a lower cost than conventional sources of protein.

10.5.1   *Future Developments*

The development of fortified foods from agro-industrial by-products, as discussed, is a rich opportunity for profit maximization, incorporation of nutrients, and diversification of the production chain. The inclusion of protein products from the rational use of agro-processing, in particular vinasse, will result in significant reduction in the environmental impacts of this activity, as well as add value to sugarcane production. Economic feasibility analysis of the SCP production process from vinasse is recommended.

The products that can be obtained from vinasse are animal feed, forage yeast, fertilizer, and methane gas. And the concept of vinasse utilization implies the recovery and sale of these marketable products or their use by the distillery itself.

References

ABNT (Associação Brasileira de Normas Técnicas) (2004) Resíduos sólidos - classificação – NBR 10.004 : 2004. Rio de Janeiro: ABNT

ANA (Agência Nacional de Águas) (2009) Manual de conservação e reuso de água na agroindústria sucroenergética. Brasilia: ANA

Anupama, Ravindra P (2000) Value-added food: single cell protein. Research review paper. Biotechnol Adv 18:459–479. https://doi.org/10.1016/S0734-9750(00)00045-8

Bacha U, Nasir M, Khalique A, Anjum AA, Jabbar MA (2011) Comparative assessment of various agro-industrial wastes for *Saccharomyces cerevisiae* biomass production and its quality production: as a single cell protein. J Anim Plant Sci 21(4):844–849

Bermúdez-Savón RC, Hoyos-Hernandez JA, Rodríguez-Perez S (2000) Evaluación de la disminución de la carga contaminante de la vinaza de destilería por tratamiento anaerobio. Rev Int Contam Ambie 16(3):103–107

Bernal AP, Santos IFS (2015) Estimativa do potencial energético a partir da digestão anaeróbia da vinhaça na cidade de Araraquara. Rev Bras Energ Renov 4:53–64

BNDES (2008) Bioetanol de cana-de-açúcar: energia para o desenvolvimento sustentável / organização BNDES e CGEE. – Rio de Janeiro: BNDES

Borzani W, Schmidell W, Lima UA, Aquarone E (2001) Industrial biotechnology Foundations. Blucher, São Paulo

Bourdichon F, Casaregola S, Farrokh C, Frisvad JC, Gerds ML, Hammes WP, Harnett J, Huys G, Laulund S, Ouwehand A, Powell IB, Prajapati JB, Seto Y, Ter Schure E, Van Boven A, Vankerckhoven V, Zgoda A, Tuijtelaars S, Hansen EB (2012) Food fermentations: Microorganisms with technological beneficial use. Int J Food Microbiol 154(3):87–97. https://doi.org/10.1016/j.ijfoodmicro.2011.12.030

Braga EAS, Aquino MD, Malveira KQ, Neto JC, Duarte Alexandrino CD (2012) Avaliação da biodegradabilidade das águas de lavagem provenientes da etapa de purificação do biodiesel produzido com óleo extraído das vísceras de tilápia. REGA 9(2):35–45

Calloway DH (1974) The place of single cell protein in man's diet. In: Davis P (ed) Single cell protein. Academic Press, New York

Campos CR, Mesquita VA, Silva CF, Schwan RF (2014) Efficiency of physico-chemical and biological treatments of vinasse and their influence on indigenous microbiota for disposal into the environment. Waste Manag 34:2016–2046. https://doi.org/10.1016/j.wasman.2014.06.006

CETESB Companhia Ambiental do Estado de São Paulo (2015) NTC P4.231 stillage – criteria and procedures for agricultural soil application, 3rd edn. CETESB, São Paulo

Chará JD, Suárez JC (1993) Utilización de vinaza y jugo de caña como fuente energética em patos Pekín alimentados con grano de soya y azolla como fuente proteica. Livestock Res Rural Dev 5(1):1–5

Cheftel JC, Cuq JL, Lorient D (1989) Proteínas alimentarias. Zaragoza: Acribia

Cherubini F (2010) The biorefinery concept: using biomass instead of oil for production energy and chemicals. Energy Convers Manag 51:1412–1421. https://doi.org/10.1016/j.enconman.2010.01.015

Chia MA, Lombardi AT, Melao MDGG, Parrish CC (2013) Lipid composition of *Chlorella vulgaris* (Trebouxiophyceae) as a function of different cadmium and phosphate concentrations. Aquat Toxicol 128(1):171–182. https://doi.org/10.1016/j.aquatox.2012.12.004

Christen P, Domenech F, Páca J, Rvah S (1999) Evaluation of four *Candida utilis* strains for biomass, acetic acid and ethyl acetate production from ethanol. Bioresour Technol 68(2):193–195. https://doi.org/10.1016/S0960-8524(98)00142-4

Christofoletti CA, Escher JP, Correia JE, Marinho JFU, Fontanetti CS (2013) Sugarcane vinasse: environmental implications of its use. Waste Manag 33:2752–2761. https://doi.org/10.1016/j.wasman.2013.09.005

Coca M, Barrocal VM, Lucas S, Gonzáles-Benito G, Gárcia-Cubero MT (2015) Protein production in *Spirulina platensis* biomass using beet vinasse-supplemented culture media. Food Bioprod Process 94:306–312. https://doi.org/10.1016/j.fbp.2014.03.012

CONAB – Companhia Nacional de Abastecimento (2017). Acompanhamento da safra brasileira de cana-de-açúcar) SAFRA 2017/18 – Brasília: Conab, 2017; 4(2):1–73. Available at: http://www.conab.gov.br/OlalaCMS/uploads/arquivos/. Accessed 30 Aug 2017

CONAB – Companhia Nacional de Abstecimento (2016) Acompanhamento da safra brasiliera. Cana-de-Açúcar. 3 - safra 2016/17 (3), dez. ISSN: 2318-7921. Available at: http://www.conab.gov.br. Accessed 19 Aug 2017

Corazza RI(1998) Reflexões sobre o papel das políticas ambientais e de ciência e tecnologia na modelagem de opções produtivas mais limpas numa perspectiva evolucionista: um estudo sobre o problema da disposição da vinhaça. Available at: https://www.race.nuca.ie.ufrj.br/eco/trabalhos/mesa3/6.doc. Accessed 13 Oct 2012

Cortez L, Magalhães P, Happy J (1992) Principais subprodutos da indústria canavieira e sua valorização. Rev Bras Energ 2(2):1–17

Crochet SL (1967) Blackstrap molasses is a major economic factor in catle operation at U.S. Sugar Corp. Sugar J 29(8):40–43

Cruz LFLS, Duarte CG, Malheiros TF, Pires EC (2013) Technical, economic and environmental viability analysis of the current vinasse use: ferti-irrigation, concentration and bio-digestion. Rev Bras Cienc Amb 29:111–127

Da Costa DA, de Souza CL, Saliba EOS, Carneiro JC (2015) By-products of sugar cane industry in ruminant nutrition. Int J Adv Agric Res 3:1–9

De Oliveira DWF, França IWL, Felix AKN, Martins JJL, Giro MEA, Melo VMM, Gonçalves LRB (2013a) Kinetic study of biosurfactant production by *Bacillus subtilis* LAMI005 grown in clarified cashew apple juice. Colloids Surf B: Biointerfaces 101:34–43. https://doi.org/10.1016/j.colsurfb.2012.06.011

De Oliveira MC, Silva DM, Carvalho CAFR, Alves MF, Dias DMB, Martins PC, Bonifácio NP, Souza Júnior MAP (2013b) Effect of including liquid vinasse in the diet of rabbits on growth performance. Rev Bras Zootec 42(4):259–263

Diaz M, Semprún A, Gualtieri M (2003) Producción de proteína unicelular a partir de desechos de vinaza. Rev Fac Farm 45(2):23–26

Dos Santos RR, Araújo OQF, de Medeiros JF, Chaloub RM (2016) Cultivation of *Spirulina maxima* in medium supplemented with sugarcane vinasse. Bioresour Biotechnol 204:38–34. https://doi.org/10.1016/j.biortech.2015.12.077

Ferreira L (2009) Biodegradação de vinhaça proveniente do processo industrial de cana- de- açúcar por fungos. Piracicaba, Tese de Doutorado. Escola Superior de Agricultura "Luis de Queiroz". Universidade de São Paulo

Ferreira GM (2012) Concentração de vinhaça a 55 °Brix integrada a usina sucroenergética. Simpósio Internacional e Mostra de Energia Canaviera, 10, Piracicaba, SP, Brasil

Ferreira LFR, Aguiar M, Pompeo G, Messias TG, Monteiro RR (2010) Selection of vinasse degrading microorganisms. World J Microbiol Biotechnol 26:1613–1621. https://doi.org/10.1007/s11274-010-0337-3

Freire WJ, Cortez LAB (2000) Vinhaça de cana-de-açúcar. Guaíba: Agriculture

Gava AJ (1998) Princípios de Tecnologia de Alimentos. São Paulo: Nobel

Gorni M, Berto DA, Moura MP, Camargo JCM (1987) Utilização da vinhaça concentrada na alimentação de suínos em crescimento e terminação. Bol Ind Anim 44(2):271–279

Hamstra RS, Schoppink PJ (1998) Process for the fractioning and recovery of valuable compounds from vinasse produced in fermentations. US Patent 5,760,078 A

Hassuda S (1989) Impactos da infiltração da vinhaça de cana no Aquífero Bauru. São Paulo, Tese de Mestrado. Instituto de Geociências. Universidade de São Paulo

Hawksworth DL, Sutton BC, Ainsworth GC (1983) Dictionary of the fungi, 7th edn. Commonwealth Mycological Institute Kew, Surrey

Hidalgo K, Rodríguez B, Valdivié M, Febles, M (2009) Utilización de la vinaza de destilería como aditivo para pollos en ceba. Rev Cubana Cienc Agrícola 43(3):281–284

ICIDCA (Instituto Cubano de Pesquisa dos Derivados da Cana-de-açúcar) (2000) Manual de los Derivados de la Caña de Azúcar, 3rd ed. ICIDCA, La Habana

Ielchishcheva I, Bozhkov A, Goltvianskiy A, Kurguzova N (2016) The effect of lipid components of corn vinasse on the growth intensity of yeast *Rhodosporidium diobovatum* IMB Y-5023. Int J Curr Microbiol App Sci 5(10):467–477. https://doi.org/10.20546/ijcmas.2016.510.053

Janke J, Leite AF, Batista K, Silva W, Nikolausz M, Nelle M, Stinner W (2016) Enhancing biogas production from vinasse in sugarcane biorefineries: effects of urea and trace elements supplementation on process performance and stability. Bioresour Technol 217:10–20. https://doi.org/10.1016/j.biortech.2016.01.110

Jardim WF, Canela MC (2004) Thematic Dossier v.1: Fundamentals of chemical oxidation in wastewater treatment and remediation of soils. Available at: http://lqa.iqm.unicamp.br/cadernos/caderno1.pdf. Accessed 22 Aug 2015

Jong E, Jungmeier G (2015) Biorefinery concepts in comparison to petrochemical refineries. In: Pandey A, Hofer R, Larroche C, Taherzadeh M, Nampoothiri M (eds) Industrial biorefineries and white biotechnology. Elsevier, Amsterdam

Kondo K, Saito T, Kajiwara S, Takagi M, Misawa NA (1995) Transformation system for the yeast *Candida utilis*: use of a modified endogenous ribosomal protein gene as a drug-resistant marker and ribosomal DNA as an integration target for vector DNA. J Bacteriol 177(24):7171–7177. https://doi.org/10.1128/jb.177.24.7171-7177

Lâcorte MCF, Bose MLV, Ripoli TCT (1989) Performance of feedlot with ration based on auto hydrolysed sugar cane bagasse, yeast and vinasse. An ESALQ 46(2):433–452

Laime EMO, Fernandes PD, Oliveira DCS, Freire EA (2011) Technological possibilities for the disposal of vinasse: a review. R Trop: Ci Agr Biol 5(3):16–29

Leme RM, Seabra JEA (2017) Technical-economic assessment of different biogas upgrading routes from vinasse anaerobic digestion in the Brazilian bioethanol industry. Energy 119:754–766. https://doi.org/10.1016/j.energy.2016.11.029

Martelli HL, Souza NO (1978) Obtaining of *Candida utilis* biomass growing in cane vinasse. Rev Bras Technol 9:157–164

Martinez-Hernandez E, Campbell G, Sakhukhan J (2013) Economic value and environmental impact (EVEI) analysis of biorefinery systems. Chem Eng Res Des 9:1418–1426. https://doi.org/10.1016/j.cherd.2013.02.025

Mohsenzadeh A, Zamani A, Taherzadeh MJ (2017) Bioethylene production from ethanol: a review and techno-economical evaluation. Chem Bio Eng Rev 4(2):75–91. https://doi.org/10.1002/cben.201600025

Moraes BS, Zaiat M, Bonomi A (2015) Anaerobic digestion of vinasse from sugarcane ethanol production in Brazil: challenges and perspectives. Renew Sust Energ Rev 44:888–903. https://doi.org/10.1016/j.rser.2015.01.023

Moraes BS, Petersen SO, Zaiat M, Sommer SG, Triolo JM (2017) Reduction in greenhouse gas emissions from vinasse through anaerobic digestion. Appl Energy 189:21–30. https://doi.org/10.1016/j.apenergy.2016.12009

Mosier NS, Ladisch MR (2009) Modern biotechnology: connecting innovations in microbiology and biochemistry to engineering fundamentals. Wiley, Hoboken

Nair RB, Taherzadeh MJ (2016) Valorization of sugar-to-ethanol process waste vinasse: a novel biorefinery approach using edible ascomycetes filamentous fungi. Bioresour Technol 221:469–476. https://doi.org/10.1016/j.biortech.2016.09.074

Nascimento D (2017) Biogás, biometano e o setor sucroenergético. Rev Canavieiros, X, 127:73–75

Nasseri AT, Rasoul-Amini S, Morowvat MH, Ghasemi Y (2011) Single cell protein: production and process. Am J Food Technol 6(2):103–116. https://doi.org/10.3923/ajft.2011.103.116

Nitayavardhana S, Khanal SK (2010) Innovative biorefinery concept for sugar-based ethanol industries: production of protein-rich fungal biomass on vinasse as an aquaculture feed ingredient. Bioresour Technol 101:9078–9085. https://doi.org/10.1016/j.biortech.2010.07.048

Nitayavardhana S, Issarapayup K, Pavasant P, Khanal SK (2013) Production of protein-rich fungal biomass in an airlift bioreactor using vinasse as substrate. Bioresour Technol 133:301–306. https://doi.org/10.1016/j.biortech.2013.01.073

Omar S, Sabry S (1991) Microbial biomass and protein production from whey. J Islamic World Acad Sci 4(170):172

Otero-Rambla MA, Almazan-Del Olmo OA, Bello-Gil D, Saura-Laria G, Martinez-Valdivieso JA (2010) Potassium removal from distillery slops by *Candida utilis* propagation. Proc Int Soc Sugar Cane Technol 27:1–7

Paananen H, Lindroos M, Nurmi J, Viljava T (2000) Process for fractioning vinasse. US Patent 6,022,394 A

Parajó JC, Santos V, Domínguez H, Vázquez M, Alvarez C (1995) Protein concentrates from yeast cultured in wood hydrolysates. Food Chem 53(2):157–163. https://doi.org/10.1016/0308-8146(95)90782-3

Pastore GM, Bicas LJ, Junior MRM (2013) Biotecnologia de Alimentos. Atheneu, São Paulo

Pedraza GX (1989) Cultivation of *Spirulina maxima* for protein supplementation. Livest Res Rural Dev 1(1):1–8

Piekarski PRB (1983) Valor nutritivo da vinhaça concentrada e do melaço na alimentação de bovinos em confinamento. Viçosa. 49p. (Mestrado - Universidade Federal de Viçosa)

Pires JF, Ferreira GMR, Reis KC, Schwan RF, Silva CF (2016) Mixed yeasts inocula for simultaneous production of SCP and treatment of vinasse to reduce soil and fresh water pollution. J Environ Manag 182:455–463. https://doi.org/10.1016/j.jenvman.2016.08.006

Poveda MMR (2004) Análise econômica e ambiental do processamento da vinhaça com aproveitamento energético. Dissertação de Mestrado, Instituto de Energia e Ambiente, Universidade de São Paulo, São Paulo, Brazil

Rasmussen M, Kambam Y, Khanal SK, Pometto AL, van Leeuwen J (Hans) (2007) Thin stillage treatment from dry-grind ethanol plants with fungi. ASABE annual international meeting of American Society of Agricultural and Biological Engineers, June 17–20, Minneapolis, MN, USA

Reis CER, Hu B (2017) Vinasse from sugarcane ethanol production: better treatment or better utilization? Front Energy Res 5:1–7. https://doi.org/10.3389/fenrg.2017.00007

Robinson T, Singh D, Nigam P (2001) Solid state fermentation: a promising microbial technology for secondary metabolite production. Appl Microbiol Biotechnol 55:284–289. https://doi.org/10.1007/s002530000565

Rodriguez B, Canela AA, Mora LM, Motta WF, Lezcano P, Euler AC (2011) Mineral composition of torula yeast (*Candida utilis*), developed from distillery vinasse. Cuban J Agric Sci 45(2):151–153

Sarria P, Preston TR (1992) Reemplazo parcial del jugo de caña con vinaza y uso del grano de soya a cambio de torta en dietas de cerdos de engorde. Livestock Res Rural Develop 4(9). http://www.lrrd.org/lrrd4/1/sarria.htm

Sartori SB, Ferreira LFR, Messias TG, Souza G, Pompeo GB, Monteiro RTR (2015) Pleurotus biomass production on vinasse and its potential use for aquaculture feed. Mycology 6(1):28–34. https://doi.org/10.1080/21501203.2014.988769

Satyawali Y, Balakrishnan M (2008) Wastemater treatment in molasses-based alcohol distilleries for COD and color removal: a review. J Environ Manag 86:481–497. https://doi.org/10.1016/j.jenvman.2006.12.024

Schoeninger V, Coelho SRM, Silochi RMQH (2014) Cadeia produtiva da cachaça. Rev Energ Agr 29(4):292–300. https://doi.org/10.17224/EnergAgric.2014v29n4p292-300

SEBRAE (Serviço Brasileiro de Apoio às Micro e Pequenas Empresas) (2005) O novo ciclo da Cana: Estudo sobre a competitividade do sistema agroindustrial da cana-de-açúcar e prospecção de novos empreendimentos. Brasília: IEL/NC, SEBRAE

Serzedello A (1986) Biomassas microbianas e algumas de suas aplicações. In: Simpósio Anual da Academia de Ciências do Estado de São Paulo (ACIESP), 11, São Paulo Anais..., São Paulo, 51:273–289

Sgarbieri VC (1996) Proteínas em alimentos protéicos; propriedades, degradações, modificações. Sao Paulo: Varela

Silva MAS, Griebeler NP, Borges LC (2007) Use of stillage and its impact on soil properties and groundwater. Rev Bras Eng Agric Environ 11(1):108–114

Silva CF, Arcuri SL, Campos CR, Vilela DM, Alves JGLF, Schwan RF (2011) Using the residue of spirit production and bio-ethanol for protein production by yeasts. Waste Manag 31:108–114. https://doi.org/10.1016/j.wasman.2010.08.015

Silva MAS, Kliemann HJ, De-Campos AB, Madari BE, Borges JD, Gonçalves JM (2013) Effects of vinasse irrigation on effluent ionic concentration in Brazilian Oxisols. Afr J Agric Res 8(45):5664–5672. https://doi.org/10.5897/AJAR12.1441

Silva ALL, Costa JL, Gollo AL, Santos JD, Forneck HR, Biasi LA, Soccol VT, Carvalho JC, Soccol CR (2014) Development of a vinasse culture medium for plant tissue culture. Pak J Bot 46(6):2195–2202

SivaRaman H, Pandle AV, Prablune AA (1984) Growth of Candida utilis on distillery effluent. Biotechnol Lett 6(11):759–762. https://doi.org/10.1007/BF00133070

Smith JE (2009) Biotechnology. Cambridge University Press, NewYork

SOPRAL (Sociedade de Produtores de Açúcar e Álcool - Brasil) (1986) Avaliação do Vinhoto como Substituto do Óleo Diesel e Outros Usos. Coleção SOPRAL. São Paulo; 10

Souza ELL, Macedo IC (2010) Ethanol and bioelectricity: the sugar cane in the future of the energy matrix. Unica, São Paulo

Souza RP, Ferrari-Lima AM, Pezoti O, Santana VS, Gimenes ML, Fernandes-Machado NRC (2016) Photodegradation of sugarcane vinasse: evaluation of the effect of vinasse pre-treatment and the crystalline phase of TiO_2. Acta Sci Technol 38(2):217–226. https://doi.org/10.4025/actascitechnol.v28i2.27440

Srividya AR, Vishnuvarthan VJ, Murugappan M, Dahake PG (2015) Single cell protein: a review. Int J Pharm Res Scholars 2(4):472–485

Suman G, Nupur M, Anuradha S, Pradeep B (2015) Single cell protein production: a review. Int J Curr Microbiol App Sci 4(9):251–262

Sydney EB (2014) Economic process to produce biohydrogen and volatile fatty acids by a mixed culture using vinasse from sugarcane ethanol industry as nutrient source. Bioresour Technol 159:380–386. https://doi.org/10.1016/j.biortech.2014.02.042

Tuse D (1984) Single cell protein: current status and future prospects. Crit Rev Food Sci Nutr 19(4):273–325. https://doi.org/10.1080/10408398409527379

Utami I, Redjeki S, Astuti DH (2016) Biogas production and removal COD-BOD and TSS from wastewater industrial alcohol (vinasse) by modified UASB bioreactor. MATEC Web of Conferences, 58:1–5, BISSTECH 2015. https://doi.org/10.1051/matecconf/20165801005,

Vadivel R, Minhas PS, Suresh KP, Singh Y, Nageshwar RDVK, Nirmale A (2014) Significance of vinasses waste management in agriculture and environmental quality – review. Afr J Agric Res 9(38):2862–2873. https://doi.org/10.5897/AJAR2014.8819

Villas-Boas SG, Esposito E, de Mendonca MM (2003) Bioconversion of apple pomace into a nutritionally enriched substrate by Candida utilis and Pleurotus ostreatus. World J Microbiol Biotechnol 19:461–467. https://doi.org/10.1023/A:1025105506004

Yang HH, Thayer DW, Yang SP (1979) Reduction of endogenous nucleic acid in single cell protein. Appl Environ Microbiol 38(1):143–147

Zhang PJ, Zhao Z-G, Yu S-J, Guan Y-G, Li D, He X (2012) Using strong acid-cation exchange resin to reduce potassium level in molasses vinasses. Desalination 286:210–216. https://doi.org/10.1016/j.desal.2011.11.024

Index

Printed in the United States
By Bookmasters